Short-Term Bioassays in the
Analysis of Complex
Environmental Mixtures II

ENVIRONMENTAL SCIENCE RESEARCH

Recent Volumes in this Series

Short-Term Bioassays in the Analysis of Complex Environmental Mixtures II

Edited by

MICHAEL D. WATERS
SHAHBEG S. SANDHU
JOELLEN LEWTAS HUISINGH
LARRY CLAXTON

and

STEPHEN NESNOW

U. S. Environmental Protection Agency
Research Triangle Park, North Carolina

PLENUM PRESS • NEW YORK AND LONDON

Library of Congress Cataloging in Publication Data

Main entry under title:

Short-term bioassays in the analysis of complex environmental mixtures II.
  (Environmental science research ; v. 22)
  ''Proceedings of the Second Symposium on the Application of Short-term Bioassays in the Fractionation and Analysis of Complex Environmental Mixtures, held in Williamsburg, Virginia, March 4-7, 1980''—Verso t.p.
  Bibliography: p.
  Includes index.
  1. Pollution—Toxicology—Congresses. 2. Toxicity testing—Congresses. 3. Biological assay—Congresses. 4. Environmental chemistry—Congresses. I. Waters, Michael D. II. Symposium on the Application of Short-term Bioassays in the Fractionation and Analysis of Complex Environmental Mixtures. (2nd : 1980 : Williamsburg, Va.) III. Series.
RA566.S47   615.9'07   81-17839
ISBN 978-1-4684-4123-9   ISBN 978-1-4684-4121-5 (eBook)   AACR2
DOI 10.1007/978-1-4684-4121-5

This report has been reviewed by the Health Effects Research Laboratory, U.S. Environmental Protection Agency, and approved for publication. Mention of trade names or commercial products does not constitute endorsement or recommendation for use.

Proceedings of the Second Symposium on the Application of Short-Term Bioassays in the Fractionation and Analysis of Complex Environmental Mixtures, held in Williamsburg, Virginia, March 4-7, 1980.

[The proceedings of the First Symposium on the Application of Short-Term Bioassays in the Fractionation and Analysis of Complex Environmental Mixtures were published under the title *Application of Short-Term Bioassays in the Fractionation and Analysis of Complex Environmental Mixtures.*]

Plenum Press, New York
Softcover reprint of the hardcover 1st edition 1981
A Division of Plenum Publishing Corporation
233 Spring Street, New York, N.Y. 10013

FOREWORD

More than one hundred short-term bioassays are now available
for detecting the toxicity, mutagenicity, and potential
carcinogenicity of chemicals. These bioassays were developed and
validated with individual compounds, and their principal
application was perceived to be in evaluating the health hazard of
such materials. However, man is rarely exposed to single
chemicals; his exposure to hazardous chemicals is more commonly a
multifactorial phenomenon. Although chemical analysis can be used
to detect known hazardous compounds, it would be a staggering and
expensive task to analyze large numbers of samples for all known or
suspected hazardous constituents. Furthermore, the biological
activity of a complex mixture cannot be reliably predicted from
knowledge of its components. On the other hand, bioassays alone
cannot tell us which components of complex mixtures are responsible
for the biological activity detected. Thus, cost effectiveness and
technical feasibility dictate stepwise and perhaps iterative
application of both chemical and biological methods in evaluating
the health effects of complex environmental mixtures.

Through the coupling of reliable biological detection systems
with methods of chemical fractionation and analysis, it is
frequently possible to isolate the individual chemical species that
show biological activity. Initially, complex mixtures may be
separated and bioassayed in carefully defined chemical fractions.
The results of such short-term screening bioassays then may be used
to guide the course of further fractionation and to determine the
need for more stringent and comprehensive biological testing.

Another approach to the screening of complex environmental
mixtures for health effects involves the use of in situ bioassays.
The biological effects of environmental chemicals are influenced by
a combination of environmental factors that cannot be completely
reproduced in the laboratory. By maintaining test organisms at
sites to be monitored, one can rapidly identify potentially
hazardous environmental mixtures that warrant further
investigation.

These are the proceedings of the Second Symposium on the
Application of Short-term Bioassays in the Fractionation and

Analysis of Complex Environmental Mixtures, held in Williamsburg, VA, March 4 through 7, 1980, and sponsored by the U.S. Environmental Protection Agency. The first symposium of this series, also held in Williamsburg, in February, 1978, combined accounts of the latest methods for collection and chemical analysis of complex samples with discussions of current research involving the use of short-term bioassays in conjunction with fractionation and analysis of such mixtures. The emphasis of the present proceedings is on the application of these methods in testing a variety of media, including ambient air, drinking water and aqueous effluents, terrestrial systems, mobile-source emissions, and stationary-source emissions and effluents. The critical problem of human health hazard and risk assessment is also addressed.

We hope that this volume will help to consolidate our knowledge of the techniques and applications of chemical analysis and bioassay of complex environmental mixtures and that it will provide direction for further research in this area.

Michael D. Waters
Shahbeg S. Sandhu

ACKNOWLEDGMENTS

We would like to thank Wendy A. Martin and Karen L. Spear of Kappa Systems, Inc., for coordinating the symposium. Our thanks are extended to Olga Wierbicki, Susan Dakin, Leslie Silkworth, Barbara Elkins, and Priscilla Skidmore of Northrop Services, Inc., for editing the proceedings and preparing the final manuscript. Our sincere appreciation also goes to Dolores Nesnow for indexing the proceedings.

CONTENTS

_____

*Invited paper.
†Contributed paper based on poster presentation.

CONTENTS

KEYNOTE ADDRESS

KEYNOTE ADDRESS

Vilma Hunt
Office of Health Research
U.S. Environmental Protection Agency
Washington, District of Columbia

I am delighted to be here today addressing this symposium on the application of short-term bioassays to the study of complex environmental mixtures. I can think of no other area that is more representative of the significant advances that have been made in the last decade in the environmental health sciences field.

The U.S. Environmental Protection Agency (EPA) has been faced since its inception with the responsibility to control harmful environmental compounds. One of the highest priorities of EPA's research and development program is the protection of health through the identification and control of toxic substances. Our responsibility in the Office of Health Research is to identify, qualitatively and quantitatively, the harmful health effects associated with environmental agents. In our effort to fulfill this responsibility, we have repeatedly been slowed by the limitations of the available testing procedures, as well as by a relative lack of understanding of health effects themselves.

One of the major problems we have faced and still face is the complexity of the agents of concern. Complex mixtures--whether industrial effluents or emissions, or ambient environmental media--are generally entities composed of hundreds of compounds. Further, these mixtures are often poorly defined and of continuously changing chemical composition. As such, complex mixtures present special challenges in quantitation and assessment that require considerable effort to overcome.

We are also faced with limitations in our understanding of the effects we must test for. The effects of concern--cancer, mutagenesis, teratogenesis, and other toxic impacts--are often

3

chronic and slow to develop. The science behind these effects is
not well understood. And, significantly, we are often looking for
effects that are expected to be of low probability for an
individual, but of large impact for the exposed population.

The magnitude of the environmental regulatory task that EPA
and other regulatory agencies face is underscored by the
astronomical numbers that come up whenever environmental assessment
is discussed. There are currently more than 4,000,000 known
chemical compounds; thousands more are being discovered each year;
70,000 are in common use, produced and distributed by some 115,000
industries and firms. Billions of gallons of industrial effluents
are discharged into our lakes, rivers, and oceans each year. The
1977 emission of criteria air pollutants to the atmosphere was 190
million tons, a quantity that does not include the other
unregulated and potentially hazardous particles, gases, and
aerosols emitted each year. In addition, several billion tons of
unwanted solid waste--some harmful, some innocuous--are disposed of
each year.

If, to fulfill our health effects assessment responsibility,
we had "only" to establish the toxic potential of some 70,000
compounds, we would be faced with a staggering assignment. When
consideration is given to the fact that these compounds appear in
the environment in diverse combinations, in a variety of media, and
often in miniscule quantities that are difficult to collect and
analyze, the staggering assignment suddenly appears overwhelming.

Our efforts to assess toxic potential include a number of
undertakings in laboratory, clinical, and epidemiological research.
Historically, EPA and other government regulatory agencies have
favored established whole-animal methods as the standard of
reference to establish carcinogenicity and other toxic effects.
Our resources, however, to conduct whole-animal studies are
limited. Indeed, the world laboratory capacity for conducting
these experiments has been estimated at 500 compounds per year, a
small number when compared with the number of compounds needing
assessment. Additionally, the long-term nature of whole animal
studies poses other problems for regulators faced with the need to
make timely regulatory decisions.

Data derived from epidemiological studies are, of course,
those most adequate from a regulatory standpoint. These data are
scarce, however, and we are again faced with fiscal limitations in
collecting sufficient information on large numbers of compounds,
not to mention the inherent difficulties in collecting meaningful
information for large populations exposed to a myriad of compounds.

Recognizing our limitations in whole-animal and
epidemiological studies, we have devoted particular attention and

considerable resources to the research and development of short-
term tests over the last several years. Our interest in short-term
bioassays was sparked by recognition of their potential value to a
health effects assessment program. Our investment has been
rewarded by highly encouraging results.

As a rapid, effective, and inexpensive means to identify the
impact of complex mixtures, short-term testing can play a critical
role in the monitoring of the environmental media for presumptive
health hazards. Today, any program aimed at identifying and
reducing production and release of large numbers of hazardous
agents must take advantage of short-term bioassays to set
priorities for further evaluation by conventional toxicological,
clinical, and epidemiological investigations. By efficiently using
short-term bioassays and through the development of approaches that
combine the use of various bioassay systems, we have begun to
screen large numbers of potentially harmful compounds in a
systematic and effective manner.

As this symposium reflects, the application of short-term
bioassays to the assessment of complex mixtures has been developed
hand-in-hand with the application of state-of-the art analytical
chemistry techniques. The iterative application of chemical and
biological analytical tools has greatly expanded the number of
environmental pollutants for which biological hazards have been
identified. Results of many of these studies will be reported on
during the next several days. However, the number of assessed
pollutants is still only a small fraction of those requiring
assessment. The lack of information on chemical composition and
biological activity continues to constitute a major barrier to the
assessment of human health hazards from complex environmental
mixtures.

At the first symposium on The Application of Short-term
Bioassays in the Fractionation and Analysis of Complex
Environmental Mixtures, Dr. Michael Waters of our Research Triangle
Park Laboratory dedicated the meeting "to the concept that the
joint application of state-of-the art biological and chemical
analytical techniques is the appropriate approach in environmental
research." I am happy to note that the multidisciplinary approach
being pursued in this area is representative of a widespread
evolution in the approach to environmental problems on many fronts.
We are moving away from the segmented, narrow approaches of the
past to joint undertakings that direct the activities of different
disciplines toward specific, unified goals.

The past two years have seen great strides in the short-term
bioassay field and in the analytical area. The symposium last year
was devoted to the "nuts and bolts" of this relatively new field.
The majority of the presentations dealt with the bioassay

techniques, the sampling methods, and the analytical procedures.
This year, the symposium is organized to report on the application
of these techniques--a reflection of the development of this field
from solely an object of research to a means for research.

In particular, the topics to be considered over the next
several days include the utilization and application of short-term
testing to different media: ambient air, drinking water and
effluents, terrestrial media, mobile-source emissions, and
stationary-source emissions and effluents. The Friday morning
session is devoted to an examination of the role of short-term
testing in hazards assessment, a topic of concern and interest to
all of us in the post-Love Canal era when we are confronted with
resolving the buried mistakes of the past.

In addition to an increase in the application and utilization
of short-term procedures, the past two years have also been marked
by the validation of many of the short-term bioassay procedures.
The validation of short-term testing was essential if short-term
testing was to find a meaningful role in environmental health
assessment. Considerable effort has been devoted to developing the
needed data, and although much work remains to be done, the initial
results are optimistic. EPA recently established the Gene-Tox
program to further the evaluation and validation of short-term
bioassays. This program, which is evaluating 27 different
short-term systems, will play an important role in identifying
aspects of short-term tests that require further development and
validation. Information from the Gene-Tox evaluations will be used
to direct future research programs.

In conclusion, the coupling of short-term bioassays with
state-of-the-art chemical analysis techniques is an exciting and
rapidly evolving field--one that offers the potential to resolve
questions of nontoxicity quickly and to provide a scientific basis
for properly allocating our resources among many studies of
potentially hazardous agents. If the past is an augury of the
future, the challenges that face this still young and rapidly
growing field will be met. I am enthusiastic about the recent
developments in this field and the developments that are
forthcoming. The importance of this research to the future of
environmental health assessment cannot be overstated.

Thank you.

SESSION 1

AMBIENT AIR

BIOASSAY OF PARTICULATE ORGANIC MATTER FROM AMBIENT AIR

Joellen Lewtas Huisingh
Health Effects Research Laboratory
U.S. Environmental Protection Agency
Research Triangle Park, North Carolina

INTRODUCTION

The influence of industrialization and consequent increased concentration of urban particulate matter on the incidence of cancer has long been a concern (Kotin and Falk, 1963; Carnow and Meier, 1973). The first bioassays used to evaluate complex ambient air samples were whole-animal carcinogenesis bioassays (Leiter et al., 1942; Hueper et al., 1962). In these studies, organic extracts of urban particulate matter were found to be carcinogenic in rodents. Such organic extracts have also been shown to transform rodent embryo cells in culture (Freeman et al., 1971; Gordon et al., 1973). Carcinogenic polycyclic aromatic hydrocarbons (PAH), such as benzo(a)pyrene, were detected in these extracts; however, these compounds did not account for all of the carcinogenic activity reported.

The development of the Ames _Salmonella_ _typhimurium_ mutagenesis bioassay (Ames et al., 1975) provided a simpler, more sensitive, and faster bioassay for potential carcinogenic activity that could be applied to air samples collected by conventional techniques. The initial applications of this bioassay to ambient air particulate organic matter (Tokiwa et al., 1976; Pitts et al., 1977; Talcott and Wei, 1977) stimulated research in the following areas:

1) improvement in sample collection, extraction, and bioassay methodology;

2) characterization and identification of potential classes of carcinogens and specific carcinogens present in ambient air particles; and

3) evaluation of emission sources and atmospheric conditions responsible for the observed mutagenicity in urban air particulate.

This overview addresses these areas of research and summarizes our current understanding of the mutagenicity of particulate organic matter found in ambient air.

ADVANCES IN SAMPLE COLLECTION AND EXTRACTION

The first reported studies on the mutagenicity of particulate organic matter from ambient air used high-volume samplers to collect air particles on glass fiber filters (Tokiwa et al., 1976; Pitts et al., 1977; Talcott and Wei, 1977). This sampler collects both respirable particles (< 5 µm) and non-respirable large particles. In these studies, the organics were extracted from the particles with either methanol (Tokiwa, 1976), acetone (Talcott and Wei, 1977), or a mixture of methanol, benzene, and dichloromethane (1:1:1)(Pitts et al., 1977). Since high-volume samplers are widely used in air-monitoring programs to determine total suspended particulate (TSP) levels, these studies provided comparative mutagenicity data for different sites.

Although high-volume samplers provide the simplest, and for many investigators the only, method available for collection of air particles, this method presents several serious disadvantages. The most consequential disadvantage is that respirable particles are collected simultaneously with larger particles. In cases where the smaller respirable particles are considerably more mutagenic than the larger particles, the larger particles dilute the overall mutagenicity of the sample, thereby biasing the analysis of the particle composition to which the human lung is exposed.

Other potential disadvantages are due to the large volume of air being drawn continuously over collected particles. Samples collected by this method may lose more volatile organics by evaporation. The organics present on the particles are also potentially subject to reactions with nitrogen dioxide (NO_2), ozone (O_3), or peroxyacetyl nitrate (PAN), which are all present in urban air. Pitts et al. (1978a, b) have shown that PAH (e.g., benzo(a)pyrene and perylene) directly coated onto glass fiber filters reacted with NO_2, O_3, and PAN, as well as ambient photochemical smog, to form several direct-acting mutagens (mutagens that do not require an exogenous microsomal activation system). Although these reactions have not been shown to occur to

PAH adsorbed on the surface of air particles, potential surface reactions during filtration still must be taken into consideration when interpreting studies using filtration for sample collection.

A recent modification of the standard high-volume sampler, the size-selective inlet (SSI) high-volume sampler, collects only particles < 15 μm (the definition of "inhalable particles"). This sampler excludes the larger particles, which are not normally inhaled.

A sampling device that does collect particles in separate size fractions is the cascade impactor (Andersen, 1966). These samplers use a series of plates with either holes or slots offset at each stage to collect separate size fractions ranging from < 2 μm to > 7 μm. The cascade impactor is generally attached to a high-volume sampler that collects the smallest particles by filtration. Teranashi et al. (1977), using an Andersen high-volume cascade impactor, found that the organics from particles < 1.1 μm, collected on the backup filter, were significantly more mutagenic than the organics from the larger particles. Pitts et al. (1978b) reported similar findings using a Sierra high-volume cascade impactor in downtown Los Angeles. This method, while providing a size-fractionated sample for bioassay, still employs filtration to collect the smallest particles, which contain most of the mutagenic components.

In order to collect larger quantities of size-fractionated air particulate matter for biological studies, the U.S. Environmental Protection Agency (EPA) had a Massive Air Volume Sampler (MAVS) designed and fabricated by Henry and Mitchell (1978) that does not require filtration. The MAVS employs two impactors that collect 3.5- to 20-μm and 1.7- to 3.5-μm particles, followed by an electrostatic precipitator (ESP) that collects the particles < 1.7 um. In our initial studies with the MAVS at a Los Angeles freeway site, we found the particles with mean diameters < 1.7 μm to be significantly more mutagenic than the larger particles (Figure 1). However, when the ESP was charged to maximize collection efficiency, as much as 0.05 ppm of O_3 was measured at the blower outlet of the sampler (Mitchell et al., 1978). Jungers et al. (1980) evaluated the effect of O_3 under the MAVS operating conditions on both the mutagenicity and chemical composition of the particulate organic matter collected in the ESP. These studies found that under these operating conditions, the O_3 did not significantly affect the mutagenicity or the PAH content of the organics.

A wide range of extraction methods have been used to remove the organics from air particles for bioassay. Solvents employed range from the nonpolar solvents cyclohexane (Møller and Alfheim, 1980) and benzene (Teranashi et al., 1977) to acetone (Talcott and Wei, 1977) and the more polar solvent methanol (Tokiwa et al.,

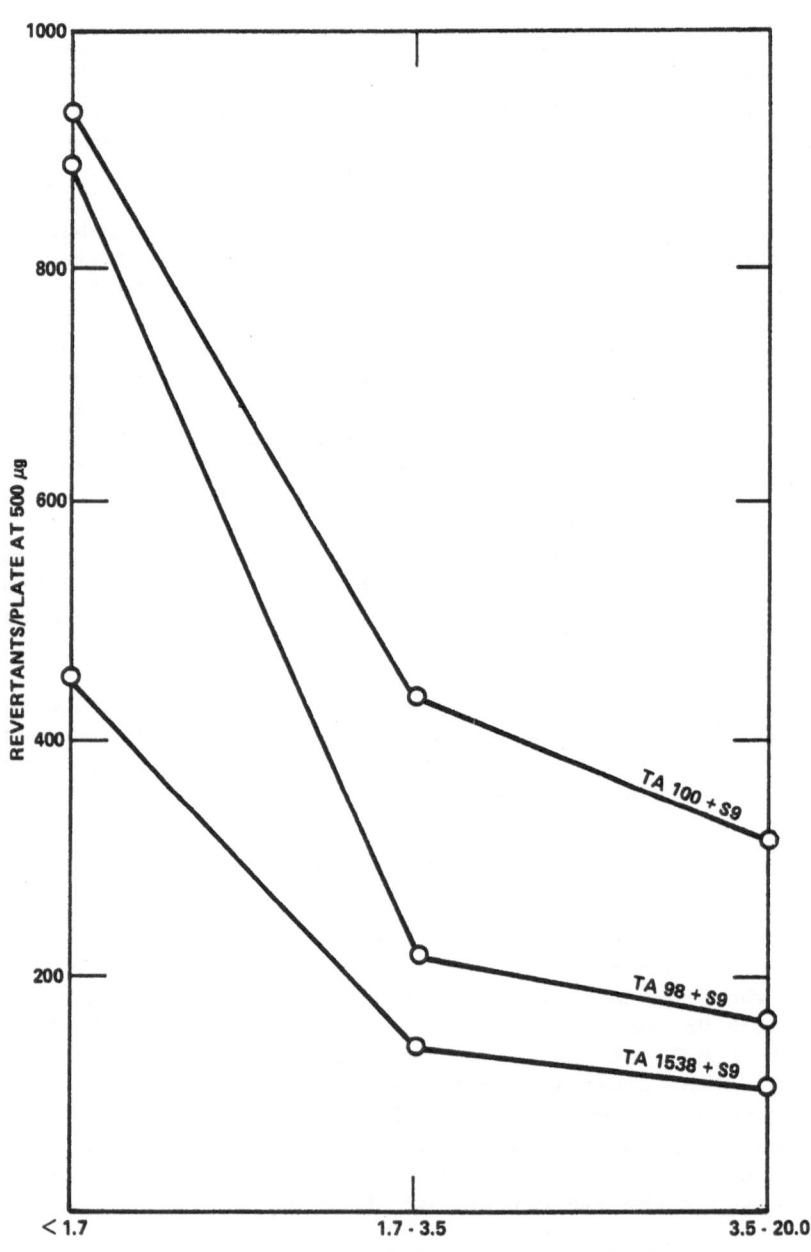

Figure 1. Mutagenicity of air particulate at the Los Angeles
         freeway site (upwind) as a function of particle size.

1976). Since different solvents will preferentially remove
different constituents from the particles, the method of extraction
can significantly alter the resulting composition and mutagenicity
of the organics. Jungers et al. (1980), after evaluating seven
solvent systems, found that dichloromethane extraction resulted in
the most mutagenic extractible organics with a minimum of inorganic
anions present.

CHARACTERIZATION AND IDENTIFICATION OF POTENTIAL CARCINOGENS

     The application of the Ames S. typhimurium plate-incorporation
assay (Ames et al., 1975) with multiple tester strains, used in the
presence and absence of a metabolic activation system, provides an
initial characterization of potential carcinogens present. The
organics from air particles generally show mutagenic activity only
in the tester strains susceptible to frameshift mutations (e.g.,
TA1538, TA1537, TA98, and TA100, but not TA1535). Mutagenic
activity is usually observed in the absence of a metabolic
activation system, indicating the presence of direct-acting
mutagens. When metabolic activation is added, certain air samples
show significant increases in mutagenicity (Talcott and Wei, 1977;
Pitts et al., 1977), while other samples show no increase in
activity, and even occasionally a decrease (Talcott and Wei, 1977).
Upon fractionation, however, air samples generally have been shown
to contain both direct-acting mutagens and mutagens that require
the addition of metabolic activation. Table 1 shows the response
of five S. typhimurium tester strains to a typical air sample, with
and without metabolic activation, using the < 1.7-μm particles
collected with a MAVS in Birmingham, AL.

Table 1. Specific Mutagenic Activity of Air Particulate Extract
          (< 1.7 μm)

Tester Strain	Revertants/100 μg Organics	
	Without S-9 Activation	With S-9 Activation
TA1538	81	121
TA1537	59	34
TA98	121	188
TA100	176	245
TA1535	–	–

Bioassay-directed chemical fractionation is an increasingly
powerful tool for the identification of potential carcinogens in
complex mixtures. This technique has been employed to characterize
and eventually identify potential carcinogens in cigarette smoke
(Swain et al., 1969), synthetic fuels (Epler, 1980), and diesel
emissions (Huisingh et al., 1978). Until the recent development
and improvement of air particle collection techniques, such studies
on particulate organic matter in air were severely limited by the
amount of sample collected. In spite of these difficulties,
Teranashi et al. (1977) fractionated particulate organic matter
from Kobe, Japan, into acidic, basic, aliphatic, aromatic, and
oxygenated fractions. The acidic, aromatic, and oxygenated
fractions accounted for most of the mutagenic activity of the total
sample. Studies by Kolber et al. (1980), who used a somewhat
different fractionation method, also showed significant mutagenic
activity in those three fractions.

EVALUATION OF EMISSION SOURCES AND ATMOSPHERIC CONDITIONS

Initial comparative studies (Tokiwa et al., 1976; Pitts
et al., 1977) showed that air particulate was more mutagenic at
industrial and urbanized sites than at rural sites. Recently,
Flessel et al. (1980) compared mutagenicity among sites in Contra
Costa County, CA, with differing amounts of industrialization and
cancer rates. These studies all indicate a higher mutagenic
activity in the more urbanized or industrialized sites.

At any one ambient sampling site, the mutagenicity appears to
vary significantly over time. A major parameter affecting airborne
mutagenicity, identified by Commoner et al. (1978), in a year-long
study at a Chicago school site, was the wind direction. In this
study, a plot of wind direction versus relative mutagenic activity
showed that wind directions of either northwest or east resulted
in air particle samples with the greatest mutagenicity. Møller
and Alfheim (1980) reported on the mutagenicity of airborne
particles from two locations in Oslo over a three-month period.
They observed higher mutagenicity in February (i.e., during the
heating season) than in March and April. They also reported
significant meteorological effects. The mutagenicity, when
calculated as revertants per cubic meter of air, was highest on
cold clear days with little wind. When revertants per milligram of
particulate matter was calculated and compared with meteorological
conditions, mutagenicity was found to be high on days with rain or
snow, when the total concentration of particles in the air was low.
Although studies have been conducted to examine the effect of
ultraviolet light (Gibson et al., 1978) and other atmospheric gases
and oxidants including O_3, NO_2, and PAN (Pitts et al., 1978a, b) on
PAH, the role of these in the mutagenicity of particulate organic
matter in ambient air is still uncertain.

In certain cases, studies can be designed to identify specific emission sources that contribute to the mutagenicity of the ambient particulate organic matter. At the Los Angeles freeway site (discussed above), samplers were situated both upwind and downwind from the freeway. Figure 2 shows the comparative mutagenic activity of organics from particles < 1.7 μm collected over the same period. The particles collected downwind from the automobiles and trucks on the freeway were significantly more mutagenic than those collected upwind. Claxton and Huisingh (1980) have shown that the organics from a gasoline catalyst automobile are significantly mutagenic. It is clear from these studies and others (Huisingh et al., 1978; Löfroth, 1980; Alfheim and Møller, 1980) that both gasoline and diesel engine exhaust from automobiles, buses, and trucks contribute to the mutagenicity of ambient air particles.

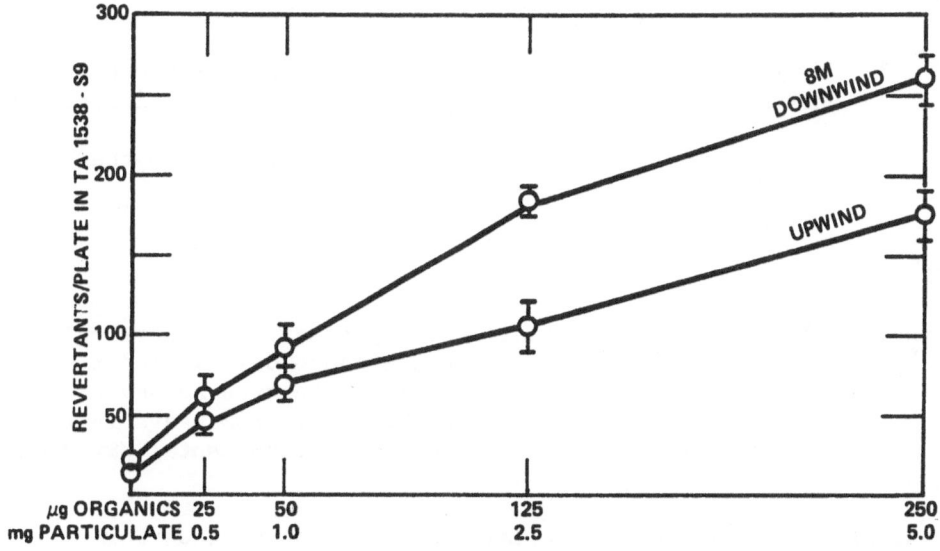

Figure 2. Mutagenicity of air particulate (< 1.7 μm) collected by MAVS upwind and downwind of the Los Angeles freeway.

It is clear that a variety of combustion sources could contribute organic mutagens to the ambient air. Claxton and Huisingh (1980) compared the mutagenic activity of organics from particles emitted from residential heaters as well as diesel and

gasoline vehicles. A study by Møller and Alfheim (1980) of two
locations in Oslo, Norway, suggested that the mutagenicity they
observed was due in part to both automotive traffic and residential
heaters, with residential heating probably contributing more in the
winter months. Industrial sources of mutagens that may be
significant are coke oven emissions (Claxton, 1980) and coal
combustion emissions (Chrisp et al., 1978).

SUMMARY

The mutagens present in ambient air particulate possess the
following characteristics:

1) They show both direct-acting and indirect-acting mutagenic
   activity. The proportions of these two classes of
   activity vary with the sample location.

2) They show mutagenic activity primarily in the tester
   strains that respond to frameshift mutagens.

3) They appear to be present in higher concentrations in the
   smallest particles (< 2 μm) than in larger particles.

4) They appear to result from specific emission sources (such
   as combustion sources).

REFERENCES

Alfheim, I., and M. Møller. Mutagenicity of airborne particulate
    matter in relation to traffic and meteorological conditions.
    Presented at the U.S. Environmental Protection Agency Second
    Symposium on the Application of Short-term Bioassays in the
    Fractionation and Analysis of Complex Environmental Mixtures,
    Williamsburg, VA.

Ames, B.N., J. McCann, and E. Yamasaki. 1975. Methods for
    detecting carcinogens and mutagens with the Salmonella/
    mammalian microsome mutagenicity test. Mutation Res.
    31:347-364.

Andersen, A.A. 1966. A sampler for respiratory health hazard
    assessment. Am. Ind. Hyg. Assoc. J. 27:160-165.

Carnow, B.W., and P. Meier. 1973. Air pollution and pulmonary
    cancer. Arch. Environ. Hlth. 27:207.

Chrisp, C.E., G.L. Fisher, and J.E. Lambert. 1978. Mutagenicity
     of filtrates from respirable coal fly ash. Science 199:73-75.

Claxton, L. 1980. Mutagenic and Carcinogenic Potency of
     Extracts of Diesel and Related Environmental Emissions:
     Salmonella typhimurium Assay. In: Proceedings of the
     International Symposium on Health Effects of Diesel Engine
     Emissions, December 1979. EPA 600/9-80-057ab.
     U.S. Environmental Protection Agency: Cincinnati, OH.

Claxton, L., and J.L. Huisingh. 1980. Comparative mutagenic
     activity of organics from combustion sources. In: Pulmonary
     Toxicology of Respirable Particles. C.L. Sanders, F.T. Cross,
     G.E. Dagle, and J.A. Mahaffey, eds. CONF-791002. U.S.
     Department of Energy: Washington, DC. pp. 453-465.

Commoner, B., P. Madyastha, A. Bronson, and A.J. Vithayathil.
     1978. Environmental mutagens in urban air particulates. J.
     Toxicol. Environ. Hlth. 4:59-77.

Epler, J.L. 1980. The use of short-term tests in the isolation
     and identification of chemical mutagens in complex mixtures.
     In: Chemical Mutagens: Principles and Methods for Their
     Detection, Vol. 6. F.J. de Serres, ed. Plenum Press: New
     York. pp. 239-270.

Flessel, C.P., J.J. Wesolowski, S. Twiss, J. Cheng, J. Ondo, N.
     Monto, and R. Chan. 1980. Integration of the Ames bioassay
     and chemical analyses in an epidemiological cancer incidence
     study. Presented at the U.S. Environmental Protection Agency
     Second Symposium on the Application of Short-term Bioassays in
     the Fractionation and Analysis of Complex Environmental
     Mixtures, Williamsburg, VA.

Freeman, A.E., P.J. Price, R.J. Bryan, R.J. Gordon, R.V. Gilden,
     G.J. Kelloff, and R.J. Hueber. 1971. Transformation of rat
     and hamster embryo cells by extracts of city smog. Proc. Nat.
     Acad. Sci. USA 68:445-449.

Gibson, T.L., V.B. Smart, L.L. Smith. 1978. Non-enzymatic
     activation of polycyclic aromatic hydrocarbons as mutagens.
     Mutation Res. 49:153-162.

Gordon, R.J., R.J. Bryan, J.S. Rhim, C. Demoise, R.G. Wolford, A.E.
     Freeman, and R.J. Huebner. 1973. Transformation of rat and
     mouse embryo cells by a new class of carcinogenic compounds
     isolated from city air. Int. J. Cancer 12:223-227.

Henry, W.M., and R.I. Mitchell. 1978. Development of a Large
     Sample Collector of Respirable Particulate Matter.
     EPA-600/4-78-009. U.S. Environmental Protection Agency:
     Research Triangle Park, NC.

Hueper, W.C., P. Kotin, E.C. Tabor, W.W. Payne, H. Falk, and E.
     Sawicki. 1962. Carcinogenic bioassays on air pollutants.
     Arch. Pathol. 74:89-116.

Huisingh, J., R. Bradow, R. Jungers, L. Claxton, R. Zweidinger,
     S. Tejada, J. Bumgarner, F. Duffield, V.F. Simmon, C. Hare,
     C. Rodriguez, L. Snow, and M. Waters. 1978. Application of
     bioassay to the characterization of diesel particle emissions.
     Part I. Characterization of Heavy Duty Diesel Particle
     Emissions. Part II. Application of a mutagenicity bioassay
     to monitoring light duty diesel particle emissions.
     Application of Short-term Bioassays in the Fractionation and
     Analysis of Complex Environmental Mixtures. M.D. Waters,
     S. Nesnow, J.L. Huisingh, S.S. Sandhu, and L. Claxton, eds.
     Plenum Press: New York. pp. 381-418.

Jungers, R., R. Burton, L. Claxton, and J. Lewtas Huisingh. 1980.
     Evaluation of collection and extraction methods for
     mutagenesis studies on ambient air particulate. Presented at
     the U.S. Environmental Protection Agency Second Symposium on
     the Application of Short-term Bioassays for the Fractionation
     and Analysis of Complex Environmental Mixtures, Williamsburg,
     VA.

Kolber, A., T. Wolff, T. Hughes, E. Pellizzari, C. Sparacino, M.
     Waters, J. Lewtas Huisingh, and L. Claxton. 1980. Collection
     chemical fractionation, and mutagenicity bioassay of ambient
     air particulate. Presented at the U.S. Environmental
     Protection Agency Second Symposium on the Application of
     Short-term Bioassays in the Fractionation and Analysis of
     Complex Environmental Mixtures, Williamsburg, VA.

Kotin, P., and H.L. Falk. 1963. Atmospheric factors in
     pathogenesis of lung cancer. Adv. Cancer Res. 7:475-514.

Leiter, J., M.B. Shimkin, and M.J. Shear. 1942. Production of
     subcutaneous sarcomas in mice with tars extracted from
     atmospheric dusts. J. Natl. Cancer Inst. 3:155-165.

Löfroth, G. 1980. Comparison of the mutagenic activity in carbon
     particulate matter and in diesel and gasoline exhaust.
     Presented at the U.S. Environmental Protection Agency Second
     Symposium on the Application of Short-term Bioassays for the
     Fractionation and Analysis of Complex Environmental Mixtures,
     Williamsburg, VA.

Mitchell, R.I., W.M. Henry, and N.C. Henderson. 1978.
    Fabrication, Optimization, and Evaluation of a Massive Air
    Volume Sampler of Sized Respirable Particulate Matter.
    EPA-600/4-78-031. U.S. Environmental Protection Agency:
    Research Triangle Park, NC.

Møller, M., and I. Alfheim. 1980. Mutagenicity and
    PAH-analysis of airborne particulate matter. Atmos. Environ.
    14:83-88.

Pitts, J.N., D. Grosjean, J.M. Mischke, V.F. Simmon, and D. Poole.
    1977. Mutagenic activity of airborne particulate organic
    pollutants. Toxicol. Lett. 1:65-70.

Pitts, J.N., K.A. Van Cauwenberghe, D. Grosjean, J.P. Schmid, D.R.
    Fitz, W.L. Belser, G.B. Knudson, and P.M. Hynds. 1978a.
    Atmospheric reactions of polycylic aromatic hydrocarbons:
    Facile formation of mutagenic nitro derivatives. Science
    202:515-519.

Pitts, J.N., K.A. Van Cauwenberghe, D. Grosjean, J.P. Schmid, D.R.
    Fitz, W.L. Belser, G.B. Knudson, and P.M. Hynds. 1978b.
    Chemical and microbiological studies of mutagenic pollutants
    in real and simulated atmospheres. In: Application of
    Short-term Bioassays in the Fractionation and Analysis of
    Complex Environmental Mixtures. M.D. Waters, S. Nesnow, J.L.
    Huisingh, S.S. Sandhu, and L. Claxton, eds. Plenum Press:
    New York. pp. 353-379.

Swain, A.P., J.E. Cooper, and R.L. Stedman. 1969. Large-scale
    fractionation of cigarette smoke condensate for chemical and
    biological investigations. Cancer Res. 29:579-583.

Talcott, R., and E. Wei. 1977. Airborne mutagens bioassayed in
    Salmonella typhimurium. J. Natl. Cancer Inst. 58:449-451.

Teranashi, K., K. Hamada, N. Tekeda, and H. Watanabe. 1977.
    Mutagenicity of the tar in air pollutants. Proc. 4th Int.
    Clean Air Congress, Tokyo. pp. 33-36.

Tokiwa, H., H. Takeyoshi, K. Morita, K. Takahashi, N. Soruta, and
    Y. Ohnishi. 1976. Detection of mutagenic activity in urban
    air pollutants. Mutation Res. 38:351-359.

Waters, M.D. 1980. An overview of the use of short-term
    bioassays in evaluation of atmospheric pollutants. In:
    Sampling and Analysis of Toxic Organics in the Atmosphere.
    ASTM STP 721. American Society for Testing and Materials:
    Philadelphia. pp. 156-165.

COLLECTION, CHEMICAL FRACTIONATION, AND MUTAGENICITY BIOASSAY OF AMBIENT AIR PARTICULATE

Alan Kolber, Thomas Wolff, Thomas Hughes, Edo Pellizzari,
and Charles Sparacino
Research Triangle Institute
Research Triangle Park, North Carolina

Michael Waters, Joellen Lewtas Huisingh, and Larry Claxton
Health Effects Research Laboratory
U.S. Environmental Protection Agency
Research Triangle Park, North Carolina

INTRODUCTION

Our industrial society has created thousands of synthetic
xenobiotics to support our modern lifestyle. Many of these
substances, and their by-products, enter our atmosphere in the form
of vapor-phase and particulate pollutants that can be ingested by
respiration and skin contact. The insults to human health and the
ecosystem from these airborne organic pollutants are due mainly to
polycyclic organic matter, specific industrial emissions such as
halogenated hydrocarbons, and end-use chemicals such as pesticides
(Fishbein, 1976). Polycyclic organics associated with air
particulates are believed to result primarily from incomplete
combustion of organic matter (Møller and Alfheim, 1980). These
carcinogenic components, (benzo(a)pyrene (B[a]P), benz(a)
anthracene) and other polycyclics have been well characterized
chemically (Shubik and Hartwell, 1969). As population and
industrial activity increase, the growing health hazards associated
with ambient air pollution must be further assessed and evaluated.

Air pollutants are chemically complex environmental mixtures,
whose compositions vary geographically with local industrial
activity and weather conditions. Seasonal variations in chemical
composition and mutagenicity of organic extracts of air particulate
have been documented in Oslo, Stockholm, and New York City (Møller
and Alfheim, 1980; Löfroth, 1980; Daisey et al., 1979). All three
communities exhibited higher mutagenicity during the winter months,
and in each case, a major contributing factor was polynuclear
aromatic hydrocarbons (PAH). These indirect-acting mutagens, which
require metabolic activation, were produced from combustion of
heating oil.

21

Ambient-air particulate matter has been estimated at
56×10^{13} g/yr, total global load, including natural pollutants
such as dusts, agricultural particulate, and others. The man-made
(anthropogenic) global load is estimated to be 28×10^{13} g/yr.
Vapor-phase pollutant organics are believed to represent 60 to 90%
of the total organic load; the remainder is organics adsorbed onto
particulate matter (Hidy and Brock, 1970; Duce, 1978).

Chemical monitoring and analysis of air pollutants have been
conducted for many years, but only within the last decade has
biotesting technology improved enough to permit simultaneous
chemical and biological characterization. However, chemical-
analytical expertise exceeds the present biological capabilities,
both qualitatively and quantitatively (Hughes et al., 1980).

In this study, air particulate was collected, size-
fractionated, solvent-extracted and fractionated into chemical
classes, which were then characterized by GC/MS/computer analysis.
These chemical class fractions were then biotested for mutagenic
activity using the Ames/Salmonella bacterial mutagenicity assay.
Vapor-phase organic pollutants were also collected and tested.
However, no quantitative method was available to adequately measure
the mutagenicity of the vapor-phase components; consequently, we
are presently developing a bioassay protocol capable of quantifying
the mutagenicity of vapor-phase substances. It should be noted
that mutagenicity or carcinogenicity is by no means the only
significant health hazard that could be presented by air
pollutants. As additional in vitro bioassay capabilities are
developed, other potential toxicity parameters will be examined,
such as neurotoxicity and lung toxicity.

METHODS

Collection of Air Particulate and Vapors

Ambient air particulate was collected by the Maxisampler, a
high-volume sampling device constructed by the Battelle Corporation
after the design by Henry and Mitchel (1978). This device can
sample 20,000 m of air in a 24-h period and can collect particulate
matter within the respirable range in three size fractions (< 1.7
μm, 1.7 to 3.5 μm, and > 3.5 μm), using impactor plates and
electrostatic precipitation. A mechanical diagram of the sampler
is shown in Figure 1. After collection, the plates were sealed in
a transportable container; and the particulate material was removed
and characterized morphologically by scanning electron microscopy.
Figure 2 is a typical scanning electron micrograph of ambient air
particulate. Air particulate was sampled at five U.S. locations:
Elizabeth, N.J.; Upland, CA; Lake Charles, LA; Houston and

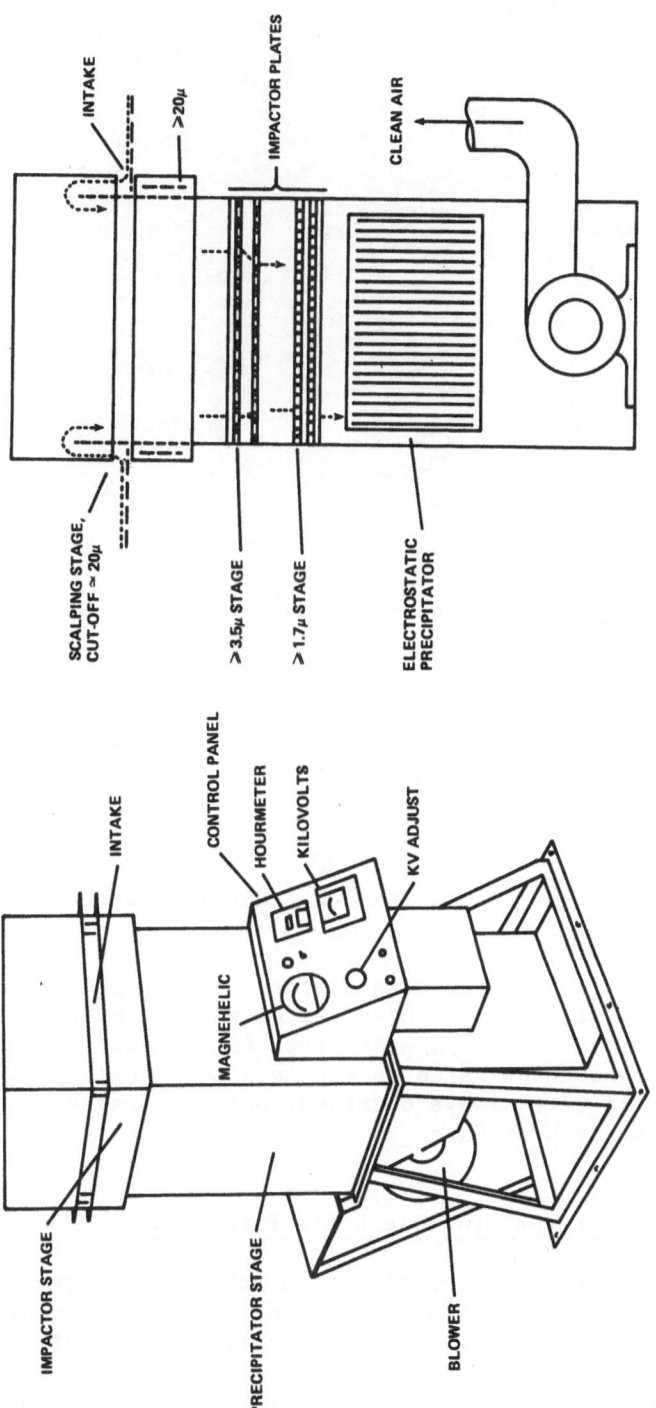

Figure 1. Schematic of the Battelle massive air volume sampler.

Figure 2.   Scanning electron micrograph of ambient air particulate
            from Lake Charles, LA (> 3.5 μm).

Beaumont, TX. Table 1 summarizes the sampling variables for
Elizabeth, N.J.

     Vapor-phase substances were sampled with cartridges containing
Tenax polymeric sorbent. Each cartridge was sealed and later
thermally desorbed into cryogenic traps (see Figure 3). This
frozen sample was then tested in the Ames/Salmonella assay.
Negative results were always obtained, but this may be due to an
inadequate testing procedure.

Chemical Fractionation of Ambient Air Particulate

     An initial chemical fractionation scheme developed by
Pellizzari et al. (1978) that generated 13 polar and nonpolar
chemical classes was utilized to chemically fractionate the size-
fractionated air particulate. However, the chemical differences
among the fractions were small, and dividing the crude particulate
into 13 fractions resulted in sample sizes of less than 5 mg, which

Table 1. Sampling Variables for Particulate Organics in Ambient Air in Elizabeth, NJ

Sample Type	Date	Temp. °C (°F)	Sample Vol. (1)	Duration of Sampling (min)	Time	Remarks	
						Relative Humidity	Wind Dir/Vel. kts (m/s)
Vapor-phase organics	9/19-20/78	17 (62)	646	1,390	1515-1425	90%	NE/5-10 kts (2.5-5)
	9/20-21/78	23 (74)	1,211	1,455	1445-1500	91%	SW/10 kts (5)
	9/21-22/78	20 (68)	1,044	1,265	1515-1220	90%	NE/10 kts (5)
	9/22-23/78	20 (68)	1,093	1,335	1245-1100	88%	NE/ 5 kts (2.5)
	9/23-24/78	16 (60)	1,279	1,690	1120-1530	75%	SE/10 kts (5)
Particulates	9/19-22/78		81,211,000	4,560	0805-1200		
	9/22-25/78		78,892,000	4,430	1210-1400		
	9/25-28/78		70,246,000	4,060	1405-0745		

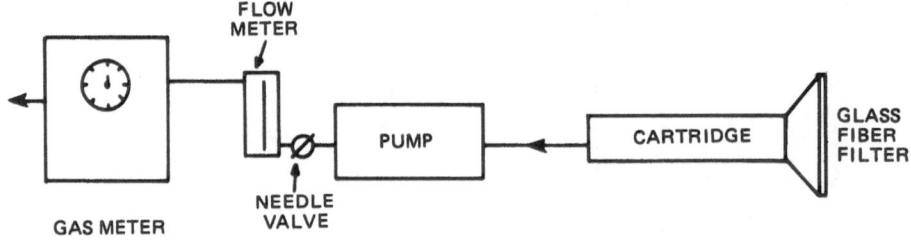

VAPOR COLLECTION SYSTEM

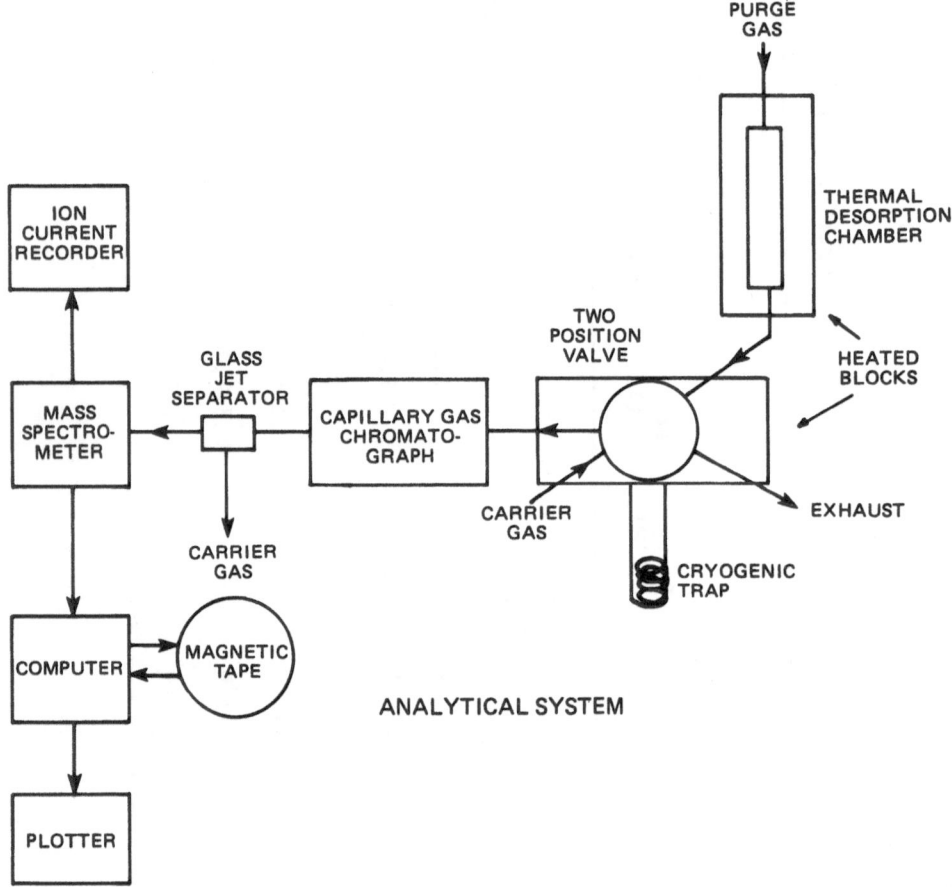

ANALYTICAL SYSTEM

Figure 3. Vapor collection and analytical systems for analysis of organic vapors in ambient air.

limited adequate bioassay and chemical identification. The scheme
was modified to generate six chemical classes: acids, bases, PAHs,
polar neutrals, nonpolar neutrals, and insolubles (see Figure 4).

        Air particulate (~1.0 g) was subjected to ultrasonic treatment
in 100 ml cyclohexane (Burdick-Jackson) for 30 min and then
filtered through a Teflon (DuPont) filter (0.5-μm pore size). The
filtrate was evaporated to dryness using a rotary evaporator; the
solids retained by the filter were dissolved in 100 ml methanol
(Burdick-Jackson), and sonicated and filtered through the Teflon
filter; and the filtrate was taken to dryness. The filtrates were
combined, dried, weighed, and redissolved in methylene chloride
(CH_2Cl_2). The solution was spiked with internal standards to allow
for quantification and to serve as a quality control parameter for
the overall partition/analysis scheme. The standards used were
quinoline-d7 (organic base), phenol-d5 (organic acid), and
anthracene-d10 (PAH).

        This CH_2Cl_2 solution was extracted twice with equal volumes
each of 10% sulfuric acid (H_2SO_4) and then once with 20% H_2SO_4.
The aqueous phases were combined and washed with CH_2Cl_2, and this
CH_2Cl_2 phase was combined with the original CH_2Cl_2 solution. The
aqueous phase was cooled (ice bath), adjusted with 25% sodium
hydroxide (NaOH) to pH 10, and extracted three times with CH_2Cl_2 to
generate the "organic bases." The aqueous phase was discarded.
The original CH_2Cl_2 solution was extracted three times with 5%
NaOH, and the aqueous phases were combined and washed with CH_2Cl_2.
The CH_2Cl_2 solution then was combined with the original CH_2Cl_2
solution. The NaOH phases were placed in an ice bath and acidified
to pH 3 with 20% H_2SO_4 and extracted three times with CH_2Cl_2 to
generate "organic acids." The remaining aqueous phase was
discarded.

        The original CH_2Cl_2 phase was evaporated to dryness,
reconstituted in cyclohexane, and filtered through a Teflon filter
(0.5 μm). The cyclohexane filtrate was extracted three times with
an equal volume of methanol:water (4:1) solution; the methanol:
water extract was then concentrated, and the water extracted three
times with ethyl acetate. The solvent was removed to generate
"polar neutrals." The cyclohexane phase was extracted three times
with an equal volume of nitromethane; the nitromethane phases were
then combined and evaporated to dryness to generate the PAHs.
The cyclohexane phase was evaporated to dryness to generate the
"nonpolar neutrals" fraction.

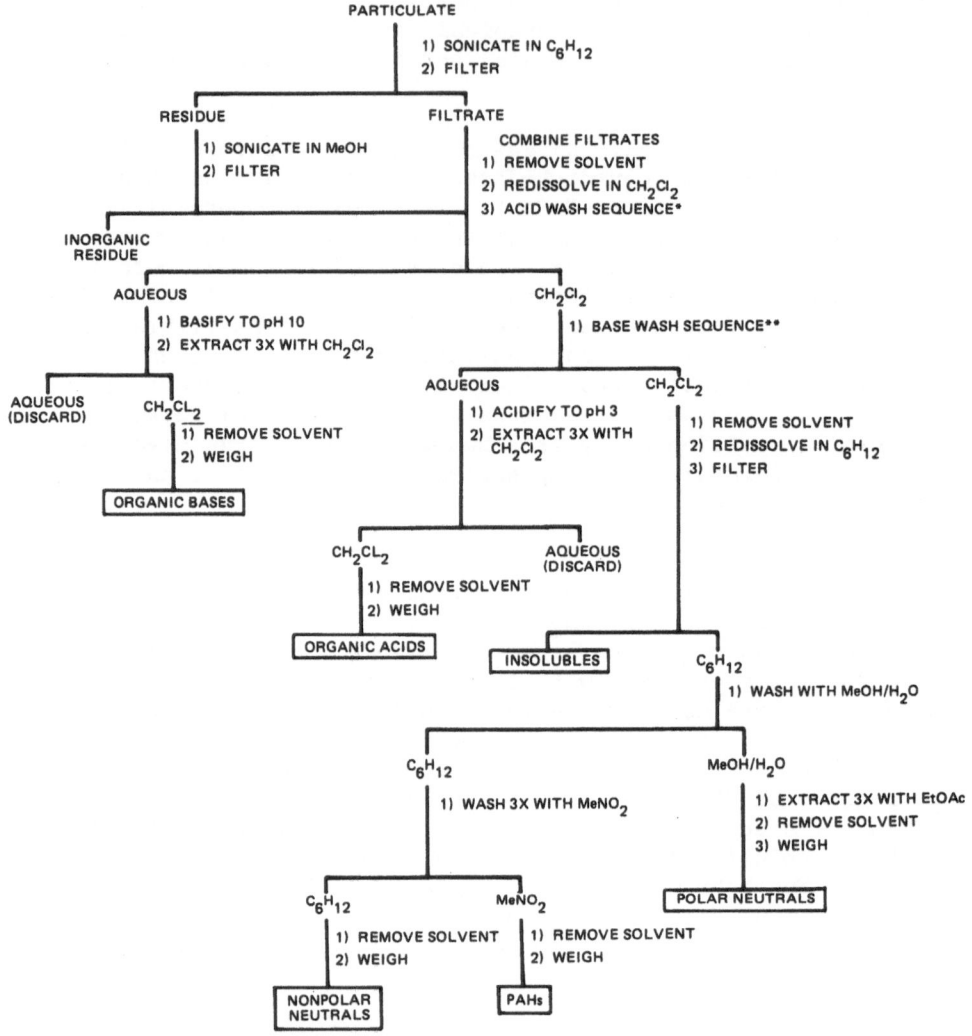

*Acid Wash Sequence: 2 X with 10% H_2SO_4, 1 X with 20% H_2SO_4
**Base Wash Sequence: 3 X with 1 N NaOH.

Figure 4.   Fractionation scheme for ambient air particulate
            organics.

This procedure is a modification of a procedure described by
Lee et al. (1976). The partition scheme has been validated for
efficiency and presence of spillover of chemicals from one class to
another, using mixtures of deuterated known substances
representative of the individual chemical classes. Known mutagens

were included in the mixture, and the recovery of mutagenic activity before and after fractionation was monitored by the Ames/Salmonella bacterial mutagenicity bioassay.

Mutagenicity Bioassay

The plate-incorporation Ames/Salmonella mutagenicity assay was performed as described by Ames et al. (1975), with and without S-9 rat liver microsomal activation preparation. All assays were conducted with standard mutagens (positive controls) and solvent (negative) controls (samples were exchanged into dimethylsulfoxide [DMSO]). For revertant selection, minimal Vogel-Bonner medium E, supplemented with 1.6% Difco bacto agar and 2% dextrose, was used for base agar layers (25 ml, poured automatically). A 2.5 ml soft agar overlay containing bacteria, minimum amounts of histidine and biotin, and sample was routinely used.

To prepare S-9 from rat liver, the following procedure was used: Male Charles River rats, strain CD_1 (200 ± 20 g), served as the source of liver material. The animals were housed in suspended cages for one week before induction. Food and water were given ad libitum. Induced rat liver (from at least three rats) was obtained from rats injected intraperitoneally on day one with 500 mg/kg of Aroclor 1254 in Mazola Corn Oil (0.5 ml of a 200 mg/ml for a 200 g rat). On day five, the animals were sacrificed and their livers removed. These were immersed in cold, sterile 1.15% KCl, washed two times, and blotted dry with sterile paper. The livers were weighed, minced, brought to 200 mg/ml with 0.25 $\underline{M}$ sucrose, and homogenized (on ice) with four strokes of a cold Potter-Elvenhjem apparatus with a Teflon pestle. The homogenate was centrifuged for 10 min at 9,000 x g, the lipoprotein layer aspirated off, and the protein concentration of the liver supernatant adjusted (after protein assay) to 30 mg protein/ml with 0.25 $\underline{M}$ sucrose stock solution. The stock solution was checked for sterility, quick-frozen in small aliquots (2 to 5 ml), and stored for a maximum of two months at -80°C. For the assay, an aliquot of stock solution was slow-thawed and adjusted to an appropriate concentration of protein in 0.25 $\underline{M}$ sucrose. NADPH was added at 320 µg/plate. Each batch was checked against known positive controls using the plate-incorporation assay. The optimal S-9 protein concentration for mutagenicity was determined by testing with B(a)P and 7-12 dimethylbenzanthracene (DMBA).

Agar diffusion well technique. A modification of the bacterial mutagenicity test was developed to screen for chemical fractions of air particulate organics when sample size was limited. In this test, the sample(s) and S-9 microsomal preparation were placed in wells cut in the agar. Soft agar then was added to fill the well, and after solidification, the bacterial overlay

in 2.5 ml soft agar was poured over the entire surface. Results
were interpreted as follows: a direct acting mutagen was observed
as a ring of colonies around the sample well; a promutagen was
observed as a line of colonies between the sample and S-9 wells,
and toxicity resulted in a clear zone around the sample well.

This qualitative bacterial mutagenicity test was verified with
known mutagens, and its sensitivity was shown to be similar to that
of the spot test. Optimum size, number, configuration, and spacing
of wells, as well as amount of S-9 rat liver microsomal preparation
required per well, were determined. A configuration suited to
general screening included a center S-9 well and four surrounding
sample and control wells, as illustrated in Figure 5. Advantages
of this test over the spot test included ability to 1) measure
multiple parameters (dosage, activation requirements, toxicity, and
mutagenicity) on a single plate, thereby greatly extending the
amount of information obtainable from the available sample; 2)
determine positive, negative, and solvent controls on the same
plate as the test compound; 3) conduct dose-response and multiple-
compound testing on a single plate; 4) eliminate adsorption of
test compound by the filter disc; 5) eliminate runoff of liquids
onto the agar; and 6) reduce potential loss of mutagenic activity
by chemical hydrolysis.

A priority scheme for biological testing was also developed,
and was based on total size of each chemical-class fraction:

1) For > 10 mg of sample, the plate-incorporation method was
    used, with five nontoxic dose levels (at least in
    duplicate) in the following order: TA98, TA100, TA1535,
    TA1537, and TA1538. Toxicity was tested at the highest
    dose only. Preferred initial dose ranges were 1000, 500,
    250, 100, and 10 µg/plate. The assay was performed first
    with, and then without, Aroclor-induced S-9 (3.0 mg/
    plate).

2) For < 10 mg of sample, the agar well diffusion test was
    used at two concentrations (500 and 100 µg/plate), except
    for PNA fractions which were tested using poured plates
    (500 µg/plate, with Aroclor-induced S-9 and TA98.
    Duplicate plates were assessed in the following order:
    TA98, TA1535, TA1537, TA1538, and TA100. Sample was
    diluted to one concentration for all tests; to obtain
    lower test concentrations, a smaller sample volume per
    well was used. Toxicity was tested at 500 µg/plate with
    TA98 and TA1535, using 1000 colonies/plate.

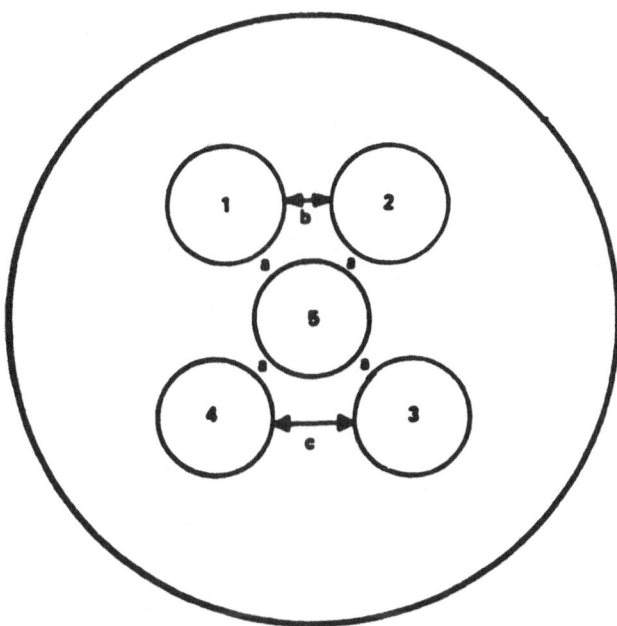

Figure 5.   Well testing configuration for bacterial mutagenesis
            screening. The wells contain: 1) positive control
            promutagen (e.g., 100 µg of 2-anthramine in 100 µl
            solvent); 2) DMSO, solvent control (100 µl); 3) low
            concentration of sample (100 µg sample in 20 µl DMSO);
            4) high concentration of sample (500 µg sample in 100
            µl DMSO); 5) Aroclor-induced S-9 microsomal preparation
            ([3.0 mg protein] + NADPH [480 µg in 300 µl]). The
            distances are: a) 2.5 mm; b) 5.0 mm; c) 10.0 mm.
            The wells are 18.0 mm in diameter.

RESULTS

Collection, Extraction, and Fractionation

     Sampling of 81×10^6 l ambient air required 76 h and generated
1.347 g of particulate, of which 180.1 mg were extractable
organics, or 2.3 ng organic/l ambient air. Table 2 shows the total
particulates collected for each size fraction from each sampling
trip. Extractable organics varied from 2.5 to 15.3% of the
particulate mass, and the smaller size fractions did not
consistently contain the greatest amounts of organics (the 53%
extractable organics for Beaumont, TX [< 1.7 µm] is probably
incorrect, due to a weighing error). Table 3 illustrates the
distribution of the extracted organic after chemical fractionation;
12 to 15% loss of organic matter was common during fractionation.

Table 2. Total Particulate Samples Collected

Sample Location	Particulate Size Range (µm)	Particulate Weight (mg)	Total Organics Extracted (mg)	% Extracted
Upland, CA	<1.7	666	33	5.0
	1.7 to 3.5	641	36	5.6
	>3.5	286	18	6.3
Houston, TX #1[a]	<1.7	337	19	5.6
	>1.7	790	86	10.9
Lake Charles, LA #1	<1.7	747	25	3.3
	>3.5	113	12	10.6
Elizabeth, NJ	<1.7	1347	180	13.4
	1.7 to 3.5	1302	120	9.2
	>3.5	1435	99	6.9
Lake Charles, LA #2	<1.7	2618	401	15.3
	1.7 to 3.5	488	15	3.1
	>3.5	548	27	4.9
Beaumont, TX	<1.7	285	151	53.0
	1.7 to 3.5	882	31	3.5
	>3.5	1701	42	2.5
Houston, TX #2	<1.7	3043	110	3.6
	1.7 to 3.5	1079	–	–
	>3.5	1353	–	–

[a]Particles > 1.7 µm were combined to generate sufficient sample for fractionation and biotesting.

Table 3. Fractionation of Organics Extracted from Ambient Air Particulate Samples

Sample Location	Size Range (μg)	Total Organics Extracted (mg)	Amount of Organics in Each Fraction (mg)					% Recovery
			Acid	Base	Nonpolar Neutrals	PAHs	Polar Neutrals	
Upland, CA	<1.7	32.7	2.9	1.0	6.5	1.6	15.8	85.0
	1.7 to 3.5	17.5	1.8	0.6	4.7	2.0	6.3	88.0
	>3.5	35.9	1.1	0.7	9.4	2.5	16.0	82.7
Houston, TX #1	<1.7	19.3	1.0	0.6	4.0	3.2	9.9	96.9
	>1.7	86.4	0.8	1.2	3.2	2.4	33.2	47.2
Lake Charles, LA #1	<1.7	24.6	7.2	0.1	6.1	2.1	3.7	78.0
	>1.7	11.9	1.7	0.1	3.9	2.3	2.4	87.4
Elizabeth, NJ	<1.7	180.1	8.5	0.5	86.2	11.1	34.6	78.2
	1.7 to 3.5	119.7	10.4	0.4	47.6	9.6	26.1	78.6
	>3.5	99.4	5.5	0.2	52.6	6.0	13.4	78.2
Lake Charles, LA #2	<1.7	401.2	170.2	16.3	49.2	7.1	152.6	98.6
	1.7 to 3.5	15.2	3.9	1.9	4.9	1.5	1.3	88.8
	>3.5	26.8	16.2	3.1	5.3	1.0	0.8	98.5
Beaumont, TX	<1.7	150.6	15.5	6.3	50.3	8.5	6.7	58.0
	1.7 to 3.5	30.7	6.8	0.6	12.2	2.2	8.9	100.0
	>3.5	42.1	10.4	0.4	21.3	4.8	3.8	96.7
Houston, TX #2	<1.7	110.2	20.4	2.8	20.9	12.2	38.0	85.6
	1.7 to 3.5	ND[a]	ND	ND	ND	ND	ND	ND
	>3.5	ND	ND	ND	ND	ND	ND	ND

[a]ND = not determined.

Mutagenicity

Significant mutagenicity was observed for various chemical
fractions of the extracted particulate organics from some sites. A
sample was considered mutagenic when it generated at least two
times the number of revertants over spontaneous background for that
bacterial strain. The PAH fraction was mutagenic at all five
sites. The polar neutral and organic acids were mutagenic at four
of the five sites, and the organic bases at three of the sites.
The mutagenicity data on fractions available in quantities too low
for the plate incorporation data were determined using the well
test. Both indirect- and direct-acting mutagens were present
in the complex mixtures and chemical fractions. An example of the
mutagenicity data obtained during this study is given in Table 4.

Chemical Analysis

Chemical fractions from each of the organic extracts of the
five geographical sites were analyzed by gas chromatography–mass
spectrometry (GC/MS). Compounds identified from the analysis were
matched against known toxic chemicals from the U.S. Health,
Education, and Welfare Chemical Registry and other sources. Table
5 lists known mutagenic, carcinogenic, neoplastic, and teratogenic
compounds identified. Many of these toxic compounds were found in
the extracted organics from all sites for which chemical analysis
was performed. Toxic PAH compounds were best represented.

DISCUSSION

This study was designed to develop and validate an integrated
multidisciplinary approach to the study of genotoxic effects of
ambient air pollutants. Engineering principles were employed to
develop and test the massive air volume sampler used to collect and
size-fractionate ambient air particulate. Analytical chemists
developed and validated the cyclohexane–methanol extraction scheme
for particulate organic matter and the acid–base extractive
fractionation scheme to separate the crude organic extract into
substituent chemical classes, which were then submitted to the
biologist for mutagenesis testing. Such multidisciplinary
approaches to the identification of bioactive chemical substituents
of various complex environmental mixtures, including organics
extracted from ambient air particulate matter, have been employed
in recent studies, including those of Epler et al. (1978), Rao et
al. (1980), Tokiwa et al. (1977), and others.

Table 4. Mutagenesis Data of Particulate Fractions from Lake Charles, LA (<1.7 μm)[a]

Sample Fraction	Doses (μg/plate)	MA Protein Concentration (mg)	TA100 +MA[b]	TA100 -MA	TA98 +MA[b]	TA98 -MA	TA1538 +MA[b]	TA1538 -MA	TA1537 +MA[b]	TA1537 -MA
Total organics	1000	3.0	507* ± 17	450* ± 21	127* ± 28	210* ± 70	152* ± 28	125* ± 12		
	500	3.0	327 ± 15	341 ± 30	74 ± 24	169* ± 12	60 ± 5	80 ± 0		
Polar neutrals[c]	500	3.0	401 ± 18	387 ± 2	129* ± 51	241* ± 6	117 ± 6	191* ± 12	67 ± 2	21 ± 11
	100	3.0	186 ± 63	256 ± 73	94 ± 23	155* ± 9	56 ± 8	74 ± 12	24 ± 10	23 ± 9
Acids[c]	500	3.0	326 ± 9	221 ± 91	82 ± 15	77 ± 17	59 ± 21	67 ± 11	62 ± 17	19 ± 10
	100	3.0	150 ± 95	168 ± 57	63 ± 20	89 ± 22	61 ± 5	52 ± 12	47 ± 21	39 ± 4
Bases	500	1.5	163 ± 106	94 ± 63	105* ± 17	51 ± 16	69 ± 1	51 ± 8		
	100	1.5	197 ± 71	276 ± 25	77 ± 21	66 ± 19	56 ± 3	46 ± 0		
	500	3.0			103* ± 40					
	100	3.0			53 ± 24					
	500	6.0			58 ± 19					
	100	6.0			62 ± 17					
Nonpolar neutrals	500	3.0	398 ± 13	224 ± 21	122* ± 7	107* ± 45	69 ± 5	53 ± 3		
	100	3.0	365 ± 7	279 ± 25	93 ± 9	88 ± 10	60 ± 7	37 ± 0		
Polynuclear aromatic hydrocarbons	500	1.5			1007* ± 159					
	100	1.5			394* ± 53					
	500	3.0			755* ± 144					
	100	3.0			283* ± 58					
	500	6.0			257* ± 198					
Positive control	100		2404* ± 218	280 ± 13	2309* ± 12	1848* ± 41	2086* ± 5	1260* ± 55	39 ± 20	174* ± 38
	10		2197* ± 682	1241* ± 215	2074* ± 31	1117* ± 47	1529* ± 13	623* ± 27	69 ± 20	20 ± 6
	1		1061* ± 14	627* ± 177	635* ± 38	109* ± 17	683* ± 30	339* ± 78	74 ± 6	20 ± 2
Solvent control	0.1 ml		218 ± 16	205 ± 9	51 ± 3	53 ± 5	64 ± 5	44 ± 3	38 ± 6	24 ± 6

[a] Plate-incorporation assays; results reported in mean ± standard deviation computed with triplicate plates.
[b] MA = metabolic activation.
[c] The polar neutrals and acids were tested in TA1535; however, the spontaneous revertants were not within acceptable limits.
* Mutagenic response is at least twofold that of the spontaneous revertants.

Table 5. Mutagenic, Carcinogenic, Neoplastic, and Teratogenic Compounds[a] Identified in Ambient Air

Compounds	Upland CA	Lake Charles LA (1)	Lake Charles LA (2)	Elizabeth, NJ	Beaumont, TX	Houston TX (1)	Houston TX (2)	Muta-genic	Carcino-genic	Neo-plastic	Terato-genic
Acenaphthene				+						+	
Allyl-phenol	+									+	
Amino-anthracene	+							+		+	
Amino-anthraquinone	+	+								+	
Aniline						+			Indefinite		
Anthanthrene	+			+					+	+	
Anthracene	+	+	+	+		+			+	+	
Anthraquinone	+	+				+				+	
Benz-anthracene	+	+	+	+		+		+	+	+	
Benzo-fluoranthene		+	+	+		+			+	+	
Benzo-perylene			+	+					+	+	
Benzo-phenanthrene	+	+	+	+		+		+	+		
Benzo-pyrene	+	+	+	+		+		+	+		
Bromophenol	+	+								+	
Butylbenzylnitrosamine						+		+	+		+
Caffeine						+				+	+
Chlorophenol	+										+
Chrysene	+	+	+	+		+		+			
Cresol	+					+		+	±	+	
Dicyclohexylamine										+	
Dimethyl-anthracene		+				+				+	
Dimethyl-phenanthrene	+			+						+	
Fluoranthene	+		+	+		+				+	
Methylbenz-anthracene	+	+		+				+	+	+	
Methylbenz-phenanthrene	+	+	+	+				+	+	+	

[a]U.S. Department of Health, Education, and Welfare, 1977.

(continued)

Table 5. Mutagenic, Carcinogenic, Neoplastic, and Tertogenic Compounds[a] Identified in Ambient Air (continued)

Compounds	Upland CA	Lake Charles LA (1)	Lake Charles LA (2)	Elizabeth NJ	Beaumont, TX	Houston TX (1)	Houston TX (2)	Muta-genic	Carcino-genic	Neo-plastic	Terato-genic
Methyl-chrysene	+									+	
Methylene-phenanthrene	+	+		+						+	
Methyl-phenanthrene	+	+								+	
Methyl-pyrene	+	+	+	+		+				+	
Methyl-stearate	+	+		+		+				+	
Naphthacene	+	+							+		
Naphthalene	+	+									
Perylene	+	+	+	+		+			+	+	
Phenanthrene	+	+	+	+		+					
Phenol	+	+				+			+	+	
Phenyl-benzene				+						+	
Phenylene-pyrene				+					+		
Phenyl-β-naphythlamine	+			+		+		+	+	+	
Phenylphenol		+								+	
Progesterone	+	+		+					Suspected	+	
Propenyl-phenol	+										+
Pyrene	+	+	+	+						+	
Quinoline	+							±	+		
Styrene	+	+						+	±		
(p-Tetramethyl-butyl)phenol											
Tetrahydrobenzo-phenanthridine						+				+	
Toluidine						+			+	+	
Trimethyl-phenanthrene				+						+	
Triphenylethylene						+				+	

[a] U.S. Department of Health, Education, and Welfare, 1977.

Several problems were encountered in the development and validation of this technical approach to the investigation of the bioactivity of air particulate matter. Technical considerations concerning the specificity of the chemical fractionation scheme have been discussed previously (Pellizzari et al., 1978). The chemical fractionation scheme separates organic acids and bases by treatment with aqueous base and aqueous acid, followed by partitioning into organic solvent. Löfroth (1980) contends that such treatment may generate artifacts, especially in the organic base (nitrogen-containing compounds), due to reaction of substances with the aqueous acid or base during the extraction. Epler et al. (1978; Rao et al., 1980) recommended separation of the acid and base components with Sephadex LH-20 to avoid this problem; however, they found no significantly different results using LH-20 vs. acid-base extraction.

Sample size was a chronic problem in this study. The massive air volume sampler did not collect the amounts of particulate expected. In a 10-day sampling period, often less than 3 g particulate matter, including all three size fractions, was collected. Only 5 to 15% of the particulate represents extractable organics, and this material, when partitioned between five fractions, usually did not provide enough sample for quantitative dose-responsive plate-incorporation mutagenicity assays. For triplicate determinations in all five Salmonella strains, using five doses, both with and without metabolic activation (as suggested by deSerres and Shelby, 1979), the assay would require 50 mg. Enough material must also be available for chemical analysis. As can be seen from Table 3, none of the samples were of adequate size for complete bioassay and chemical analysis.

Despite the limited data available, the results of air particulate research by various investigators at different sites show certain similarities to this study. For example, Pitts et al. (1977) collected air particulate at rural and urban sites in California and extracted the organics using sonication with equal parts of methanol, benzene, and dichloromethane. The particulate collected from the rural site was not mutagenic, but material from all urban locations was mutagenic without S-9 addition, exhibiting a linear dose response for Salmonella strains TA98, TA1538, and TA1537. Thus, the mutagenic components present were probably not polynuclear aromatic hydrocarbons (PAH), which require activation. In another case, mutagenic activity requiring S-9 metabolic activation was demonstrated in a California air sample; this activity was then quantitatively related to B(a)P concentration (Flessel, 1980).

In Japan, ambient air was sampled at six residential and industrial sites (Tokiwa et al., 1977). The organic fraction from ambient air particulate (extracted with methanol) collected in

Ohmuta (an industrial city) was mutagenic, exhibiting a linear dose
response from 100 to 400 μg with <u>Salmonella</u> strain TA98. A sample
collected from a residential area (Fukuoka City) required from 370
to 1970 μg/plate to generate a positive mutagenic response. The
mutagenicity was approximately three times greater for the sample
collected from the industrial city, and a linear dose response was
observed for both samples. These researchers identified the
compounds in the methanol extract using an alumina column and GC/MS
analysis. A majority of the compounds identified were PAHs such as
benzopyrenes, anthracenes, fluorenes, and dibenzoisomers of these
compounds. These Japanese studies revealed that air particulate
near industrial sites possessed higher mutagenic potential than
that from rural sites and that both S-9-activated (PAH) and direct-
acting (non-PAH) mutagenic components were present in air
particulates, agreeing with the California studies of Pitts et al.
(1977) and Flessel (1977).

Talcott and Wei (1977) collected air particulate on glass
fiber filters and extracted the filter with acetone in a Soxhlet
apparatus. The extracted material exhibited the highest activity
with TA100 and required S-9 metabolic activation. Direct-acting
mutagens were also detected with TA98 and TA1537. Activity was
again attributed to at least two types of mutagens: a PAH and a
non-PAH fraction. The view that PAH compounds are not the sole (or
major) mutagenic factor associated with air particulate was further
reinforced by the studies of Dehnen et al. (1977). Here, mutagenic
activity was predominantly produced by compounds other than PAHs,
since the observed mutagenicity did not require S-9. Dehnen used a
chemical fractionation scheme to divide his crude air particulate
sample into cyclohexane and methanol-extractable fractions. An
alumina column was employed to further fractionate the cyclohexane
extract into a purified cyclohexane fraction (containing PAH-type
compounds) and a 2-propanol fraction (containing azo-heterocyclic
compounds). The highest mutagenic response was found in the
2-propanol and methanol extracts; neither fraction contained
PAH-like compounds.

Other investigators (Tokiwa et al., 1977) have combined a
chemical fractionation scheme with a mutagenesis-detection system.
After solvent extraction, fractionation of the crude organic
extract into chemical classes permitted identification of the
mutagenically active chemical compounds within each chemical class.
In some cases, fractionation appeared to reduce or remove toxic
effects that otherwise would have prevented expression of
mutagenicity. Thus, the mechanism of synergism or antagonism of
the individual components in the crude mixture is open to
investigation. Pelroy et al. (1978) observed such effects with
shale oil and its fractions, as did Rao et al. (1980) with fossil
fuels and their fractions.

Pellizzari et al. (1978) initially tested a West Virginia ambient air particulate for mutagenicity using a fractionation scheme and the qualitative spot test (Ames et al., 1975) with the standard Ames Salmonella tester strains (TA98, TA100, TA1535, TA1537, and TA1538), both with and without a metabolic activation system (S-9). Mutagenic activity was observed for the organic bases, organic acids, and aromatics. Each of the five tester strains gave a positive response with at least one of the active fractions, and only the aromatics required metabolic activation (Hughes et al., 1978; Pellizzari et al., 1978). In this West Virginia sample, 2-nitro-4,5-dichlorophenol, a direct-acting mutagen (Pellizzari et al., 1978), was identified in the polar neutrals, and fluoranthene, pyrene, benz(a)anthracene, and B(a)P were identified in the aromatic fraction. All of these compounds have been identified either as co-carcinogens (Van Duuren, 1976) or as mutagens requiring metabolic activation (McCann et al., 1975).

These preliminary results indicated that the Ames assay could detect mutagenic activity in small sample amounts of ambient air (as a complex mixture), and that the fractionation and analysis scheme was capable of identifying broad chemical mutagenic classes within these mixtures. It was concluded that chemical fractionation of the crude complex sample was both useful and necessary to accomplish adequate biotesting and chemical identification of signature mutagenic components in ambient air particulate.

ACKNOWLEDGMENTS

This study was supported by U.S. Environmental Protection Agency Contract No. 68-02-2724.

REFERENCES

Ames, B., J. McCann, and E. Yamasaki. 1975. Methods for detecting carcinogens and mutagens with the Salmonella/mammalian microsome mutagenicity test. Mutation Res. 31:347-364.

Chrisp, C., G. Fisher, and J. Lammert. 1978. Mutagenicity of filtrates from respirable coal fly ash. Science 199:73-75.

Daisey, T., I. Hawryleck, T. Kneip, and F. Mukai. 1979. Mutagenic activity in organic fractions of airborne particulate matter. In: Conference on Carbonaceous Particles in the Atmosphere. U.S. Dept. of Commerce: Washington, DC. pp. 187-192.

Dehnen, W., R. Pitz, and R. Tomingas. 1977. The mutagenicity of airborne particulate pollutants. Cancer Lett. 4:5-12.

DeSerres, F., and M.D. Shelby. 1979. Recommendations on data production and analysis using the Salmonella/microsome mutagenicity assay. Mutation Res. 64:159-165.

Duce, R.A. 1978. Speculations on the budget of particulate and vapor-phase non-methane organic carbon in the global troposphere. Pure and Appl. Geophys. 116:244-273.

Epler, J.L., B.R. Clark, C.-h. Ho, M.R. Guerin, and T.K. Rao. 1978. Short-term bioassay of complex organic mixtures: Part II, mutagenicity testing. In: Application of Short-term Bioassays in the Fractionation and Analysis of Complex Environmental Mixtures. M.D. Waters, S. Nesnow, J.L. Huisingh, S.S. Sandhu, and L. Claxton, eds. Plenum Press: New York. pp. 269-289.

Fishbein, L. 1976. Atmospheric mutagens. In: Chemical Mutagens: Principles and Methods for their Detection. A. Hollaender, ed. Plenum Press: New York. pp. 219-319.

Flessel, P. 1977. Mutagenic activity of particulate matter in California Hi-Vol samples. Presented at the Third Interagency Symposium on Air Monitoring Quality Assurance, San Francisco, CA.

Henry, W., and R. Mitchell. 1978. Development of a large sampler for the collection of respirable matter. EPA 600/4-78-009. U.S. Environmental Protection Agency: Research Triangle Park, NC.

Hidy, G., and J. Brock. 1970. An assessment of the global sources of tropospheric aerosols. In: Proceedings of the Second International Clean Air Congress. H.M. Englund and W.T. Beery, eds. Academic Press: New York. pp. 1088-1097.

Hughes, T., L. Little, L. Claxton, M. Waters, E. Pellizzari, and C. Sparacino. 1978. Microbial mutagenesis testing of air pollution samples. In: Application of Short-term Bioassays in the Fractionation and Analysis of Complex Environmental Mixtures. M.D. Waters, S. Nesnow, J.L. Huisingh, S.S. Sandhu, and L. Claxton, eds. EPA 600/9-78-027. U.S. Environmental Protection Agency: Research Triangle Park, NC. (abstr.) p. 582.

Hughes, T.J., E. Pellizzari, L. Little, C. Sparacino, and A. Kolber. 1980. Ambient air pollutants: Collection, chemical characterization and mutagenicity testing. Mutation Res. 76:51-83.

King, L.C., and J. Lewtas Huisingh. 1980. Evaluation of the release of mutagens from diesel particles in the presence of serum. Presented at the U.S. Environmental Protection Agency Second Symposium on the Application of Short-term Bioassays in the Fractionation and Analysis of Complex Environmental Mixtures, Williamsburg, VA.

Lee, M., M. Novotny, and K. Bartle. 1976. GC/MS and nuclear magnetic resonance determination of polynuclear aromatic hydrocarbons in airborne particulates. Anal. Chem. 48:1566.

Löfroth, G. 1980. Comparison of the mutagenic activity in diesel and gasoline engine exhaust and in carbon particulate matter. Presented at the U.S. EPA Second Symposium on the Application of Short-term Bioassays in the Fractionation and Analysis of Complex Environmental Mixtures, Williamsburg, VA.

McCann, T., E. Choi, E. Yamasaki, and B.N. Ames. 1975. Detection of carcinogens as mutagens in the Salmonella/microsome test: assay of 300 chemicals. Proc. Nat. Acad. Sci. USA 72:5135–5139.

Møller, M., and I. Alfheim. 1980. Mutagenicity and PAH-analysis of airborne particulate matter. Atmos. Exper. 14:83–88.

National Academy of Sciences. 1972. Particulate Polycyclic Organic Matter, Biological Effects of Atmospheric Pollutants. Baltimore Press: Washington, DC.

Pellizzari, E., L. Little, C. Sparacino, T. Hughes, L. Claxton, and M. Waters. 1978. Integrating microbiological and chemical testing into the screening of air samples for potential mutagenicity. In: Application of Short-term Bioassays in the Fractionation and Analysis of Complex Environmental Mixtures. M.D. Waters, S. Nesnow, J.L. Huisingh, S.S. Sandhu, and L. Claxton, eds. Plenum Press: New York. pp. 331–351.

Pelroy, R., and M. Peterson. 1978. Mutagenicity of shale oil components. In: Application of Short-term Bioassays in the Analysis of Complex Environmental Mixtures. M.D. Waters, S. Nesnow, J.L. Huisingh, S.S. Sandhu, and L. Claxton, eds. Plenum Press: New York. pp. 463–475.

Pitts, J., D. Grosjean, T. Mischke, V. Simmon, and D. Poole. 1977. Mutagenic activity of airborne particulate organic pollutants. Toxicol. Lett. 1:65–70.

Rao, T.K., J.L. Epler, M.R. Guerin, B.R. Clark, and C.-h. Ho.
    1980. Mutagenicity of nitrogen compounds from synthetic crude
    oils: collection, separation, and biological testing.
    Presented at the U.S. EPA Second Symposium on the Application
    of Short-term Bioassays in the Fractionation and Analysis of
    Complex Environmental Mixtures, Williamsburg, VA.

Shubik, P., and J. Hartwell. 1969. Survey of compounds which have
    been tested for carcinogenic activity. Supp. 2. PHS 149-2.
    U.S. Public Health Service: Washington, DC.

Talcott, R., and W. Harger. 1979. Mutagenic Activity of Aerosol-
    size Fractions. EPA-600/3-79-032. U.S. Environmental
    Protection Agency: Research Triangle Park, NC.

Talcott, R., and E. Wei. 1977. Airborne mutagens bioassayed in
    Salmonella typhimurium. J. Natl. Cancer Inst. 58:449-451.

Tokiwa, H., K. Morita, H. Tokeyoshi, K. Takahashi, and Y. Ohnishi.
    1977. Detection of mutagenic activity in particulate air
    pollutants. Mutation Res. 48:237-248.

U.S. Department of Health, Education, and Welfare. 1977. Registry
    of Toxic Effects of Chemical Substances, Vol. 1 and 2. U.S.
    Depart. of Health, Education, and Welfare: Washington, DC.

Van Duuren, B. 1976. Tumor promoting and co-carcinogenic agents
    in chemical carcinogenesis. In: Chemical Carcinogenesis.
    ACS Monograph 173. C. Searle, ed. American Chemical Society:
    Washington, DC. pp. 24-51.

EVALUATION OF COLLECTION AND EXTRACTION METHODS FOR MUTAGENESIS STUDIES ON AMBIENT AIR PARTICULATE

R. Jungers, R. Burton, L. Claxton, and J. Lewtas Huisingh
Environmental Monitoring Systems Laboratory and
Health Effects Research Laboratory
U.S. Environmental Protection Agency
Research Triangle Park, North Carolina

INTRODUCTION

The extractable organics associated with air particles have been shown to be carcinogenic (Hueper et al., 1962; Leiter et al., 1942) and mutagenic in a short-term microbial bioassay (Lewtas Huisingh, 1980; Teranishi et al., 1977). The identification and characterization of the potentially hazardous chemical components associated with ambient air particles requires that efficient collection and extraction methods be developed and validated. It is important that the original chemical composition of the particle-bound organics be maintained through collection and extraction to establish that the bioassay results are directly relatable to the chemistry of the original organics as they exist on particles in the atmosphere. The particles of major interest are those that are in the inhalable (< 15 μm) or respirable (< 5 μm) size range and therefore can be trapped in the human respiratory tract. Several studies have shown that the organics associated with ambient air particles are more mutagenic in the smaller (respirable) particle size range (Huisingh, 1980; Teranashi et al., 1977).

In the past, most conventional ambient air samplers were not designed to collect particles in size-separated stages, nor could they collect the quantities of particulate matter desired for integrated biological and chemical studies. A new sample collection instrument, the Massive Air Volume Sampler (MAVS), has been designed to separate collected particles into three size ranges (0 to 1.7 μm, 1.7 to 3.5 μm, and 3.5 to 20 μm) and to collect large amounts of particles. The objective of this study is to evaluate current ambient air particle sampling and extraction methods for short-term mutagenesis bioassay applications.

COMPARISON OF AVAILABLE COLLECTION METHODS

A great number of field air particle samplers with varied applications have been designed and used. Filtration and impaction are the two techniques most commonly employed to collect particles for the determination of particle concentration and chemical composition. The conventional air particle collection instrument, recommended in the Federal Register for determining total suspended particulate (TSP), is the standard high-volume sampler. Recently, the size selective inlet (SSI) high-volume sampler has been introduced to collect size-selected inhalable particles less than 15 µm (McFarland and Rodes, 1979). In addition, the dichotomous sampler (virtual impactor) is being employed when elemental chemical analysis is desired (McFarland and Rodes, 1979). All three of these instruments employ filtration as the primary collection method. Preliminary studies have shown that these samplers do not collect sufficient amounts of particles for complete bioassay studies. Henry and Mitchell (1978) designed and constructed a massive air volume sampler (MAVS) for the U.S. Environmental Protection Agency (EPA) that would collect large quantities of size-fractionated air particles. The principles and operation of this sampler are described in the next section. The new sampler, MAVS, capable of collecting gram amounts of sized particles in a reasonable sampling period, is currently being evaluated for collecting ambient air particles to be used in bioassay studies.

A comparison of particle mass that can theoretically be collected by the TSP high-volume sampler, the SSI high-volume sampler, the dichotomous sampler, and the new MAVS is shown in Table 1 for different air pollution TSP concentrations. The comparison was made for 24-h sampling periods in ambient air with TSP particle concentrations of 60, 100, and 200 $\mu g/m^3$. As noted in the table, the mass of particles collected depends on the ambient particle concentration and, more importantly, on the flow rate of sample air through the given sampler. The flow rate of the MAVS (18.5 m^3/min) enables it to collect a greater amount of particulate matter than the other samplers.

The Massive Air Volume Sampler

Since the MAVS is a relatively new particle sampler, a brief description of its operation is given here. The sampler, shown in Figures 1, 2, and 3, has an inlet which serves as the scalping stage that permits only particles of 20 µm aerodynamic diameter or

Table 1. Comparison of Amounts of Particulate Matter
Theoretically Collected by Four Types of Samplers

Type of Sampler	Flow Rate (m^3/min)	Vol. of Air Sampled[a] (m^3)	Calculated Particulate Matter Collected at: TSP Concentrations[a]		
			60 µg/m^3 (g)	100 µg/m^3 (g)	200 µg/m^3 (g)
Dichotomous	0.017	24.0	0.0015	0.0024	0.0048
SSI high-volume	1.13	1,627.0	0.097	0.16	0.33
TSP high-volume	1.40	2,016.0	0.12	0.20	0.40
MAVS	18.5	26,640.0	1.6	2.7	5.3

[a]During 24-h sampling period.

less to enter the sampler. The sample air then enters an impaction
plate assembly containing four stainless steel Teflon-coated plates
with slots, which serve as impactor jets and collection surfaces
for the two-stage impactor. The first stage of the impactor
collects particles between 20 and 3.5 µm, and the second stage
collects particles between 3.5 µm and 1.7 µm. Immediately below
the impaction plates, the remaining particles less than 1.7 µm
pass through a particle charging field and are collected on the
third stage, which consists of vertically charged electrostatic
precipitator (ESP) plates. Both the impaction plates and the
electrostatic collector plates are Teflon-coated to minimize
substrate contamination and to facilitate removal of the particles.
The particle collector plate assemblies, after sampling, are
shipped in sealed containers to the laboratory, where the particles
are mechanically scraped from the Teflon plates. Figure 4 shows
the particles being removed inside a sealed glove box to minimize
contamination of the sample and to insure the safety of personnel.

Ozone Generation in the MAVS

The MAVS does not have the potential artifact-formation
problems associated with substantial flow of reactive gases (e.g.,
ozone and nitrogen oxides) over particles collected on reactive
filtration media, since the MAVS employs impaction and

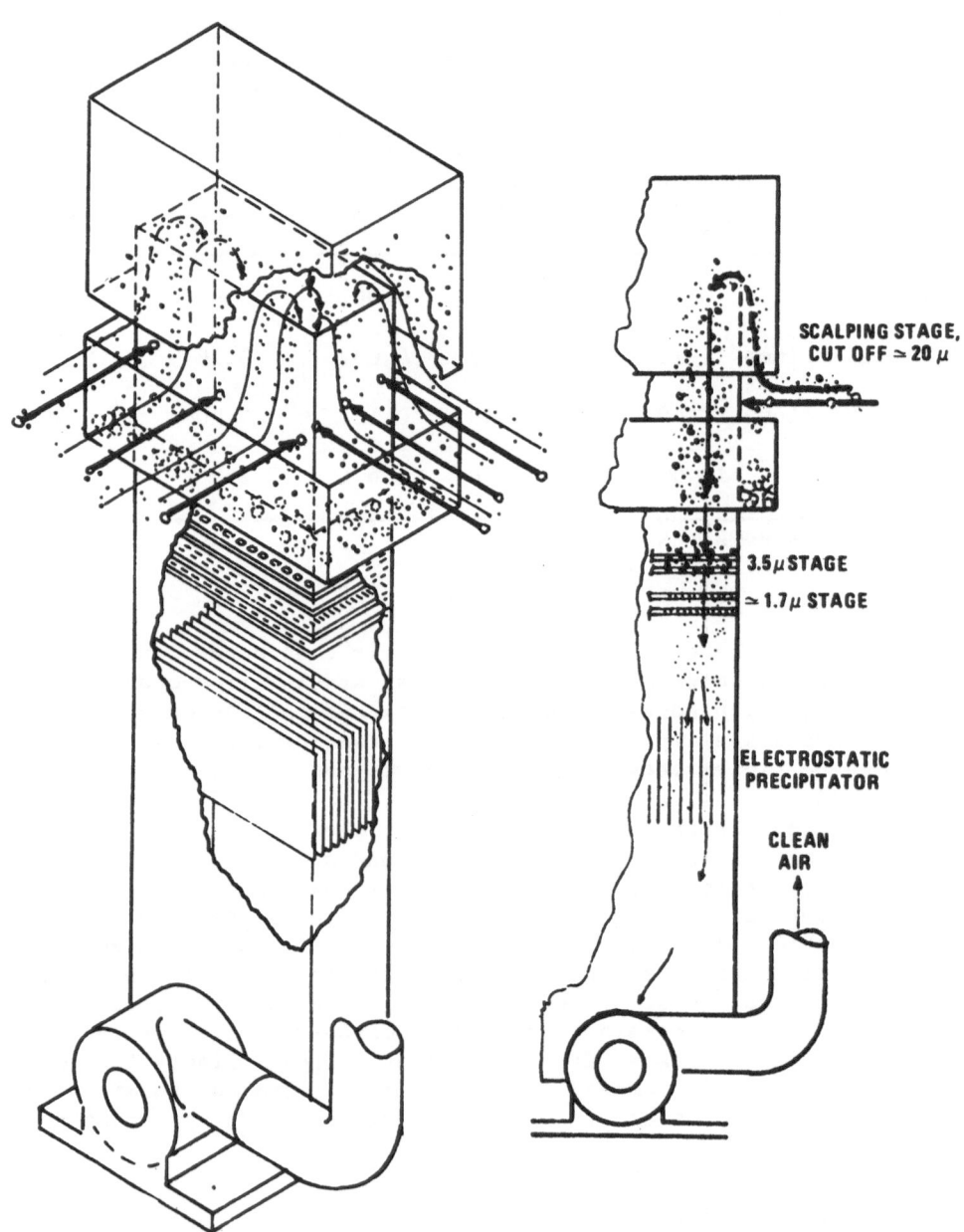

Figure 1. Massive Air Volume Sampler schematic.

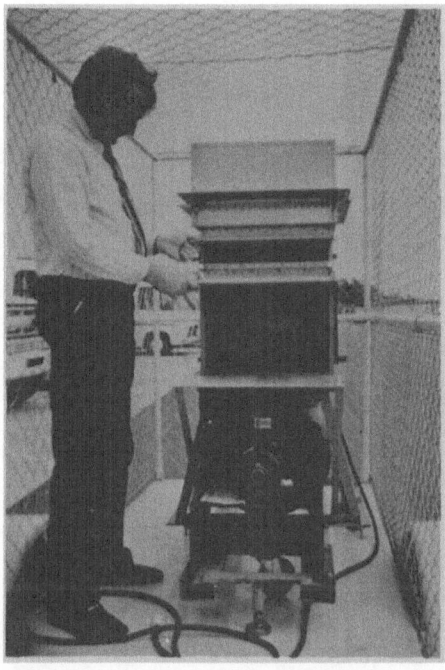

Figure 2. Massive Air Volume Sampler.

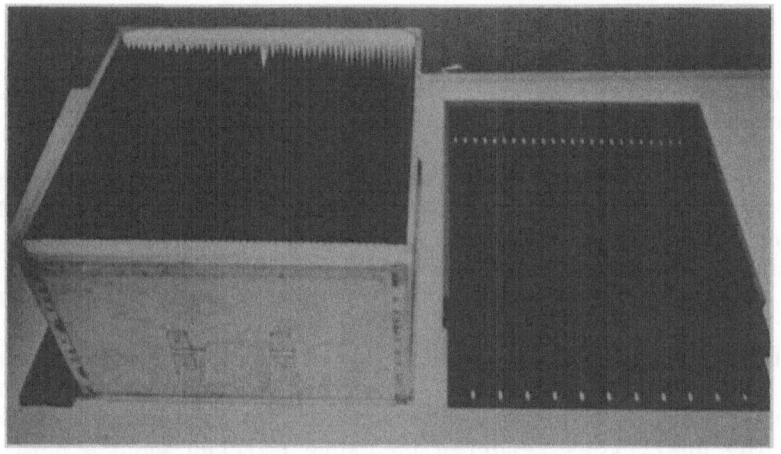

Figure 3. Impactor plate and electrostatic precipitator.

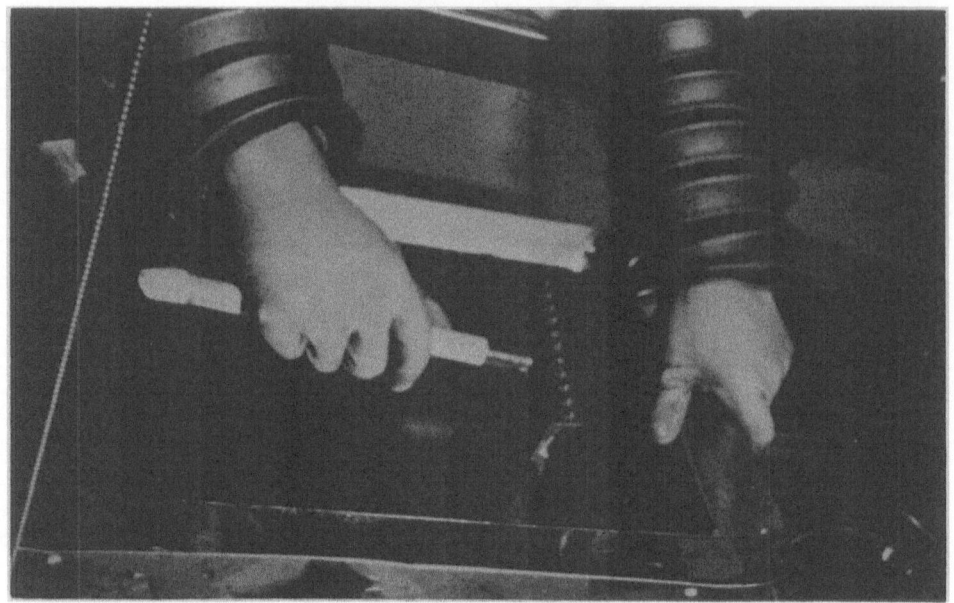

Figure 4.   MAVS plate being scraped inside glove box.

electrostatic precipitation onto Teflon-coated surfaces.   The ESP
particle-charging section of the sampler, however, generates low
levels of ozone that passes through the fine particle (< 1.7 μm)
collection stage.   Since ozone may react with organic compounds
(NSF, 1977), including polynuclear aromatic hydrocarbons (PAH), it
is important to determine whether the generated ozone reacts with
the organic compounds associated with the collected particles and
whether such reactivity would bias mutagenicity bioassay results.

Mitchell et al. (1978) reported the ozone concentrations
present in the blower outlet of the first MAVS under different
operating conditions.   After subtracting the ambient background
ozone (0.05 ppm), they reported no ozone emitted when a positive
corona was employed and 0.025 to 0.050 ppm ozone emitted when a

negative corona was employed. As the corona sign was changed from positive to negative, the particle collection efficiency increased from 66% to 80% at a corona voltage of 7,800. As the negative corona voltage was increased, both collection efficiency and ozone concentration increased.

As the MAVS has become viable as the most acceptable method for collecting large samples of size-separated ambient particles for bioassay screening studies, artifact formation due to reaction of the collected material with ozone has become an important consideration. Extractable organics from ambient air particles collected at the third stage of the MAVS, the ESP, where an ozone reaction would be most suspect, were found to be more mutagenic than the larger particles collected at the first and second stages (Huisingh, 1980). A second, more detailed measurement of ozone within the sampler, as well as at the inlet and outlet, was performed; the results are given in Table 2. The standard operating particle charging voltage of 10,000 volts was used along with negative polarity and corona. As noted in Table 2, the highest ozone concentration in this study was found at the blower outlet of the sampler, which is downstream of the third collection stage. The highest ozone concentration observed in any section of the sampler where particles are actually collected was 0.08 ppm. When the voltage was reversed to give a positive corona, no measurable amounts of ozone were generated in the sampler.

Table 2. MAVS Ozone Generation Test Results

Charging Wire Voltage[a]	Plate Voltage[a]	Location	Ozone (O_3) Generated (ppm)
10,000	7,800	Sampler inlet	--- [b]
10,000	7,800	Precipitator inlet	0.07
10,000	7,800	Between plates	0.08
10,000	7,800	Precipitator outlet	0.09
10,000	7,800	Blower outlet	0.11
10,000	5,000	Between plates	0.08

[a]All voltages are negative polarity.
[b]Room ambient concentration was 0.02 ppm O_3. All other values were corrected for this amount to determine the concentration of O_3 generated.

MAVS Ozone Field Study

A field study was developed to determine whether the concentration of ozone generated in the electrostatic precipitator collector section of the MAVS causes reaction between the ozone and the organic constituents of the collected particles and consequently biases the results of the bioassay. Three different sampling sites were selected on the basis of possible PAH presence in the air: Durham, NC (EPA Cameo Building in downtown Durham); Birmingham, AL (industrial site in north Birmingham); and Gadsden, AL (steel mill coke oven).

Since the MAVS can be operated with either positive particle charging corona (without ozone generation) or with negative particle charging corona (with ozone generation), field samples were collected by co-located samplers (one with positive charging corona and one with negative charging corona). All samples were extracted, and split blind samples were both chemically characterized and tested in the Ames Salmonella typhimurium microbial mutagenesis bioassay. Chemical characterization and bioassay results from particles charged with positive corona were compared with results from particles charged with negative corona. Samplers were operated in a normal manner with particle charging voltage of 10,000 volts and collector plate voltage of approximately 7,800 volts. As shown in Table 2, 0.08 ppm of ozone was generated in the negative charging mode and no measureable ozone in the positive charging mode.

At the Birmingham site, a sample was collected simultaneously by each of the two co-located samplers (one with positive and one with negative particle charging corona). The same sampling scheme was used in Gadsden, where the two co-located samplers were operated for two sequential sampling periods. In Durham, the same sampling scheme was used, with the sampler corona reversed in the second of the two sequential sampling periods to eliminate any possible sampler bias. The sampling scheme for all three tests is shown in Table 3.

The particles collected at the Durham field site were mechanically extracted from the collector plates, and the Birmingham and Gadsden particles were rinsed from the collector plates with dichloromethane (DCM; methylene chloride). These particle samples and solvent extract samples were sealed and blind-coded for laboratory analysis (e.g., organic extraction, chemical analysis, and bioassay preparation).

Table 3.   Ozone Field Study Sampling Scheme

| Location | Sampler | Particle Charging Corona | |
		Experiment 1	Experiment 2
Durham, NC	A	+	−
	B	−	+
Birmingham, AL	A	+	
	B	−	
Gadsden, AL	A	+	+
	B	−	−

Analysis Methodology

The air particle samples from Durham were Soxhlet-extracted with DCM after mechanical removal from the plates, and those from Birmingham and Gadsden, which were originally received in DCM, were filtered. The DCM extract from both types of samples was dried in a stream of dry nitrogen. One aliquot was used for chemical analysis, and one aliquot was prepared for bioassay by adding dimethylsulfoxide (DMSO) to obtain a concentration of 2 mg/ml.

The DCM extracts were prepared for gas chromatographic (GC) and GC/MS analysis of polycyclic aromatic hydrocarbon (PAH) using procedures developed by Bjørseth (1977). The GC analysis was performed on these extracts with a flame ionization detector (FID) and a glass capillary column with hydrogen as the carrier gas. Splitless injections were used while the column was at ambient temperature, and peak area integration was performed with an automatic digital integrator.

The GC/MS analysis was performed using a Finnigan Model 3200 MS and a Model 9500 GC with a glass capillary column. Splitless injections were made while the column remained at ambient temperatures. Electron impact (EI) spectra were obtained and processed with a Digital System 150 and INCOS data system.

Bioassay Methodologies

The S. typhimurium plate incorporation assay was performed as described by Ames et al. (1975), with minor modifications. The modifications consisted of adding the standard minimal concentration of histidine directly to the plate media instead of to the overlay and then incubating the plates for 72 rather than 48 h. The comparative samples were bioassayed simultaneously in the same experiment for each tester strain.

The studies were performed with triplicate plates at five doses with and without metabolic activation. Activation was provided by a 9000 x g supernatant of liver from Aroclor-1254-induced CD rats (Ames et al., 1975). The linear portion of the dose-response curves were used to calculate a linear regression line. The slope of the linear regression analysis is reported as revertants per microgram (rev/µg) of organics tested.

Discussion of Analytical Results

GC chromatographs were obtained under identical conditions and show a similar relative concentration of PAH components in the pairs of samples for all three sites. This similarity was shown regardless of collection date, precipitator ion potential, or whether the precipitator plates were washed or scraped.

The similarity noted in the chromatograms was the basis of a reduced number of samples being analyzed by GC/MS. With the identification of the major components determined by GC/MS, the total ion chromatogram was compared to the GC chromatograph to aid in the identification of the various components. The quantification for various PAH species and the concentration ratios for two pairs of isomeric PAH are shown in Table 4.

Some PAHs are oxidized readily, while others are relatively insensitive to oxidation (Committee on Biological Effects, 1972). Pyrene and anthracene are readily oxidized, while fluoranthene and phenanthrene are relatively inert to oxidation. Assuming that reactions with ozone proceed in the same manner in the MAVS with negative corona as in the laboratory studies, the ratio between the inert and reactive compounds should reflect any potential reactions of ozone with the reactive PAHs. Comparison of these ratios (Table 4) in the presence and absence of ozone (negative and positive corona) show no significant or consistent changes, suggesting no significant reaction of the PAHs with ozone. In fact, the values obtained were well within the range reported in other studies, including ambient air and aluminum and coke plants that were conventionally sampled (Bjørseth et al., 1978; Hoffman and Wynder, 1976).

Table 4. Comparisons of PAH Levels at Three Sampling Locations[a]

Chemical Compound	Durham		Birmingham		Gadsden	
	+	–	+	–	+	–
Phenanthrene	1470	872	116	64	776	616
					896	940
Anthracene	1960	1150	92	72	288	180
					256	524
Ratio P/A	0.75	0.76	1.23	0.90	2.69	3.46
					3.46	1.79
Fluoranthene	2190	648	208	196	924	1370
					1840	2220
Pyrene	1980	696	224	184	1250	1240
					1900	2900
Ratio F/P	1.11	0.93	0.94	1.08	0.74	1.10
					0.97	0.77

Column header spanning: PAH ($\mu g/g$)

[a]Positive corona = +; negative corona = –.

The data provide interesting information about the MAVS.
Since the Birmingham samples (#1 and #2) were collected in the same
manner, they should exhibit similar patterns. In order to study
this relationship, parent PAH profiles (PPP) were constructed. PPP
depict relative distribution of the key PAH compounds in the
sample, thus providing a convenient method of comparing samples
(see Figure 5). The profile remained relatively constant over the
sampling period, indicating that operation of the electrostatic
precipitator in the positive corona (#1) or negative corona (#2)
mode did not affect the composition of the samples.

The data indicate that no significant degradation of PAH
attributable to ozonation occurred, compared with the inert
compounds. The stability of the profiles indicates that variation
of sampling parameters had no influence on the collected organics.
Therefore, it is unlikely that the MAVS electrostatic precipitation

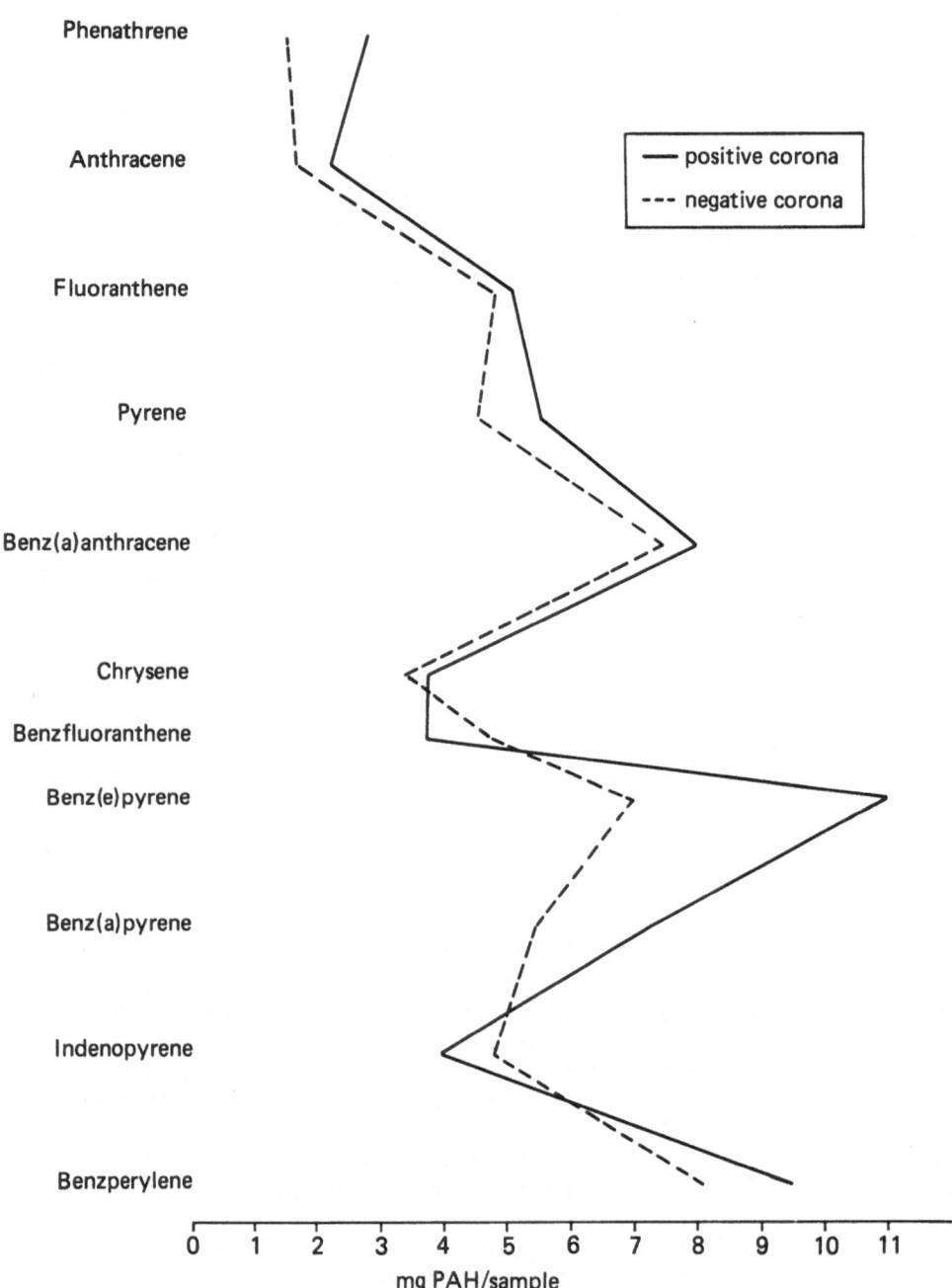

Figure 5.   Parent PAH profile of ambient air.

plates in the positive or negative corona mode caused any artifact formation due to ozonation.

Discussion of Bioassay Results

The mutagenic activity in S. typhimurium tester strain TA98 is shown in Figure 6 for each of the three sites as a function of ionization potential. At the Durham site, the slope of the mutagenic activity curve ranged from 0.37 to 0.42 rev/µg in the absence of S-9 activation and from 0.52 to 0.75 rev/µg in the presence of activation. No significant difference was observed in mutagenicity as a function either of ionization potential or of sampler.

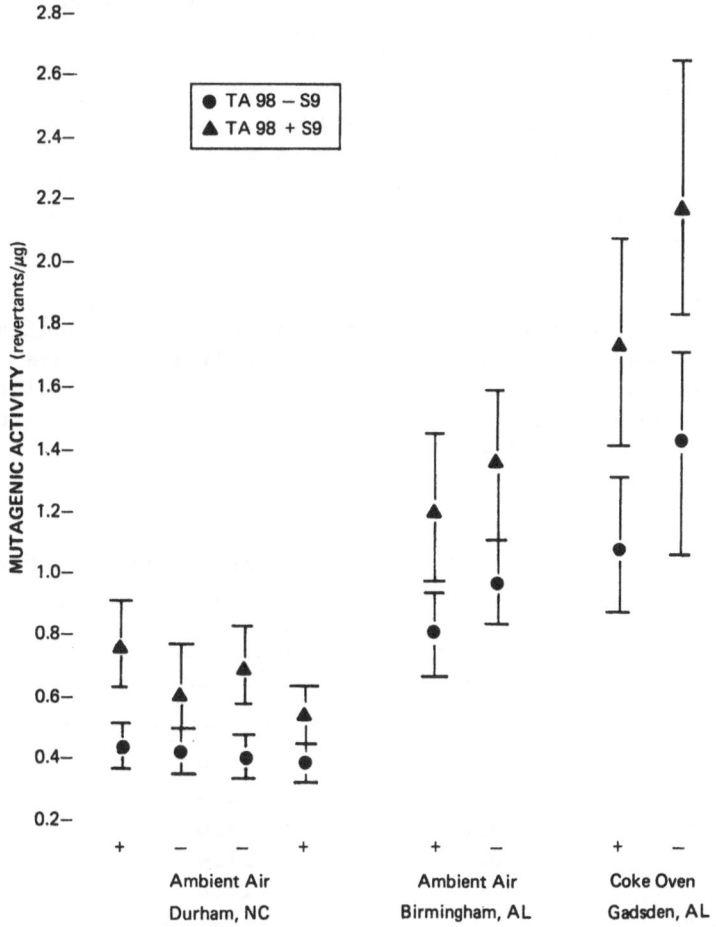

Figure 6.   Corona potential.

Although the mutagenic activity of the Birmingham and Gadsden samples was greater than the Durham samples, no significant difference was observed between the two samples collected during the same time periods with different coronas.

Evaluation of Extraction Techniques

To determine which extraction technique was the most effective in removing mutagens from ambient air particles, two extraction methods and seven solvent systems were evaluated. The two extraction methods were Soxhlet extraction for 24 h and sonication for both 30 min and 2 h. The four basic single solvents, selected for increasing polarity index, were cyclohexane (CH), dichloromethane (DCM), acetone (Ac), and methanol (MeOH), with polarity indices of 0.0, 3.4, 5.4, and 6.6, respectively. Three additional comparative solvent systems were sequential sonication of cyclohexane and methanol (CH/MeOH), 1:1:1 toluene: dichloromethane:methanol (T:DCM:MeOH), and direct suspension in dimethylsulfoxide (DMSO). These solvent systems were selected for comparison with previous work (Pitts et al., 1977; Pellizzari et al., 1979). Only the four single solvents were used in the Soxhlet extraction method, while two of the comparative solvent systems were also used in sonication.

The data in Table 5 indicate no significant difference for percent extractables in the length of time used for sonication, but acetone and methanol extracted approximately three times as much mass as cyclohexane and twice as much as DCM. The Soxhlet extraction method indicated a significant increase of percent extractables using acetone as the solvent.

Table 5. Ambient Air Particles Percent Extractables

Solvent	Soxhlet 24 h	Sonication 30 min	Sonication 2 h
CH	4.0	2.9	3.0
DCM	6.4	4.3	4.7
Ac	21.0	9.0	8.2
MeOH	8.9	11.4	11.5
CH/MeOH		11.6	11.6
T:DCM:MeOH		8.7	8.9

An aliquot of each extract was dried under nitrogen and diluted with a 1:1 mixture of DCM and cyclohexane. Fifty microliters were spotted on a TLC plate and developed in a tank containing a 1:2 mixture of DCM and ethanol. Analysis was done with a Perkin Elmer MPF-44B Fluorescence Spectrometer. The results are shown in Table 6. Regardless of method of extraction or solvent used, the benzo(a)pyrene (B[a]P) analysis is relatively constant on a weight-by-weight basis.

Table 6.   Ambient Air--Glass Fiber Filters B(a)P Analysis

Solvent	B(a)P (µg/g)		
	Soxhlet 24 h	Sonication 30 min	Sonication 2 h
CH	11.8	8.2	8.4
DCM	12.0	10.9	9.5
Ac	11.4	11.7	10.1
MeOH	11.8	11.1	10.4
CH/MeOH		13.2	12.8
T:DCM:MeOH		13.9	11.7

An aliquot of 10% of the extract was dried and prepared in aqueous solution for analysis by a Dionex Model 10 ion chromatograph (IC). The IC was equipped with a standard column and 0.003 $\underline{M}$ sodium bicarbonate and 0.0024 $\underline{M}$ sodium carbonate eluent. The sample was analyzed for fluoride, chloride, nitrate, and sulfate ions. Total anion concentration was calculated and compared for the four single solvents used in Soxhlet extraction and the two additional comparative solvent systems used in the sonication extraction techniques. Table 7 shows that the increasing polarity of the solvent is comparable to the increasing quantity of total anions extracted, regardless of extraction technique, although the increased time of sonication did extract increased amounts of total anions. Table 7 also shows that the methanol in both the Soxhlet and 30-min sonication extraction techniques and the comparative sorbent system CH/MeOH in the 30-min sonication extraction technique extracted approximately the same quantities of total anions. It appears that methanol is the basic extracting solvent of total anions in both systems.

Table 7.   Comparison of Total Anion Concentrations

	Total Anions (µg/g)		
Solvent	Soxhlet 24 h	Sonication 30 min	Sonication 2 h
CH	29.12	17.09	13.80
DCM	19.32	21.77	81.89
Ac	583.71	319.66	528.49
MeOH	992.67	878.54	1124.87
CH/MeOH		949.44	99.749
T:DCM:MeOH		443.69	500.17

Table 8 summarizes anion analysis of the four single solvents in the Soxhlet extraction technique.   The major quantity of anion extracted for acetone is nitrate ion and for methanol is sulfate ion.   Neither of these ions appreciably affect the microbial mutagenicity bioassay results.   The results of further studies incorporating sonication and the comparative solvent systems are shown in Table 9 for nitrate ion and Table 10 for sulfate ion.

It should be noted that while the comparative solvent systems do not extract significantly more nitrate ion than does the single solvent acetone, they extract far less of the sulfate ion than does the single solvent methanol.   Therefore, it appears that the two comparative solvent systems (CH/MeOH and T:DCM:MeOH) are not more effective than the Soxhlet-extracted single solvents (acetone and methanol).

Bioassay Results and Discussion

The S. typhimurium mutagenesis data for tester strain TA98 for each solvent system and extraction technique are shown in Table 11.   DCM extraction, either by Soxhlet or sonication, resulted in an extractable material that was more mutagenic than that resulting from any of the other solvents.   The least polar solvent, cyclohexane, was the least effective in extracting mutagens by either method.   Acetone and methanol solvent extraction generally resulted in less-mutagenic samples than DCM extraction, except when acetone was used in sonication.   Acetone removed considerably more extractable mass from air particles, including more inorganics.   Therefore, when the mutagenic activity was

Table 8.   Ambient Air Anion Analysis

Solvent	Anion Concentrations (μg/g)[a]			
	Fl	Cl	NO$_3$	SO$_4$
CH	0.02	2.6	7.0	19.5
DCM	0.02	2.2	2.8	14.3
Ac	2.41	83.7	477.5	20.1
MeOH	0.77	18.7	56.1	917.1

[a]Soxhlet extraction for 24 h.

Table 9.   Ambient Air--Glass Fiber Filters Nitrate Analysis

Solvent	Nitrate (μg/g)		
	Soxhlet 24 h	Sonication 30 min	Sonication 2 h
CH	7.01	6.73	1.22
DCM	2.81	3.93	3.04
Ac	477.48	265.27	290.62
MeOH	56.11	42.92	498.57
CH/MeOH		505.48	468.43
T:DCM:MeOH		318.32	397.02

Table 10. Ambient Air--Glass Fiber Filters Sulfate Analysis

	Sulfate (μg/g)		
Solvent	Soxhlet 24 h	Sonication 30 min	Sonication 2 h
CH	19.48	7.61	8.82
DCM	14.33	15.48	71.74
Ac	20.06	12.61	4.26
MeOH	917.12	825.41	516.80
CH/MeOH		331.40	425.61
T:DCM:MeOH		51.54	6.88

Table 11. Mutagenic Activity of Different Solvent Extractable Material from Air Particulate in TA98 (without S-9 activation)

	Soxhlet Extraction		Sonication	
Solvent	Rev/μg	95% Confidence Limits	Rev/μg	95% Confidence Limits
CH	0.11	0.05 - 0.16	0.09	0.06 - 0.13
DCM	0.52	0.43 - 0.62	0.48	0.37 - 0.58
Ac	0.28	0.23 - 0.33	0.45	0.38 - 0.53
MeOH	0.26	0.23 - 0.30	0.35	0.28 - 0.41
CH/MeOH			0.38	0.30 - 0.45
T:DCM:MeOH			0.37	0.26 - 0.48
DMSO			0.07	0.03 - 0.10

calculated per milligram of particle, as shown in Table 12, acetone appeared to result in greater mutagenic activity. Consequently, the use of acetone may be advantageous for certain studies. Analytical problems associated with the use of acetone plus the presence of higher concentrations of inorganic salts have resulted in our selection of a solvent other than acetone for most studies.

Table 12. Ambient Air Particle Mutagenicity in TA98
(without S-9 activation)

Solvent	% Extractable[a]	Revertants/μg Extractable	Revertants/mg Particle
CH	4.0	0.11	4.4
DCM	6.4	0.52	33.3
AC	21.0	0.28	58.8
MeOH	8.9	0.26	23.1

[a]Soxhlet extraction.

While DMSO can be used directly both to suspend the particles and administer to the bioassay without evaporation or solvent exchange, it was the least effective method for detecting the mutagenic activity present. Thus, DCM is the preferable solvent, particularly for studies in which the amount of nonmutagenic mass should be minimized.

REFERENCES

Ames, B.N., J. McCann, and E. Yamasaki. 1975. Methods for detecting carcinogens and mutagens with the Salmonella/ mammalian microsome mutagenicity test. Mutation Res. 31:347-364.

Bjørseth, A. 1977. Analysis of polycyclic aromatic hydrocarbons in particulate matter by glass capillary gas chromatography. Anal. Chim. Acta 94:21-27.

Bjørseth, A., O. Bjørseth, and P.E. Fjeldstad. 1978. Polycyclic aromatic hydrocarbons (PAH) in industrial atmospheres. I. Analysis of the PAH content in an aluminum reduction plant. Scand. J. Work Environ. Hlth 4:212-216.

Code of Federal Regulations. 1980. Title 40, part 50, appendix B. Reference Method for Determination of Suspended Particulates in the Atmosphere (High Volume Method). General Service Administration: Washington, DC. pp. 531-535.

Committee on Biological Effects of Atmospheric Pollutants. 1972. Particulate polycyclic organic matter. National Academy of Science: Washington, DC.

Henry, W.M., and R.I. Mitchell. 1978. Development of a large sample collector of respirable particulate matter. EPA-600/4-78-009. U.S. Environmental Protection Agency: Research Triangle Park, NC.

Hoffman, D., and E.L. Wynder. 1976. In: Chemical Carcinogens. C.E. Searl, ed. ACS Monograph, 173, American Chemical Society: Washington, DC. pp. 324-365.

Hueper, W.C., P. Kobin, E.C. Tabor, W.W. Payne, H. Falk, and E. Sawicki. 1962. Carcinogenesis bioassays on air pollutants. Arch. Pathol. 74:89-116.

Huisingh, J. Lewtas. 1980. Bioassay of particulate organic matter from ambient air. Presented at the U.S. Environmental Protection Agency Second Symposium on the Application of Short-term Bioassays in the Fractionation and Analysis of Complex Environmental Mixtures, Williamsburg, VA.

Leiter, J., M.B. Shimkin, and M.J. Shear. 1942. Production of subcutaneous sarcomas in mice with tars extracted from atmospheric dusts. J. Natl. Cancer Inst. 3:155-165.

McFarland, A., and C.E. Rodes. 1979. Characteristics of aerosol samplers used in ambient air monitoring. Presented at the 86th National Meeting of Chemical Engineers, Houston, TX.

Mitchell, R.I., W.M. Henry, and N.C. Henderson. 1978. Fabrication, optimization, and evaluation of a massive air volume sampler of sized respirable particulate matter. EPA-600/4-78-031. U.S. Environmental Protection Agency: Research Triangle Park, NC.

National Science Foundation, Subcommittee on Ozone and Other Photochemical Oxidants. 1977. Ozone and other photochemical oxidants. Printing and Publishing Office, National Academy of Sciences: Washington, DC.

Pellizzari, E.D., L.W. Little, C. Sparacino, T.J. Hughes,
    L. Claxton, and M.D. Waters. 1979. Integrating
    microbiological and chemical testing into the screening of
    air samples for potential mutagenicity. In: Application of
    Short-term Bioassays in the Fractionation and Analysis of
    Complex Environmental Mixtures. M.D. Waters, S. Nesnow,
    J.L. Huisingh, S.S. Sandhu, and L. Claxton, eds. Plenum
    Press: New York. pp. 382-418.

Pitts, J.N., D. Grosjean, J.M. Mischke, V.F. Simmon, and D. Poole.
    1977. Mutagenic activity of airborne particulate organic
    pollutants. Toxicol. Lett. 1:65-70.

Teranashi, K., K. Hamada, N. Tekeda, and H. Watanabe. 1977.
    Mutagenicity of the tar in air pollutants. Proceedings of the
    4th International Clean Air Congress, Tokyo, Japan.
    pp. 33-36.

INTEGRATION OF THE AMES BIOASSAY AND CHEMICAL ANALYSES IN AN EPIDEMIOLOGICAL CANCER INCIDENCE STUDY

C. Peter Flessel, Jerome J. Wesolowski, SuzAnne Twiss,
James Cheng, Joel Ondo, Nadine Monto, and Raymond Chan
Air and Industrial Hygiene Laboratory
California Department of Health Services
Berkeley, California

INTRODUCTION

The development of the Ames bioassay as an instrument for assessing public health problems involving mutagenicity and potential carcinogenicity has resembled the development of other quantitative techniques for assessing public health problems. The Ames test has developed from a qualitative assay of samples in simple matrices into a quantitative determination of complicated environmental mixtures, and its results are now being integrated with human epidemiological studies. Originally, the Ames test was used primarily to determine whether or not a compound was mutagenic (Ames, 1971). Soon quantitative methods were introduced (Ames et al., 1975), and a significant correlation between mutagenicity in the Ames test and carcinogenicity in animal bioassays emerged (McCann et al., 1975). Shortly thereafter, the test was applied to environmental mixtures in the analysis of air particulate material (Pitts et al., 1977; Talcott and Wei, 1977). Currently, the Ames test is used to detect mutagens in a variety of media and sample types. The rapidly expanding list of applications includes drinking water, cigarette smoke, auto exhaust, foods, drugs, and urine (Holstein et al., 1979). Although not a quantitative test in the sense of having well-established precision and accuracy, the Ames bioassay yields results that indicate relative mutagenicity. Thus, it is appropriate to consider its use in epidemiological cancer studies.

The application of the Ames mutagenicity test to epidemiological cancer studies is logical because mutagenicity is in some measure a composite index of total potential carcinogenicity (McCann et al., 1975). It can be argued that

67

determining the concentrations of all carcinogens in the medium
also should give, at least in theory, the information needed to
determine the carcinogenic exposure to humans. However, chemical
methods are not available to detect all possible carcinogens in a
given medium. Furthermore, even an exhaustive compilation of
carcinogens would neglect synergistic or antagonistic effects.
Thus, application of both chemical and bioassay techniques more
completely characterizes environmental samples (Bjørseth et al.,
1980).

Contra Costa County, CA, was suspected of having high rates of
respiratory-tract cancer, based on the findings of the national
survey of cancer mortality by county, 1950 through 1969 (Mason and
McKay, 1973). The northern section of Contra Costa County is
heavily industrialized, with five major petroleum refineries and
many petrochemical plants. These facts prompted the U.S.
Environmental Protection Agency (EPA) to fund an epidemiological
study of cancer incidence as related to airborne emissions in
Contra Costa County. This study is part of a larger study funded
jointly by the State of California and the Occupational Safety and
Health Administration (OSHA). The larger study includes four other
counties in the San Francisco Bay Area and is being carried out by
the California Resource for Cancer Epidemiology (RCE). The present
discussion is restricted to the Contra Costa County portion of the
project. The major objectives of the study are 1) to identify
environmental factors which may contribute significantly to cancer
incidence in the county, 2) to determine whether various groups of
workers are more likely to develop cancer than others, and 3) to
evaluate whether air pollutants have affected the observed
incidence of respiratory-tract cancer in the county.

An important component of the study is environmental
monitoring, consisting of ambient air particulate matter sampling
and subsequent chemical and biological analyses. A major goal is
to determine whether or not mutagenic activity, as measured by the
Ames assay, can be accounted for by the chemical characterization
of the samples. Environmental monitoring will provide a means of
correlating the geographic distribution of current cancer incidence
with current ambient air pollutants and to develop baseline air
pollution data for comparison with future measurements and use in
future epidemiological cancer studies. Because of the latency of
cancer onset, current incidence levels cannot be causally related
to current air quality. Analysis of past and present emission data
and air quality patterns could, however, reveal historical
pollution trends useful in determining the association, if any,
between air pollution and current cancer incidence.

This paper will discuss the design of the environmental
monitoring program and preliminary data from the chemical and

biological assays. The planned integration of this environmental
data base with the epidemiological cancer study will be described.

PROCEDURE

The Sampling Program

Fifteen high-volume particulate samplers were placed at
thirteen locations in Contra Costa County and two locations in
adjacent counties (shown in Figure 1), in order to characterize
air quality variations over the entire county. Five of the
locations are permanent stations of the Bay Area Air Quality
Management District (BAAQMD). Although these five stations monitor
for several pollutant gases as well as for particulate matter, the
present paper will discuss only particulate matter results.

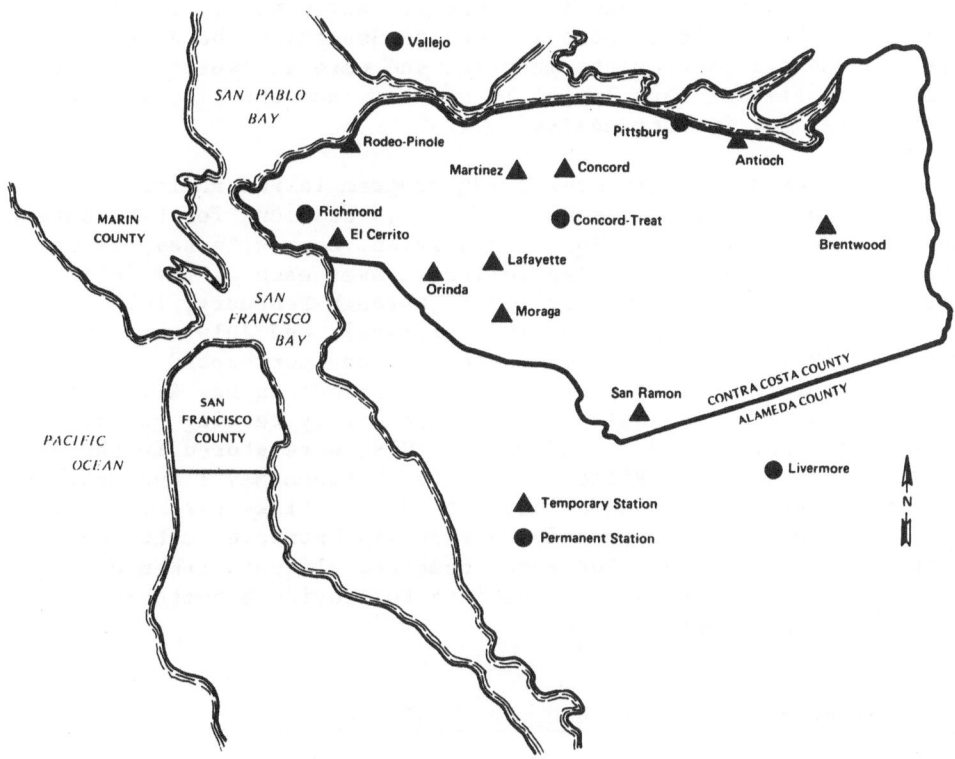

Figure 1. Locations of sampling stations in Contra Costa County,
CA.

Air particulate material was collected on 8- x 10-in. (20- x 25-cm) glass-fiber filters (EPA Grade Whatman) in standard high-volume samplers, which drew approximately 1,200 m^3 of air through a filter during each 24-h run. Filters were collected every sixth day at each of the 15 sampling stations, from November 3, 1978, through October 31, 1979. In this period, nearly 900 air-filter samples were collected and analyzed.

Particulate matter was analyzed for benzene-soluble organics (BSO), lead (Pb), total suspended particulate matter (TSP), nitrates (NO_3^-), sulfates ($SO_4^=$), specific polycyclic aromatic hydrocarbons (PAH), and mutagenic activity, using the Ames test. The BAAQMD collected the air samples and analyzed them for TSP, NO_3^-, and $SO_4^=$, and the Air and Industrial Hygiene Laboratory (AIHL) carried out the other analyses.

Logistics of Sample Analysis and Data Management

The plan for distributing filter samples for analysis and reporting results is shown in Figure 2. After weekly sample collection, the filters were weighed to determine the amounts of total suspended particulate material and were delivered to AIHL. There, the filters were logged in and cut, and the pieces were distributed for further analysis.

The crux of the air-monitoring program is the analysis of composite air samples from each of the 15 stations for PAH content and mutagenic activity. For each station, composite samples were prepared by combining samples collected over each of the following four-month periods: November, 1978, through February, 1979 (winter); March through June, 1979 (spring); and July through October, 1979 (summer). These three periods correspond to the three meteorological seasons of the San Francisco Bay Air Basin. Filter disks for PAH analysis and mutagenicity testing collected between November 1, 1978, and May 1, 1979, were stored in the dark at room temperature. Filters collected between May 1 and October 31, 1979, were stored in the dark at -20°C. Disks (47-mm diam.) were cut from each filter and individually extracted ultrasonically with organic solvents. For each location, aliquots taken during a given four-month period were combined to provide a composite sample for PAH and mutagenic analysis.

Analysis Methods for TSP, BSO, Pb, NO_3^-, and $SO_4^=$

Standard methods were used to analyze for the following five pollutants: TSP was determined gravimetrically (BAAQMD, 1977); NO_3^- colorimetrically (BAAQMD Method N-7, 1976); $SO_4^=$ turbidimetrically (BAAQMD Method S-42, 1976); BSO by Soxhlet

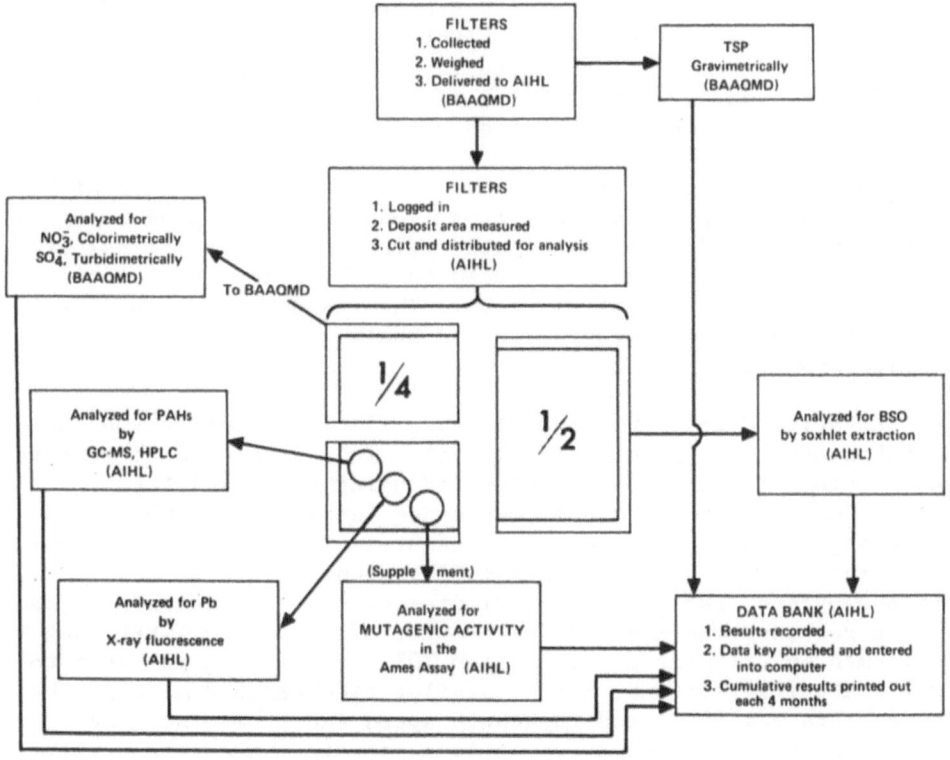

Figure 2.   Logistical plan for analysis of high-volume air filters
            collected in Contra Costa County, CA, November, 1978,
            through October, 1979.

extraction (AIHL, 1975); and Pb by wavelength dispersive X-ray
fluorescence (Moore, 1976).

Analysis Methods for PAH

     Analysis methods for PAH were modified from Bjørseth et al.
(1980).   Individual (47-mm) filter disks were placed in screw-cap
test tubes and extracted twice at 40 to 45°C for 20 minutes, first
with 8 ml and then with 6 ml of cyclohexane (MCB, OmniSolv) in an
ultrasonic bath (Bransonic Models 220 or 32).   Each cyclohexane
extract was filtered through a 0.5-μm Fluoropore filter
(Millipore).   Extracts comprising each composite were combined in a
round-bottom evaporating flask and concentrated to 6 ml in a rotary
evaporator at 45 to 50°C.   To separate PAH from interfering
material, the 6 ml of concentrated cyclohexane extract was combined
with 2 ml of toluene (MCB, OmniSolv) and extracted ultrasonically

with 12 ml of a 10:1 mixture of N,N-dimethylformamide (DMF) (MCB
Manufacturing Chemists, Inc., OmniSolv) and water in a screw-cap
test tube for 15 min. The bottom layer, containing the PAH, was
transferred by pipette to a 60-ml separatory funnel. The remaining
cyclohexane phase was re-extracted ultrasonically twice more with
6 ml of the DMF-H$_2$O mixture, and the phases containing the PAH were
combined in the separatory funnel. Following the addition of 24 ml
of distilled water to the separatory funnel, the 2 ml of toluene,
containing the PAH, separated. The toluene phase was transferred
to a centrifuge tube and evaporated to dryness in a heating block
at 45°C under a stream of nitrogen. The residue was then dissolved
in 200 to 400 µl of acetonitrile for analysis by high-pressure
liquid chromatography (HPLC).

A Varian Model 5000 high-pressure liquid chromatograph and
Microbondapak C18 (Waters Associates) column were used to separate
PAH. Column effluents were monitored using ultraviolet (UV)
absorption (at 254 nm) and fluorescence (excitation, 263 nm;
emission, 407 nm). Fluorescence measurements were used to resolve
and quantitate three carcinogenic PAH: benz(a)anthracene,
benzo(a)pyrene, and chrysene. Fluorescence measurements were made
with a Perkin-Elmer Model MPF-44A spectrofluorometer. HPLC was
performed in a linear gradient from 70% acetonitrile in water to
100% acetonitrile, in 50 min. The flow rate was 0.8 ml/min, the
temperature was 30°C, and the chart speed was 1 cm/min. The
injection was made with a sample loop operated by a rotary valve,
using a 10-µl injection volume.

The efficiency of extraction of PAH from high-volume filters
has been studied. Fluorescence measurements show that more than
95% of PAH can be recovered from a spiked filter.

Peaks observed in the HPLC chromatograms were identified by
three methods. First, peaks were tentatively identified by
comparing their retention times to those of standards. Second,
the peak height ratios of samples and standards were compared at
the wavelengths used to measure absorbance and fluorescence.
Identifications of the three PAH were confirmed by stopping the
flow during HPLC analysis and scanning the fluorescence spectra
using the optimum excitation wavelength for each compound. Peak
heights were used for quantitation of PAH.

The selectivity of the fluorescence detection method is
illustrated in Figures 3 and 4. Using fluorescence excitation and
emission wavelengths of 263 nm and 407 nm respectively, most of the
UV-absorbing PAH peaks were suppressed, but benzo(a)pyrene,
benz(a)anthracene, and chrysene were enhanced. Such specificity is
critical, because many poorly resolved peaks, including those
containing the 10 PAH standards, are visible in the chromatograms
of air samples using UV detection (Figure 4). The major peak

eluting just before phenanthrene was due to an organic contaminant in the cyclohexane used in early experiments. This artifact disappeared when the brand of cyclohexane specified above was substituted.

The Ames Test for Mutagenic Activity

Methods for extracting air particulate material from high-volume glass-fiber filters and making composite samples were adapted from Pitts et al. (1979). The solvent, a 1:1:1 mixture of methanol, dichloromethane, and toluene (MCB, OmniSolv), was prepared fresh daily and saturated with nitrogen. Extractions were carried out in an ultrasonic bath under low light. Each individual 47-mm filter disk cut from a filter was placed in a 16- x 125-mm screw-cap tube with a Teflon liner, and 4 ml of solvent was added. Tubes were sonicated for 20 min at maximum power in the ultrasonic bath containing water at 45°C. Extracts were filtered through 0.5- m Fluoropore filters (Millipore); filter disks were re-extracted using 3 ml solvent, and the filtrates were combined. The volume of each extract was adjusted to exactly 10 ml with the solvent, and the extracts were stored at -20°C.

Composite samples for mutagenic testing were prepared by combining aliquots of stored extracts in a vacuum flask, saturating the extracts with nitrogen, and reducing the volume of solvent in a rotary evaporator, under reduced pressure and at a temperature of 45°C. The composite samples were then transferred to preweighed tubes, which were placed in a heat block at 45°C; the remaining extraction solvent was removed under a stream of nitrogen. After weighing, residues were redissolved in dimethylsulfoxide for mutagenic analysis.

The method for detecting mutagens with the Salmonella/ mammalian microsome test was as described by Ames et al. (1975), with the following changes: rat liver homogenate (S-9) was prepared from rats fed commercial rodent foods; rats were anesthesized with carbon dioxide before surgery; and plates were incubated for 72 instead of 48 h. S-9 protein concentrations were determined by the method of Lowry et al. (1951).

Negative solvent (dimethylsulfoxide) and S-9 sterility controls and positive controls for each strain used were run with each experiment. The control mutagens for the five tester strains were sodium azide in TA1535; 9-aminoacridine in TA1537; 2-aminofluorene in TA1538 and TA98; and methyl methanesulfonate in TA100. The Ames assay was applied according to a two-part protocol (Pitts et al., 1979). The first step involved screening the sample in the five standard Ames tester strains both with and without metabolic activation. These data gave a qualitative estimate of

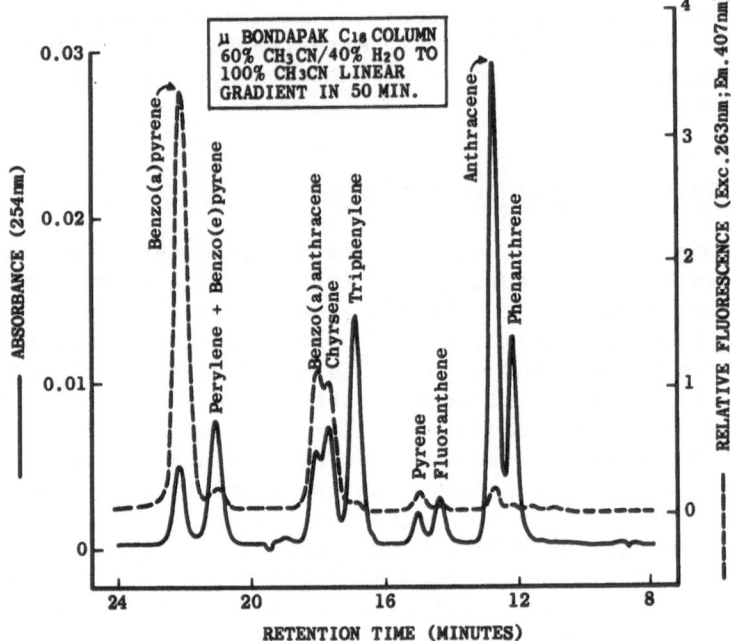

Figure 3.   High-pressure liquid chromatogram of a 10-PAH standard
            detected by fluorescence and UV absorbance.

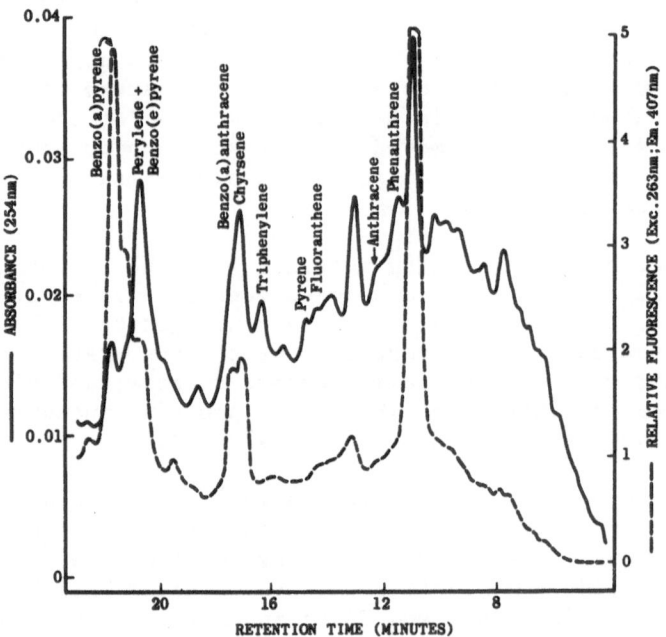

Figure 4.   High-pressure liquid chromatogram of the PAH fraction
            from air particulate material collected in Antioch, CA,
            November, 1978, through February, 1979.

the mutagenic activity and indicated the most sensitive strain and the optimum conditions of metabolic activation for subsequent quantitative analysis. For initial screening, each composite sample was assayed at one dose in the range 20 to 1000 µg/plate. Determinations were made with and without added rat liver S-9 at both low (~ 0.6 mg/plate) and high (~ 3 mg/plate) protein concentrations.

All composite samples exhibited mutagenic activity in the initial screening and were reanalyzed in the strain showing the greatest response. The dose range and conditions of metabolic activation also were chosen to maximize activity. Duplicate determinations were made at each of several doses and the result expressed as revertants per cubic meter.

An interlaboratory comparison of the mutagenicity of ambient particulates was carried out between AIHL and the Statewide Air Pollution Research Center, University of California, Riverside. High-volume filter samples collected in southern and northern California were split and analyzed for mutagenic activity in the Ames test. The determinations made by the two laboratories agreed within a factor of two.

RESULTS

As analyses are still in progress, only partial results are presented.

Table 1 gives the median and maximum values for TSP, BSO, and Pb for winter and spring. The median and maximum levels of the three pollutants were the highest in winter, when meterological inversions frequently occur. The extreme values for individual 24-h runs differed by a factor of 50. For example, the highest 24-h level of BSO was 41.3 mg/m^3, in Concord in early December, 1978, and the lowest was less than 0.8 mg/m^3 (the detection limit) at several sites in January and February, 1979. The highest Pb concentration was found in Antioch, also in early December, while the highest TSP concentration was in Brentwood in November.

To describe variations in levels of community air pollution throughout the county, maps showing the geographical distribution of the seven measured pollutants are being constructed using a computer program called SYMAP (developed by the Laboratory for Computer Graphics, Harvard University, Cambridge, MA). In this program, sampling station coordinates and associated pollutant levels are used to construct a matrix containing the pollution levels at each station. Contours are constructed by interpolation. These distributions will be used to estimate community exposure levels and will be compared with the patterns of cancer discovered

Table 1. Analysis of Air Particulate Material Collected
in Contra Costa County, CA

Pollutant	24-Hour Median Value		24-Hour Maximum Value	
	Nov. 1978–Feb. 1979	March–June 1979	Nov. 1978–Feb. 1979	March–June 1979
Total suspended particulate material	60	42	229	126
Benzene-soluble organics	4.8	1.4	41.3	32.1
Lead	0.7	0.2	2.5	0.6

in epidemiological studies. The distributions may also provide clues to pollution sources.

Thus far, computer-drawn contour maps of TSP mass, BSO, and Pb levels for winter and spring have been prepared; these are shown in Figure 5. They were constructed using average values obtained at each sampling station during the first two seasons for which composite samples were analyzed. Panels A, B, and C show the distributions of these pollutants during winter; panels D, E and F show the distributions in the spring. Concentrations of the pollutants were generally higher in winter than in spring, and seasonal variations were most pronounced for Pb and BSO levels, which changed by more than a factor of three (see Table 1). The geographical distributions of Pb and BSO were similar. For both pollutants, the highest levels were found in winter in a north-south band located in central Contra Costa County. This region corresponds roughly to the Diablo Valley, a natural pollution sink through which runs a major freeway.

At present, three PAH have been quantitated: benz(a)anthracene benzo(a)pyrene, and chrysene. Concentrations from the Antioch and Brentwood winter composite samples are listed in Table 2. Antioch is an urban-industrial site; Brentwood is a rural location. The range of concentrations of benz(a)pyrene found in these samples (0.1 to 0.8 ng/m) was comparable to, although somewhat lower than, that found in particulate samples collected in San Francisco 20 years ago (Sawicki et al., 1960). The present values were also

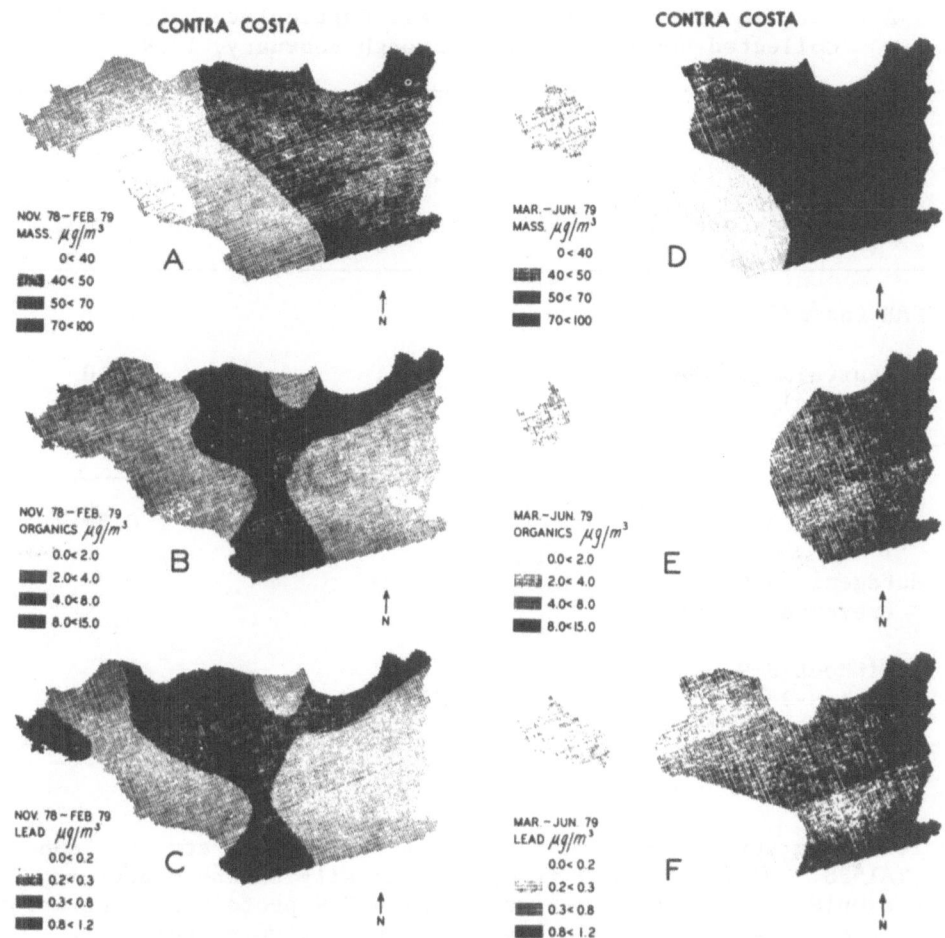

Figure 5. Computer-drawn contour maps of the geographical
distribution of levels of total mass, benzene-soluble
organics, and lead in air particulate material
collected in Contra Costa County, CA. A, B, and C
show distributions for November, 1978 through
February, 1979, and D, E, and F show March through
June, 1979, distributions.

comparable to those measured more recently in New York City (Daisey
et al., 1979) and Germany (Gusten and Heinrich, 1978).

Extracts from samples collected at the 15 stations have been
analyzed qualitatively for mutagenicity, and all showed activity in

Table 2. Polycyclic Aromatic Hydrocarbon Content and Mutagenic
Activity in Contra Costa County Air Particulate Material
Collected November, 1978, through February, 1979

	Sampling Location	
Type of Measurement	Antioch	Brentwood
PAH (ng/m^3)		
Benz(a)anthrene	0.81	0.10
Benzo(a)pyrene	0.81	0.10
Chrysene	0.90	0.16
Sum of three PAH	2.52	0.36
Mutagenic activity (revertants/m^3)		
Without S-9	6.3	3.9
With S-9	25.4	5.9

at least one strain. The most activity was seen in strains TA98
and TA1538. Adding the S-9 fraction generally enhanced activity;
some samples were most active at the high S-9 protein concentration
(3 mg/plate), while the majority were most active at the lower
concentration (0.6 mg/plate). Dose-response curves for the Antioch
and Brentwood composites (with activities given as revertants/
plate) are shown in Figures 6 and 7. Over the dose range used (up
to 20 m^3 of air), the mutagenic responses appear linear. The
dose-response curves obtained with the Antioch winter composite
(Figure 6) were somewhat atypical, in that most samples did not
show a fourfold increase in activity in the presence of S-9. The
response to S-9 in samples analyzed to date was more typically that
shown by the Brentwood sample (Figure 7), although stations in more
heavily polluted areas generally showed greater S-9 enhancement.

Seasonal variations in mutagenic activity were also observed.
Activities (expressed as revertants per cubic meter) were generally
higher for samples collected in winter than for samples collected
in spring or summer. Table 2 summarizes results from the Antioch
and Brentwood Stations. Samples from Antioch, a more urban setting
than Brentwood, showed both higher concentrations of PAH and

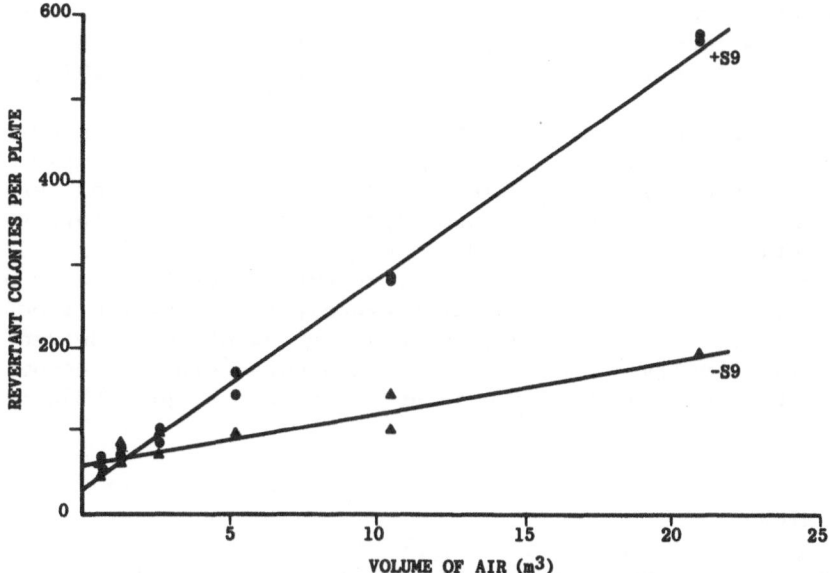

Figure 6. Ames test dose-response curves for a composite sample from Antioch, CA, November, 1978, through February, 1979, assayed in strain TA98, with and without 3 mg S-9 protein/plate.

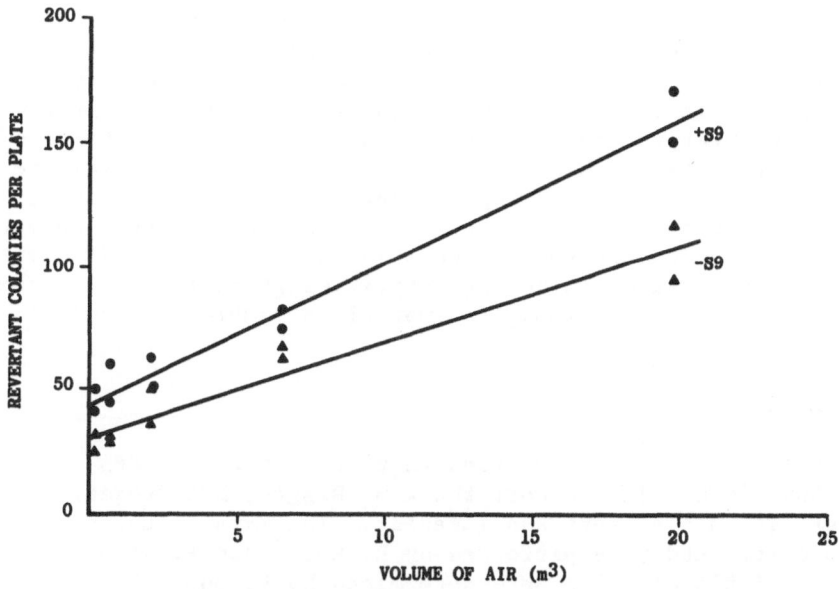

Figure 7. Ames test dose-response curves for a composite sample from Brentwood, CA, November, 1978, through February, 1979, assayed in strain TA98, with and without 0.6 mg S-9 protein/plate.

increased mutagenic activity. The activities and the urban-rural differences found in this study were similar to those found previously in California (Flessel, 1977; Pitts et al., 1977, 1979).

One can also compare chemical composition and biological activity in a given sample. For example, the Antioch sample analyzed above had a benzo(a)pyrene concentration of 0.81 ng/m^3. As the molecular weight of benzo(a)pyrene is 252, this corresponds to approximately 0.003 nmol/m^3. The specific mutagenic activity of benzo(a)pyrene is approximately 121 revertants/nmol (Ames et al., 1975). Therefore, the amount of benzo(a)pyrene measured accounts for about 0.36 revertants/m^3, or less than 2% of the observed activity. Clearly, chemicals other than benzo(a)pyrene account for most of the mutagenicity in this and other air samples (Bjørseth et al., 1980).

DISCUSSION

The cancer epidemiology project plan consists of a series of studies, both prospective and retrospective, to investigate environmental factors relevant to cancer in Contra Costa County. The study will include a census tract analysis of cancer incidence at specific sites and is designed to ultimately attempt to distinguish between the contributions of occupational and community exposures to the incidence of cancer. From the mutagencity and PAH data, community exposures in Contra Costa County will be estimated. These exposures will be examined for correlations with present and future cancer rates in various geographical areas. Although this is fundamentally a prospective study, attempts will be made to interpret current cancer incidence rates in terms of current exposures and historical pollution trends. Definitive studies will require following current Contra Costa residents through the next several decades and monitoring a larger number and variety of carcinogens and mutagens in community air. It might be worthwhile to expand monitoring activities to include data from community water, soil, food, and workplace samples.

ACKNOWLEDGMENTS

This research was supported in part by the U.S. EPA grant no. R 806396010. The authors thank W. Riggan, EPA Project Officer, for his support and interest. The X-ray fluorescence analyses for lead were performed by H. Moore and A. Alcocer (AIHL). Benzene-soluble organics were determined by F. Boo (AIHL). Nitrates, sulfates, and total suspended particulate matter were measured by S. Balestrieri (BAAQMD). We also wish to acknowledge the cooperation of M. Imada, E. Jeung, and R. Stanley (AIHL); D.

Levaggi, W. Siu, D. England, and N. Balberan (BAAQMD); and D.
Austin, W. Mandel, and S. Lum (RCE, California Department of Health
Services).

REFERENCES

AIHL Method 67. 1975. Determination of Total Organic Materials
    in Atmospheric Particulate Matter. Air and Industrial Hygiene
    Laboratory, California Department of Health Services,
    Berkeley, CA.

Ames, B. 1971. The detection of chemical mutagens with enteric
    bacteria. In: Chemical Mutagens: Principles and Methods
    for their Detection, Vol. 1. A. Hollaender, ed. Plenum
    Press: New York. pp. 267-282.

Ames, B. J. McCann, and E. Yamasaki. 1975. Method for detecting
    carcinogens and mutagens with the Salmonella/mammalian-
    microsome mutagenicity test. Mutation Res. 31:347-364.

BAAQMD Method. 1977. Total Suspended Particulate Gravimetric
    Analysis Procedure. Bay Area Air Quality Management District,
    San Francisco, CA.

BAAQMD Method N-7. 1976. Determination of Nitrate in Glass Fiber
    Hi-Vol Filters. Bay Area Air Quality Management District,
    San Francisco, CA.

BAAQMD Method S-42. 1976. Determination of Sulfate in Glass Fiber
    Hi-Vol Filters. Bay Area Air Quality Management District,
    San Francisco, CA.

Bjørseth, O., P. Flessel, N. Monto, J. Wesolowski, T. Parker, and
    P. Ouchida. (1980). Monitoring for polycyclic aromatic
    hydrocarbon (PAH) content and mutagenic activity in products
    and emissions from a gasifier demonstration project. In:
    Safe Handling of Chemical Carcinogens, Mutagens, and
    Teratogens--A Chemist's Viewpoint, Vol 2. D. Waters, ed.
    Ann Arbor Press: Ann Arbor, MI. pp. 635-651.

Daisey, J., M. Leyko, and T. Kniep. 1979. Source identification
    and allocation of polynuclear aromatic hydrocarbon compounds
    in the New York City aerosol: methods and applications. In:
    Polynuclear Aromatic Hydrocarbons. P. Jones and P. Leber,
    eds. Ann Arbor Science: Ann Arbor, MI. pp. 201-215.

Flessel, P. 1977. Mutagenicity of Particulate Matter. Presented
    at the Third Interagency Symposium on Air Monitoring Quality
    Assurance, Air, and Industrial Hygiene Laboratory, California
    Department of Health Services, Berkeley, CA.

Gusten, H., and G. Heinrich. 1979. Polycyclic aromatic
    hydrocarbons in the lower atmosphere of Karlsruhe. In:
    Polynuclear Aromatic Hydrocarbons. P. Jones and P. Leber,
    eds. Ann Arbor Science: Ann Arbor, MI. pp. 357-370.

Holstein, M., J. McCann, F. Angelosanto, and W. Nichols. 1979.
    Short-term tests for carcinogens and mutagens. Mutation Res.
    65:133-226.

Lowry, O., N. Rosebrough, A. Farr, and R. Randall. 1951. Protein
    measurement with the folin phenol reagent. J. Biol. Chem.
    193:265-271.

Mason, P., and F. McKay. 1973. U.S. Cancer Mortality by County,
    1950-1969. DHEW Publication No. (NIH)-74-615. U.S.
    Government Printing Office: Washington, D.C.

McCann, J., E. Choi, E. Yamasaki, and B. Ames. 1975. Detection
    of carcinogens as mutagens in the Salmonella/microsome test:
    Assay of 300 chemicals. Proc. Natl. Acad. Sci. USA
    72:5135-5139.

Moore, H. 1976. Application of Wavelength Dispersive X-ray
    Fluorescence Spectrometry to the Determination of Lead in
    Atmospheric Particulate Matter Collected on High-Volume Glass
    Fiber Filters. AIHL Report 183, Air and Industrial Hygiene
    Laboratory, California Department of Health Services,
    Berkeley, CA.

Pitts, J., D. Grosjean, and T. Mischke. 1977. Mutagenic activity
    of airborne particulate organic pollutants. Toxicol. Lett.
    1:65-70.

Pitts, J., K. Van Cauwenberghe, D. Grosjean, J. Schmid, D. Fitz,
    W. Belser, G. Knudson, and P. Hynds. 1979. Chemical and
    microbiological studies of mutagenic pollutants in real and
    simulated atmospheres. In: Application of Short-term
    Bioassays in the Fractionation and Analysis of Complex
    Environmental Mixtures. M. Waters, S. Nesnow, J. Huisingh, S.
    Sandhu, and L. Claxton, eds. Plenum Press: New York. pp.
    355-378.

Sawicki, E., W. Elbert, T. Hauzer, F. Fox, and T. Stanley. 1960.
    Benzo(a)pyrene content of the air of American communities.
    Am. Ind. Hyg. J. 21:443-451.

Talcott, R., and E. Wei. 1977. Airborne mutagens bioassayed in
Salmonella typhimurium. J. Natl. Cancer Inst. 58:449-451.

MUTAGENICITY OF AIRBORNE PARTICULATE MATTER IN RELATION TO TRAFFIC AND METEOROLOGICAL CONDITIONS

Ingrid Alfheim and Mona Møller
Central Institute for Industrial Research
Oslo, Norway

INTRODUCTION

It is now well established that environmental factors are major causes of cancer in man (Higginson and Muir, 1973). Epidemiological studies have shown that the incidence of lung cancer is higher in urban than in rural areas (Henderson et al., 1975). Urban air contains large amounts of particulate pollutants, which are believed to contribute to this higher lung-cancer incidence (Menck et al., 1974). Furthermore, the carcinogenic potential of organic extracts from airborne particles has been demonstrated in animal experiments (Hueper et al., 1962). These observations have made it increasingly important to identify carcinogenic compounds in ambient air. Since animal studies are expensive and time-consuming, short-term tests for mutagenicity in microbial systems are currently used to identify possible carcinogens.

The presence of mutagenic compounds in airborne particulate matter from urban and industrial areas has been documented by us and others using the Ames Salmonella test system (Dehnen et al., 1977; Löfroth, 1978; Møller and Alfheim, 1980; Pitts et al., 1977; Talcott and Wei, 1977; Teranishi et al., 1978; Tokiwa et al., 1977). Mutagenic activity of airborne particulates from a rural site was investigated in only one of these studies; all of the rural samples were reported to be inactive (Pitts, 1977). Part of the observed mutagenic activity in samples from urban air is probably due to polycyclic aromatic hydrocarbons (PAH). Several PAH compounds cause cancer in animals and are also suspected of causing cancer in man. However, these compounds require metabolic activation before they can act as mutagens in the Salmonella test

system (Ames et al., 1973). In the studies cited, all urban
samples were mutagenic both with and without metabolic activation,
indicating the presence of mutagens other than PAH.

Mutagens in ambient air originate from various combustion
sources, including residential heating and motor vehicle exhausts
(Møller and Alfheim, 1980; Löfroth, 1979; Wang, 1978), and may be
transported over long distances (Alfheim and Møller, 1979). Many
parameters will influence the mutagenicity of airborne particulate
matter from a given source, including the distance from the source,
meteorological conditions, the presence of other pollutants, and
the location of the source. In this work, we compared the
mutagenicity of urban air particulate matter at street level to
that at roof level, to determine the contribution of traffic to
the mutagenicity of urban air. We also compared the mutagenicity
of urban and rural airborne particulate matter, and we related
mutagenic activity to meteorological conditions and to the
composition of the particulate matter.

MATERIALS AND METHODS

Sampling

The urban sampling sites were all located in the center of
Oslo, Norway. Two sampling sites were in a narrow street with
heavy traffic (averaging 2000 cars/h during the day and 500 cars/h
at night). Site A, street level, was 2 m above the ground, and
site A, roof level, was 25 m above the ground. Sampling site B
was in a park with much lower traffic frequency than at site A.
Roof samples were taken at two other locations in Oslo: sampling
site C was at a junction with heavy traffic, and site D was located
in a more industrial area with less traffic. Sites A and C were
considered to be more polluted than sites B and D, based on SO_2
and soot measurements. Rural samples were collected in southern
Norway near the coast at Birkenes (site E) and in central Norway in
a mountainous area at Hummelfjell (site F).

Airborne particulate matter was collected during the winter
and spring of 1978 and 1979. Samples were collected on glass fiber
filters (Gelman type A-E) with high-volume samplers. At sites A
and B, the air was also passed through plugs of polyurethane (PUR)
to adsorb the more volatile compounds (especially volatile PAH)
(Alfheim et al., 1977). Separate day and night samples were
collected, each during two 12-h periods (~400 m^3 air). At sites C
and D, approximately 700 m^3 air passed through each filter during a
24-h period. About 2000 m^3 were sampled at the rural sites. For
particle fractionation, a Sierra High Volume Cascade Impactor
Sampler with split filters (Gelman type C-230) was connected to the
sampler.

Extractions and Mutagenicity Testing

The organic compounds were extracted from the filters with
50 ml of acetone or 50 ml of cyclohexane, for nonpolar compounds,
in a Soxhlet apparatus for 16 h. For mutagenicity testing, the
samples were either concentrated by evaporation and tested directly
or were evaporated to near dryness and dissolved in
dimethylsulfoxide. Samples from locations A and B were extracted
with acetone, and samples from location C, D, E, and F were
extracted with cyclohexane. Acetone or cyclohexane extracts from
unused filters were tested as·controls; positive controls were
2-aminoanthracene and benzo(a)pyrene (B[a]P).

Salmonella typhimurium strains TA98 and TA100 were kindly
supplied by Dr. B.N. Ames, University of California at Berkeley.
Liver homogenate fractions were prepared from male Wistar rats
injected with Aroclor 1254, 500 mg/kg i.p., five days prior to
preparation. The assay was carried out as described by Ames et al.
(1975).

Analysis of PAH

PAH were analyzed by the procedure of Grimmer and Bohnke
(1972) as modified by Bjørseth (1977). Internal standards were
added to the cyclohexane before extraction of the filters. The
extracts were shaken once with 50 ml and once with 25 ml
dimethylformamide (DMF):water (9:1). The DMF:water phases were
separated, water was added to give a DMF:water ratio 1:1, and the
samples were re-extracted with cyclohexane. The final cyclohexane
extracts were concentrated, first with a modified Vigreux column
under nitrogen and reduced pressure and then with a stream of
highly purified nitrogen at 30°C. The analysis was performed on a
Carlo-Erba Fractovap 2101 AC gas chromatograph with a glass
capillary column and flame ionization detector. The cyclohexane
extract, 2 µl, was injected splitless (Grob and Grob, 1969).

Chemical Fractionation

Acetone extracts from three samples collected at site A,
street level, were pooled, evaporated to near dryness, transferred
to diethylether, and fractionated into acidic, basic, and neutral
fractions. The acidic and basic fractions contained compounds that
were extractable from ether by aqueous sulfuric acid and sodium
hydroxide, respectively, and re-extractable into ether after
neutralization. The neutral fraction was further separated on a
silica column by elution with cyclohexane, benzene, and ether. The
cyclohexane fraction was subdivided into three fractions, on a

column of silica gel with a layer of aluminum oxide on top, by
elution with pentane containing increasing amounts of ether.

One sample from each of these seven fractions was evaporated
to dryness and tested for mutagenic activity. The most mutagenic
fractions were analyzed by glass capillary gas chromatography, as
described above, or by combined gas chromatography/mass
spectrometry (GC/MS) (Bjørseth et al., 1977).

RESULTS

Mutagenicity Testing

Preliminary results revealed very little or no mutagenic
activity in Salmonella strain TA100; therefore, strain TA98 was
used for further studies, including all those reported here. This
strain had 35 to 45 spontaneous revertants per plate. Twice the
number of spontaneous mutants was considered a significant
mutagenic response.

Mutagenic substances either act directly or require metabolic
conversion to mutagenic products. Unsubstituted PAH compounds
require activation with mammalian enzymes to be mutagenic in the
Ames test (Ames et al., 1973). The relative contributions of these
two groups of substances were estimated by testing the extracts in
both the presence and absence of liver microsomal preparations.

All extracts of airborne particulate matter collected on glass
fiber filters were mutagenic both with and without metabolic
activation. The dose response, expressed as number of mutants
per plate, was linear both with and without S-9. The standard
amount of S-9, 50 µl per plate, produced maximum activity for most
samples, but more S-9 was needed for a few extracts. Extracts from
the polyurethane filters showed no or very weak mutagenic
responses. Some samples taken at street level were fractionated
according to particle size. The results showed that only extracts
from particles less than 2.7 µm were mutagenic in both the presence
and absence of S-9.

Figure 1 shows the mutagenicity results for samples from site
A, street and roof levels, and from site B, expressed as
revertants per cubic meter of air. Daytime samples taken at street
level were about twice as mutagenic with microsomal activation as
without. (Mean values in February were 69 and 38 revertants/m^3
with and without S-9, respectively.) The activity of daytime
samples collected at site A, roof level, was approximately the same
with and without S-9; the same was true for site B daytime samples.
These samples were only 5 to 25% as mutagenic as the site A
street-level samples. For site A street-level samples, the

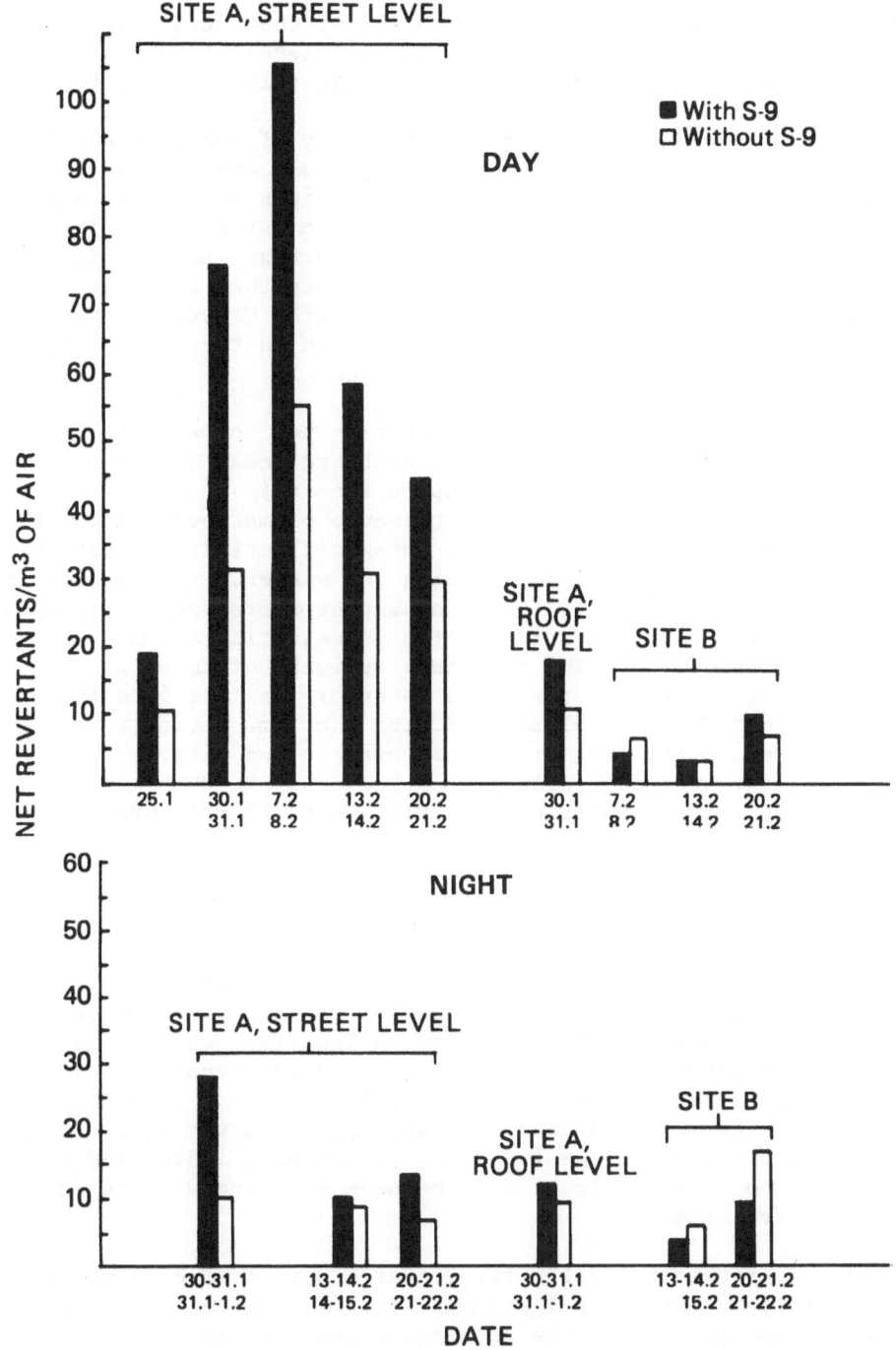

Figure 1.   Mutagenicity of samples from site A (street-level and
            roof-level samples) and site B, expressed as
            revertants per cubic meter of air.

mutagenicity at night was only 20 to 25% of that during the day.
The activity of extracts from site A, roof level, and from site B
was relatively constant from one day to the following night.

The daily variation in the mutagenicity of nonpolar extracts
sampled at sites C and D is shown in Figures 2 and 3. The
mutagenicity of the samples from both locations was of the same
magnitude (expressed as revertants per cubic meter of air in
Figure 2). However, when expressed as revertants per milligram of
particulate matter, the mutagenicity was more than twice as high
at site D than at site C (Figure 3). The site C extracts were
more mutagenic in the winter (February) than in the spring (April),
both with and without metabolic activation.

The mutagenicity of urban air samples was compared to that of
samples collected at site E, on the southern coast of Norway, far
from any source of pollution. Samples from this location were
mutagenic in both the absence and presence of metabolic activation
(S-9) (see Figure 4). The highest mutagenic activity at site E was
obtained for samples collected during the winter. Furthermore, the
activity was significantly lower in samples representing air masses
coming from the north than in samples representing air masses
coming from the south. The mutagenic activity of these samples
(expressed as revertants per cubic meter of air) was 5 to 10% of
that for urban samples collected during the same period (sites C
and D). The samples from site F, a rural inland site in the
mountains, far from any pollution source, did not show any
mutagenic activity. These nine samples were all made during the
winter.

Chemical Fractionation of Street Samples

To characterize the mutagenic compounds in urban air, a few
Oslo samples were fractionated, and the fractions were tested for
mutagenicity. The results are given in Table 1. Only 48% of the
mutagenicity assayed in the presence of S-9 was recovered after
fractionation. Without S-9, the recovery was 30%. The mutagenic
activity of the combined fractions was approximately equal to the
sum of the activity of the individual fractions, indicating that
there were no synergistic effects between substances found in
different fractions.

The mutagenicity of the fractionated street-level samples was
found mainly in the neutral aromatic fraction (N-2). Some activity
was also associated with the acidic and the neutral fractions
(N-4). The GC analysis showed that the N-2 fraction contained all
the common PAH except for the most volatile compounds. In
addition, this fraction contained unidentified polynuclear aromatic

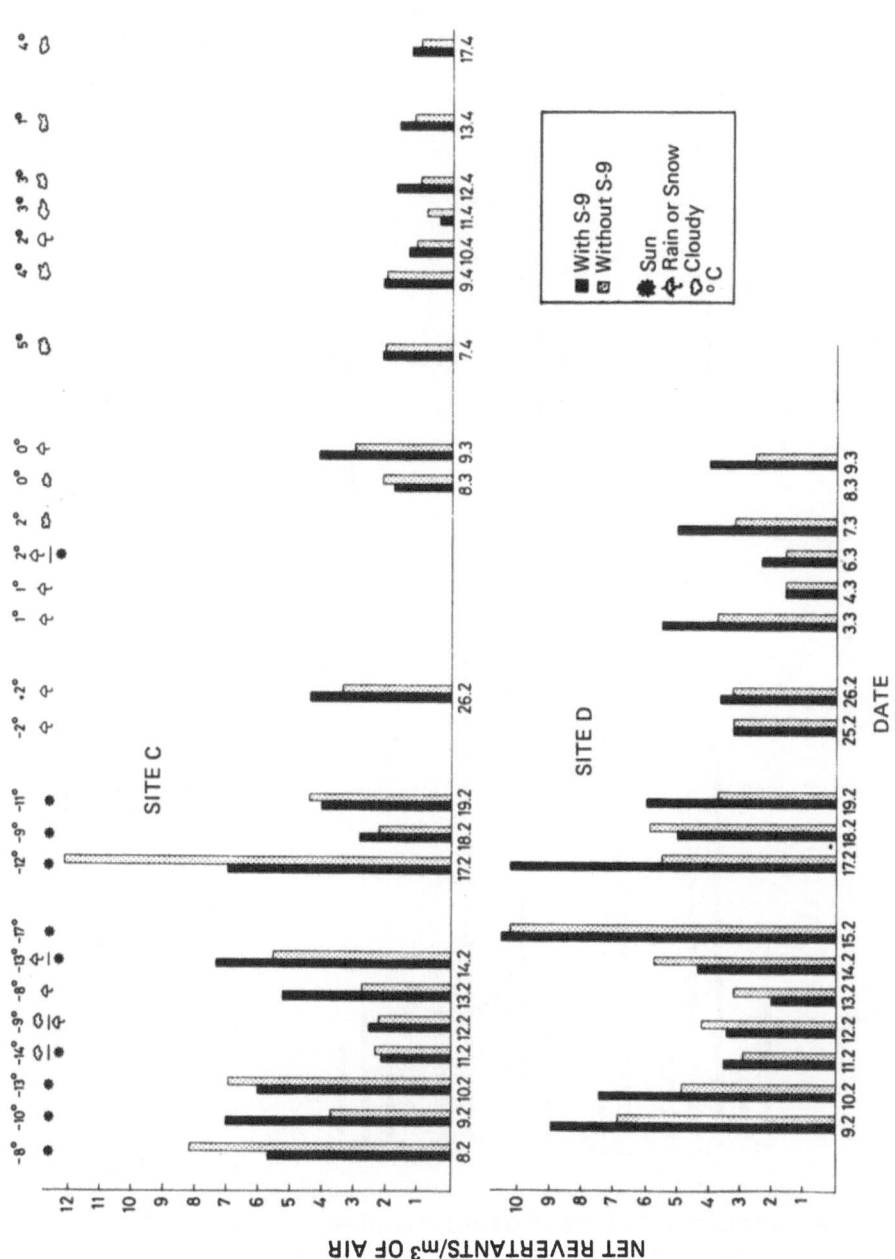

Figure 2. Mutagenicity of roof-level samples from sites C and D, expressed as revertants per cubic meter of air, and weather conditions on dates of sampling.

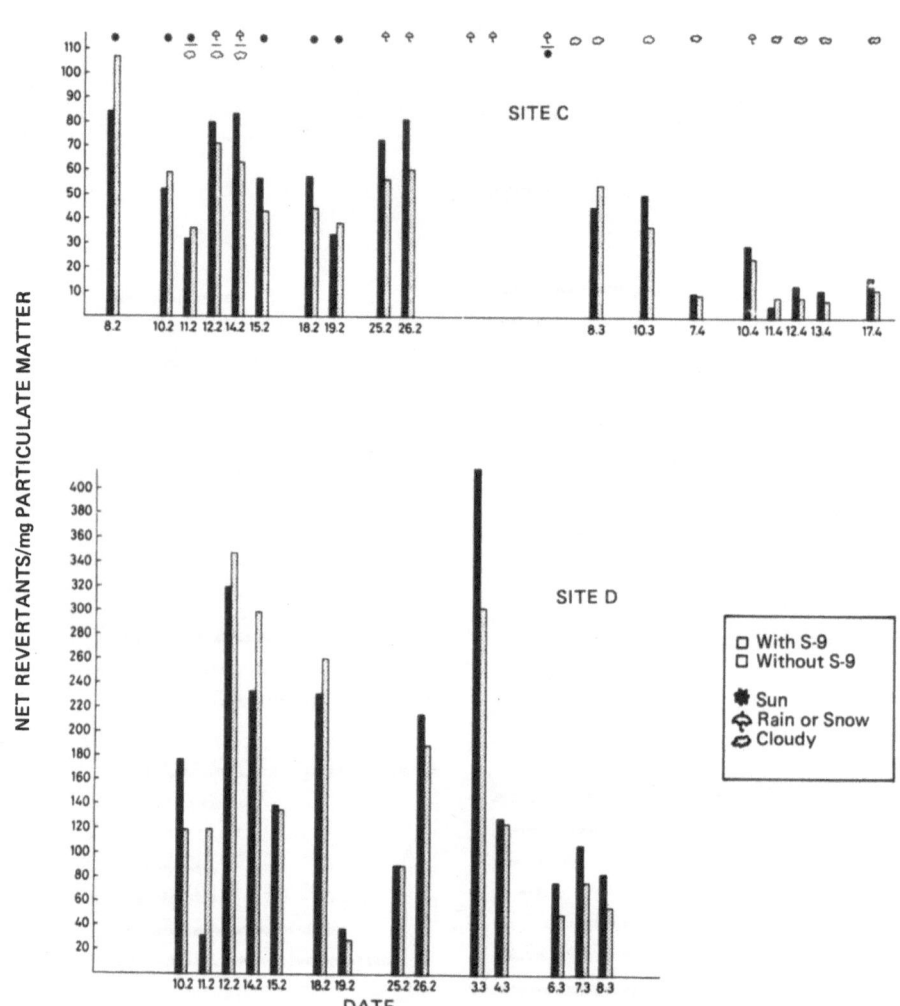

Figure 3. Mutagenicity of roof-level samples from sites C and D,
expressed as revertants per milligram particulate
matter, and weather conditions on dates of sampling.

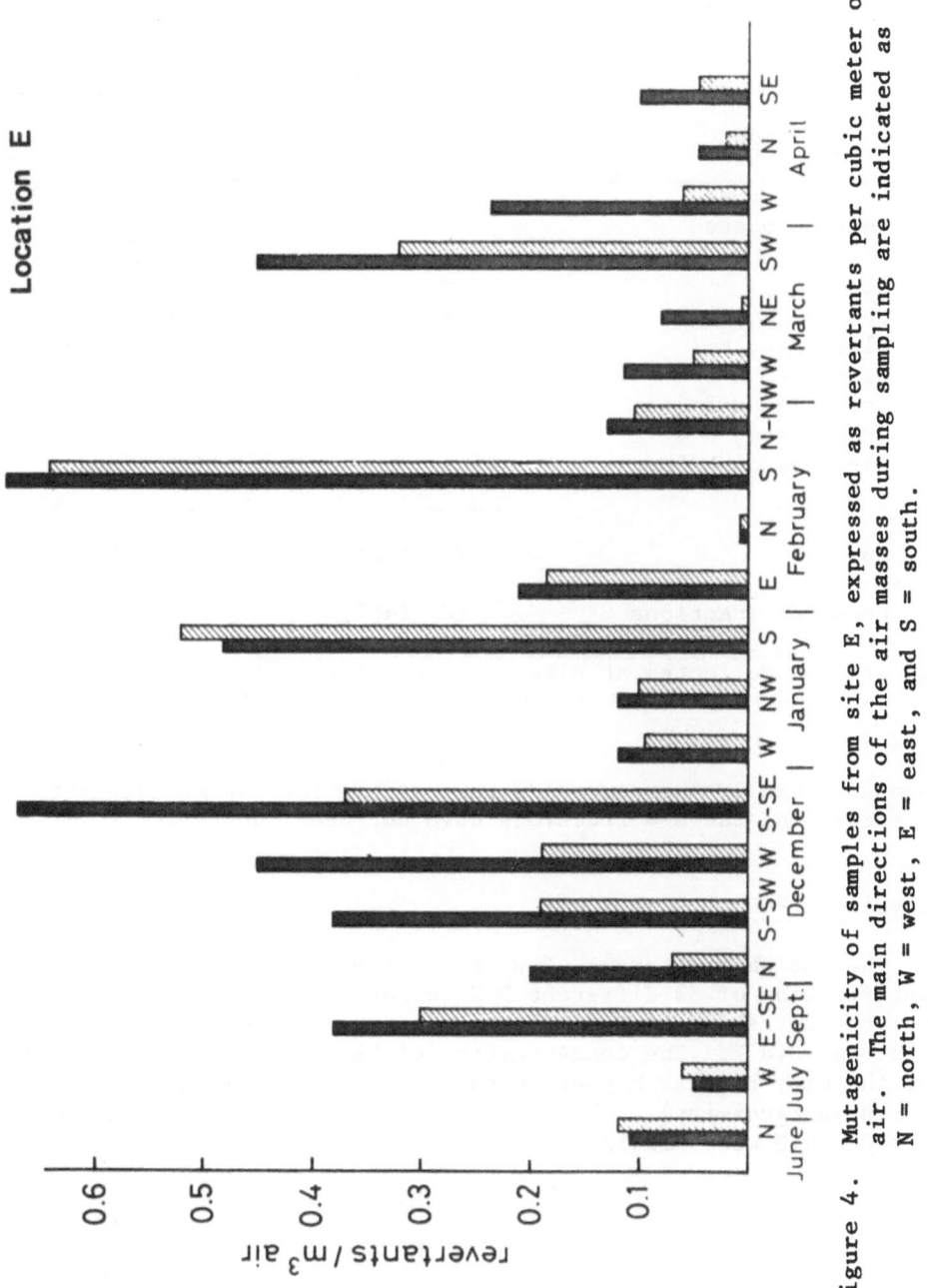

Figure 4. Mutagenicity of samples from site E, expressed as revertants per cubic meter of air. The main directions of the air masses during sampling are indicated as N = north, W = west, E = east, and S = south.

Table 1.   Distribution of Mutagenicity Among Fractions of
Extracts of Airborne Particulate Matter[a]

Fraction	Net Revertants/m^3 of Air	
	With S-9	Without S-9
Unfractionated	59	35
Acidic	5	2
Basic	1	1
Aliphatic, N-1	0	0
Aromatic, N-2	18	6
Aromatic, N-3	1	0
Oxygenated, N-4	3	2
Oxygenated, N-5	< 1	< 1
Sum	28 (48%)	11 (30%)
Combined fractions	28 (48%)	14 (39%)

[a]Day samples collected at site A, street level, in February, 1979.

compounds (PNA). The compounds responsible for the mutagenicity in
the acidic and the N-4 fractions have not been identified.

Analysis of PAH

The gas chromatograms of samples from sites C and D allowed
quantification of 33 different PAH compounds. Table 2 shows the
total concentration of PAH, together with the concentrations of
pyrene and B(a)P. The concentration of PAH in particulate matter
from the city air was higher in the winter than in the spring (as
was its mutagenicity). The PAH concentrations in samples from
sites C and D were similar.

DISCUSSION

The contribution from traffic to the mutagenicity of the air
at street level appeared to be substantial. The mutagenicity of
daytime samples at street level in the presence of S-9 was 4 to 20
times higher than for the corresponding samples from roof level or

Table 2. Concentration of PAH in Airborne Particulate Matter

Sampling Site	Date of Sampling	Concentration of PAH (ng/m^3)		
		Pyrene	B(a)P	$\int$PAH (33 comp.)
C	1978 February 20	66.0	12.0	414
	" 24	16.0	4.1	127
	" 28	3.9	2.1	72
	March 05	1.2	0.6	45
	April 04	2.1	0.9	17
	" 18	1.5	1.0	14
D	1978 February 20	40.0	7.9	319
	March 05	1.9	1.0	51
	April 04	3.5	1.4	29
C	1979 February 14	71.0	11.0	355
	" 20	5.5	3.7	60
	March 04	1.5	0.1	7
	" 10	1.1	0.6	14

from site B. Furthermore, the mutagenicity at street level varied
with traffic frequency (i.e., day vs. night), whereas activity of
samples from roof level and site B showed no such variations.

The mutagenicity of street samples was enhanced by the presence
of liver microsomes, indicating a greater contribution from gasoline
engine exhaust than from diesel exhaust. Exhaust from diesel
engines is more mutagenic in the Ames test than is exhaust from
gasoline engines. However, the activity of diesel exhaust is lower
in the presence of S-9, while the opposite is true for gasoline-
engine exhaust (Löfroth, 1979). It has been suggested that nitro-
substituted PAH contributes to the mutagenic activity of exhaust
from traffic (Wang, 1978). Most of these compounds are direct-
acting mutagens in the Ames test.

All mutagenic activity at street level was associated with
particles less than 2.7 μm in diameter. Furthermore, only traces
of PAH were found in particles greater than 2.7 μm (Alfheim et al.,
1977). This agrees with cascade impactor measurements made in
Belgium of the size distribution of 60 organic pollutants. Van
Vaeck and Van Cauwenberghe (1978) showed that 95 to 98% of the PAH,

as well as some heterocyclic polyaromatics (PNA), are found in particles less than 3 µm in diameter. Most carbocyclic acids and aliphatic hydrocarbons (90%) are also associated with particles less than 3 µm.

Fractionation of extracts from particulate matter at street level showed that most of the mutagenic activity recovered was due to PNA. A similar investigation based on roof sampling in Japan showed a somewhat different result (Teranishi et al., 1978). In this study, the mutagenicity was evenly distributed among the acidic, aromatic, and oxygenated fractions (~15% in each fraction), with slight activity in the basic fraction. Of the original activity, 76% was recovered after the fractions were combined. Fractionation of particulate matter of roof samples in Stockholm (Löfroth, 1979) gave a distribution of mutagenicity similar to that reported from Japan. The recovery in our experiments was somewhat lower, possibly because the components of particulate matter collected at street level were more labile. Earlier investigations, in which particulate matter from a heavy traffic area was fractionated and tested for carcinogenicity through skin-painting experiments on mice (Wynder and Hoffman, 1965), showed that the aromatic fraction was associated with carcinogenic properties, while the acidic and oxygenated fractions were associated with promoter effects.

The mutagenicity of roof-level samples (A, C, and D) and samples from site B probably has main sources other than traffic. Mutagenicity and PAH concentrations were both higher in the February samples than in the spring samples, which is most likely explained by the higher use of residential-heating fuels during the winter months in Oslo. The average temperatures during the sampling period were -8.3°C for February, -0.9°C for March, and +3°C for April. Using measurements at street level, we found that the amount of B(a)P could only account for ~1% of the mutagenic activity in day samples and up to 4% in night samples. Data for extracts of roof-level samples from other cities show that the mutagenicity of urban air (revertants per cubic meter) was of the same magnitude in Oslo as in Stockholm (Löfroth, 1979) and Los Angeles (Pitts et al., 1977).

Meteorological conditions may also contribute to seasonal differences in mutagenicity levels and cause daily variations in the mutagenicity of winter samples. The mutagenicity (revertants per cubic meter of air) was highest on cold, clear weekdays with little wind and stagnant air.

Daily measurements of SO_2 and soot in the air have been made at these same sampling sites by the Oslo City Department of Health. The most mutagenic samples from each sampling site also had the highest concentrations of SO_2, soot, and PAH at that site.

However, while mutagenicity (revertants per cubic meter of air) and PAH concentrations were approximately equal for the two sampling sites, the values for SO$_2$ and soot were nearly twice as high at site C as at site D. Contributions from different pollution sources may explain why SO$_2$ and soot concentrations were higher at site C than at site D, while mutagenicity and PAH concentrations were equal.

Mutagenic activity in strain TA98 (revertants per cubic milligram of particulate matter) was high on days with a low total concentration of particles in the air--mainly days with rain and snow. The mutagenicity was especially high at location D on such days. Variations in the amount of nonmutagenic particles will greatly influence the mutagenicity (revertants per milligram of particles). Currently, we consider revertants per cubic meter of air to better express the mutagenic potential of the air samples. However, this subject requires further investigation.

For most samples from the rural site in southern Norway, mutagenicity either was the same with and without metabolic activation or was higher with it. The activity with metabolic activation might be explained by the presence of PAH compounds in the samples. Such compounds have previously been demonstrated in long-range-transported aerosols collected at this site (Lunde and Bjørseth, 1977), and their amount have been shown to vary with the origin of the air masses in the same way as the mutagenic effect does. Like the mutagenicity, the concentration of PAH at site E also was 5 to 10% of that found in samples from Oslo. At site F, the PAH concentrations were up to 1% of the corresponding Oslo values.

REFERENCES

Alfheim, I., and M. Møller. 1979. Mutagenicity of long-range transported atmospheric aerosols. Sci. Tot. Environ. 13:275-278.

Alfheim, I., M. Møller, S. Larssen, and A. Mikalsen. 1977. Undersøkelse av PAH og mutagene stoffer i Oslo-luft--relasjon til trafikk. Report to Norwegian State Pollution Control Authority. 42 pp.

Ames, B.N., W.E. Durston, E. Yamasaki, and F.D. Lee. 1973. Carcinogens are mutagens: a simple test system combining liver homogenate for activation and bacteria for detection. Proc. Natl. Acad. Sci. USA 70:2281-2285.

Ames, B.N., J. McCann, and E. Yamasaki. 1975. Methods for
    detecting carcinogens and mutagens with the Salmonella/
    mammalian-microsome mutagenicity test. Mutation Res.
    31:347-364.

Bjørseth, A. 1977. Analysis of polycyclic aromatic hydrocarbons
    in particulate matter by glass capillary gas chromatography.
    Anal. Chim. Acta 94:21-27.

Bjørseth, A., G. Lunde, and N. Gjøs. 1977. Analysis of
    organochlorine compounds in effluents from bleacheries by
    neutron activation analysis and gas chromatography/mass
    spectrometry. Acta Chem. Scand. B 31:797-801.

Dehnen, W., N. Pitz, and R. Tomingas. 1977. The mutagenicity of
    airborne particulate pollutants. Cancer Lett. 4:5-12.

Grimmer, G., and H. Bohnke. 1972. Bestimmung des Gesamtgehaltes
    aller Polycyclischen Aromatischen Kohlenwasserstoffe in
    Luftstaub und Kraftfahrzeugabgas mit der Capillar-gas-
    Chromatographie. Z. Anal. Chem. 261:310-314.

Grob, K., and G. Grob. 1969. Splitless injection on capillary
    columns (Part I and Part II). J. Chromatogr. Sci. 7:584-591.

Henderson, B.E., R.J. Gordon, H. Menck, J. Soohoo, S.P. Martin, and
    M.C. Pike. 1975. Lung cancer and air pollution in
    Southcentral Los Angeles County. Am. J. Epidemiol.
    101:477-488.

Higginson, J., and C.S. Muir. 1973. In: Cancer Medicine. J.F.
    Holland and R. Frei, eds. Lea and Febiger: Philadelphia.

Hueper, W.C., P. Kotin, E.C. Tabor, W.W. Payne, H. Falk, and E.
    Sawicki. 1962. Carcinogenic bioassays on air pollutants.
    Arch. Pathol. 74:89-116.

Löfroth, G. 1978. Mutagenicity assay of combustion emissions.
    Chemosphere 7:791-798.

Löfroth, G. 1979. Salmonella/microsome assays of exhaust from
    diesel and gasoline powered motor vehicles. Presented at
    International Symposium on Health Effects of Diesel Engine
    Emissions, Cincinnati, OH.

Lunde, G., and A. Bjørseth. 1977. Polycyclic aromatic
    hydrocarbons in long-range transported aerosols. Nature
    268:518-519.

Menck, H.R., J.T. Casagrande, and B.E. Henderson. 1974.
    Industrial air pollution: possible effect on lung cancer.
    Science 183:210-212.

Møller, M., and I. Alfheim. 1980. Mutagenicity and PAH analysis
    of airborne particulate matter. Atmos. Environ. 14:83-88.

Pitts, J.N., Jr., D. Grosjean, and T.M. Mischke. 1977. Mutagenic
    activity of airborne particulate organic pollutants. Toxicol.
    Lett. 1:65-70.

Talcott, R., and E. Wei. 1977. Brief communication: airborne
    mutagens bioassayed in Salmonella typhimurium. J. Natl.
    Cancer Inst. 58:449-451.

Teranishi, K., K. Hamada, and H. Watanabe. 1978. Mutagenicity in
    Salmonella typhimurium mutants of the benzene-soluble organic
    matter derived from airborne particulate matter and its five
    fractions. Mutation Res. 56:273-280.

Tokiwa, H., K. Morita, H. Takeyoshi, D. Takahashi, and Y. Ohnishi.
    1977. Detection of mutagenic activity of particulate air
    pollutants. Mutation Res. 48:237-248.

Van Vaeck, L., and K. Van Cauwenberghe. 1978. Cascade impactor
    measurements of the size distribution of the major classes of
    organic pollutants in atmospheric particulate matter. Atmos.
    Environ. 12:229-239.

Wang, Y.Y., S.M. Rappaport, R.F. Sawyer, R.E. Talcott, and E.T.
    Wei. 1978. Direct-acting mutagens in automobile exhaust.
    Cancer Lett. 5:39-47.

Wynder, E.L., and D. Hoffman. 1965. Some laboratory and
    epidemiological aspects of air pollution carcinogenesis. J.
    Air Pollut. Contr. Assoc. 15:155-159.

DETECTION OF GENETICALLY TOXIC METALS BY A MICROTITER MICROBIAL DNA REPAIR ASSAY

Guylyn R. Warren
Chemistry Department
Montana State University
Bozeman, Montana

INTRODUCTION

For some years, our laboratory has been involved in assessing the genetically toxic effects of inorganic chemicals found in the environment near mining and smelting operations (Tindall et al., 1978; Warren et al., 1979) and of metal-containing pesticides used in Montana (Warren et al., 1976). Although many inorganic species are known or suspected carcinogens or are genetically active in many test systems (Flessel et al., 1979; Sunderman, 1978), most existing short-term biological screening methods are unsuitable for use with this class of suspected carcinogen. Only two systems, the Bacillus subtilus rec assay (Nishioka, 1975; Kanematsu et al., 1980) and an in vitro DNA synthesis fidelity assay (Loeb et al., 1979), have been useful for a large number of inorganic chemicals. Mutagenicity of some inorganic chemicals has been demonstrated in a CHO/HGPRT assay (Hsie et al., 1979) with considerable technical difficulty due to the general toxicity of inorganic chemicals. Green and Muriel (1976) have used a repair-deficient series of Escherichia coli B strains to detect mutagenicity of some chromate salts; also by this method, they found that nickel chloride salts do not cause differential lethality.

Recent discoveries about the relationship of DNA repair functions to mutagenesis in bacteria (Witkin, 1976; Kimball, 1978) and of error-prone repair mechanisms such as inducible SOS repair and another excision repair mechanism (Hanawalt et al., 1979), provide reasons for using batteries of repair-deficient organisms in a screening system, rather than using a single repair-deficient strain, such as pol A (Hyman et al., 1980). The pleiotropy of rec- mutants is a major problem in using only one double-rec-deficient

101

strain, such as Bacillus subtilis M45, for the assay. Such
rec⁻ mutants exhibit permeability changes, possibly due to cell-
wall and membrane-surface defects, as in rec A (Tomizawa and Ogawa,
1968). Differences in inhibitory effects of test chemicals might
be due to differences in their rates of penetration into the
repair-defective strain as compared with wild type rather than to
differences in DNA repair capacity.

To rapidly screen large numbers of samples, we have developed
a microbial repair assay using a series of singly- and multiply-
mutant DNA-repair-deficient strains in an E. coli K12 background.
The mutational repair defects of the K12 series have been studied
in much more detail, both genetically and biochemically, than have
those of E. coli B or B. subtilis (Witkin, 1976; Kimball, 1978;
Hanawalt et al., 1979). The strains are stable in culture and easy
to grow, and they carry several biochemical markers that permit
genetic manipulation for strain construction, and possibly
mutagenesis assays, in the same repair strains. A number of metal
salts and metal salt compounds have been tested in the K12 system
(Warren et al., in press a, b; MS). Our results correlate well
with mutagenesis as assayed by the Ames test. This paper describes
in detail the repair assay and its use to assay metal-containing
samples.

METHODS

Construction and Testing of Strains

The derivation of the repair-deficient E. coli bacterial
strains is diagrammed in Figure 1. In each case, isogenicity of
the repair defectives with the wild type AB1157 was optimized, to
minimize mutational effects other than on DNA repair. For
discussions of repair defects and phenotypes see Kimball (1978),
Hanawalt et al. (1979), and Witkin (1976).

The bacteria are stored at room temperature in Lederberg stabs
of nutrient broth (Difco) and have remained viable for at least
three years. Each month, working slants are made from single
colonies and stored at room temperature. Bacteria from these
single colonies are streaked onto minimal medium plus the amino
acids required by AB1157 and onto minimal medium alone, to verify
the strains' nutritional requirements. Every two weeks, single
colonies are isolated from the working slants to check the repair
characteristics of the strains. Two colonies of each strain are
isolated and inoculated into Mueller-Hinton (MH) broth (5 ml), and
cultures are grown overnight at 37°C. Ultraviolet-light
sensitivity of each culture is checked by the rapid method of
Greenberg (1967) as follows: A streak of each strain is placed by
capillary pipette on each of seven MH agar plates. Plates are

AB1157 (a Y10 E. coli K12, F⁻, λ⁻) wild-type for repair

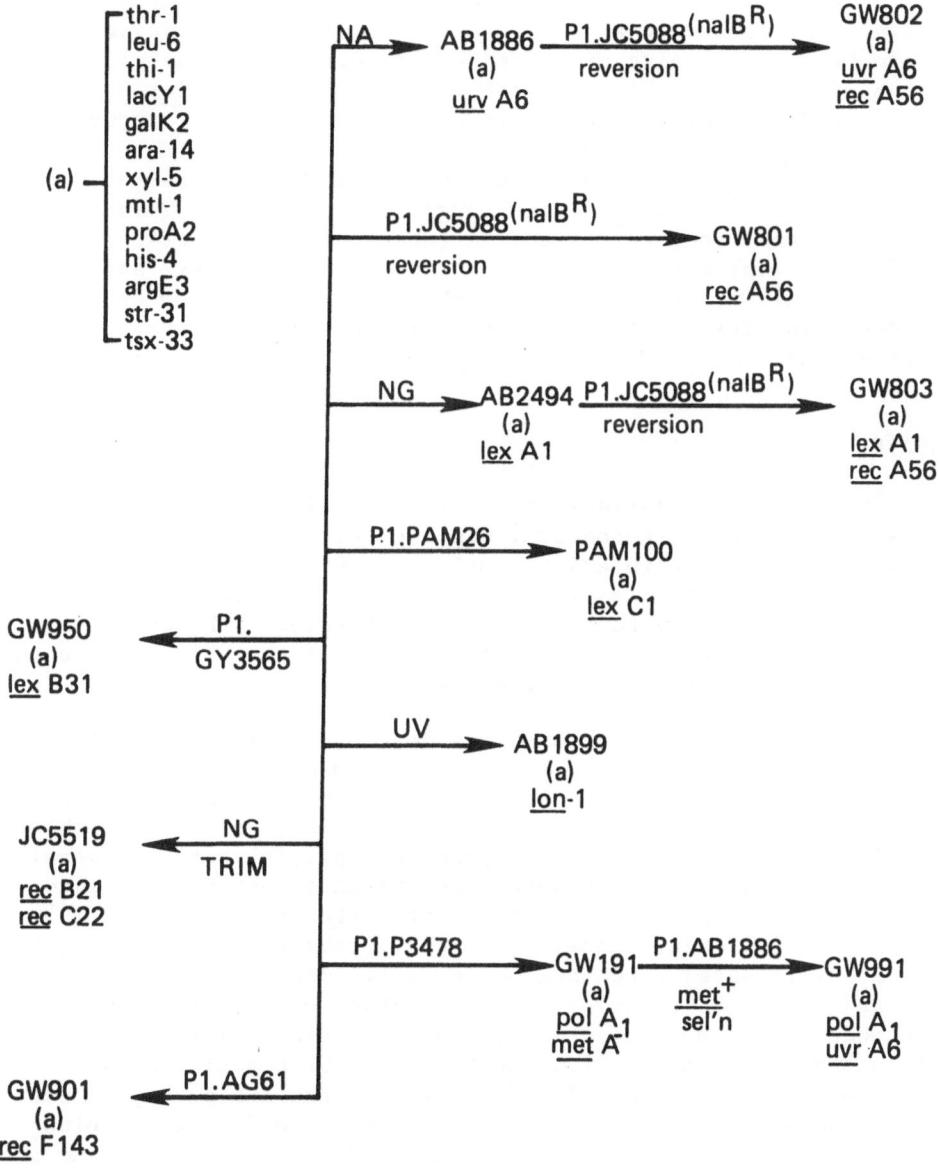

Figure 1.   Construction of a series of E. coli K12 repair-deficient strains from strain AB1157 (wild-type for repair).

immediately irradiated with known fluences of 254-nm light (for
example, 0, 20, 50, 100, 200, 300, and 400 ergs/mm^2) and then
incubated in the dark at 37°C. Results such as those shown in
Table 1 can be determined after incubation for 18 h. Any culture
not giving the expected result is rejected, and a new colony of
that strain is picked and tested. Acceptable cultures are labelled
for use and stored. Broth cultures may be used for testing for up
to two weeks if stored at 2°C. Several of the repair-deficient
strains are still being evaluated as testers, and those with the
most useful traits will be chosen. An example is the lex C mutant
strain, PAM100, which is uniquely sensitive to nickel compounds.

We are constructing a second identical isogenic series of
strains containing an rfa (deep rough) mutant locus. This series
will be tested later this year. A series of repair-deficient
strains containing the ochre mutant trp⁻ locus from WP2 and one
strain containing two of the Yanofsky trp⁻ series (Yanofsky, 1963)
for detecting frameshifts and base pair substitutions are under
construction.

ASSAYS

Initially, we developed a sensitivity disc assay in which
zones of inhibition caused by inorganics were compared between each
repair-deficient strain and the wild-type (repair-proficient)
strain. Such a method allowed differential inhibition (in many
instances, lethality) to be grossly quantitated. DNA damage of
some sort should have been the cause, because the only differences
between strains were in repair capacities.

The sensitivity disc assay was done by spreading 0.1 ml of
each broth culture directly on a complete-medium agar plate.
Sterile discs of paper were impregnated with a known concentration
of test agent. Three of these discs were placed on each freshly
spread agar surface, and plates were run in triplicate. Several
different media were tried initially, in an effort to maxmimize
zone size differences. For a group of salts of known metal
carcinogens (chromium, cadmium, mercury, cobalt, and arsenic), the
largest differences were obtained on solid medium containing 10 g/l
agar and MH broth (Difco). When 0.2% glucose was added, chromate
compounds gave a much clearer and larger differential response,
indicating a glucose requirement for rapid bacterial growth.
Addition of 0.5% sodium chloride increased the inhibitory effects
of cadmium and cobalt. The standard medium now used is MH plus
glucose and salt. The effects of these additions to MH medium are
shown in Table 2 as differences in zones of inhibition (mm).
Differences greater than 5 mm were significant at the 95% level
(Tindall, 1977).

Table 1. Verification of Repair Deficiencies by Ultraviolet Irradiation

Strain	Repair Defect[a]	Growth in Streaks on MH Plates After Irradiation with 254-nm Light[b]						
		Fluences (ergs/mm^2)						
		0	20	50	100	200	300	400
AB1157	wild type	+++	+++	+++	+++	+++	+++	+++
AB1886	uvr A6	+++	++	- (1)	- (20)	- (0)	- (1)	- (2)
AB1899	lon-1	+++	+++	+++	+++	+++	+++	- (8)
AB2494	lex A1	+++	+++	+	+-	- (15)	- (0)	- (0)
GW801	rec A56	++	+-	- (>200)	- (35)	- (0)	- (0)	- (0)
PAM100	lex C1	++	+++	+-	-+	- (20)	- (0)	- (0)
GW802	uvr A6, rec A56	+++	- (3)	- (0)	- (0)	- (0)	- (0)	- (0)
GW803	lex A1, rec A56	+++	- (>200)	- (45)	- (8)	- (3)	- (0)	- (0)
GW191	pol A1	+++	+++	+++	+++	+-	-+	- (36)

[a]Actual kill curves can be determined if necessary.

[b]+++ = maximal growth; ++ = noticeably less dense streak; + = thin streak; +- = visible lethality; -+ = spotty but >500 colonies; - = kill (no. of colonies counted).

Table 2. Effects of Additions to Growth Medium on Differential Inhibition by Metals: Difference from Strain AB1157 (Wild Type) in Zone Diameter (mm)

	Carcinogenic Metal Salts															
	K_2CrO_4[a]				Na_2TeO_3[b]				$BeSO_4$[c]				$KSbO_4(C_4H_4O_6) \cdot 1/2H_2O$[d]			
Medium[e]:	MH	MHG	MHS	MHGS	MH	MHG	MHS	MHGS	MH	MHG	MHS	MHGS	MH	MHG	MHS	MHGS
Strain																
AB1886	0	-10	3	-4	5	5	3	4	0	0	0	0	2	5	4	2
GW801	2	3	3	6	-4	1	-2	0	0	0	0	0	-1	7	6	4
AB2494	-4	-14	-2	2	12	16	16	8	0	1	0	0	-2	6	1	1
GW802	1	10	3	12	-2	3	0	2	10	10	8	0	-1	11	4	4
GW803	1	4	3	5	16	6	3	8	0	0	0	0	-7	0	2	1
GW191	0	3	2	9	10	14	15	12	0	0	1	0	-6	3	-3	-1

[a] 0.10 M in H_2O, 10 µl/disc.
[b] 0.01 M in H_2O, 15 µl/disc.
[c] 0.10 M in H_2O, 20 µl/disc.
[d] 0.10 M in H_2O, 20 µl/disc.

[e] MH = Mueller-Hinton agar (Difco).
MHG = MH agar + 0.1% glucose.
MHS = MH agar + 0.05% NaCl.
MHGS = MH agar + glucose + NaCl.

Kanematsu et al. (1980) used a cold incubation period after spreading and disc placement to enhance the inhibition by test metals in the B. subtilis rec assay (this allows the test substance to diffuse before bacterial growth starts). We have observed this effect with most chemicals, although some (mostly organic compounds) have shown weaker responses after cold incubation than without it.

Because migration through an agar medium is affected by the depth and viscosity of the agar, temperature, charge, and pH, and because all of these parameters are subject to laboratory variation, we searched for a more reproducible and sensitive method. The disc repair assay required high levels of test agent and was not quantitative, due to solubility problems. McCarroll et al. (1979) have developed a microtiter assay system to study effects of organic toxicants using repair-defective strains of E. coli B obtained from Green and Muriel (see Green and Muriel, 1976; McCarroll et al., 1980a, b). We have adapted this system to the repair assay with E. coli K12. Microtiter techniques (Cooke Engineering, 1972) can be used to quantitatively measure inhibition of growth (as in antibiotic testing) and also to quantitatively measure lethality, since colonies can be counted in each well at the end of a treatment period (McCarroll et al., 1980b). Mutagenesis testing also can be done from the same microtiter plate wells, thus generating actual mutation frequencies (mutants per survivor). Several loci could be monitored for mutagenesis at the same time.

A microtiter plate set up for an eight-strain test is shown in Figure 2. Initially, 0.05 ml of MH broth is added to each well of the microtiter plate, using a Tridak Stepper, 3-ml syringe, or Gilson Repetman. Rows 1 and 12 receive a double dose. Rows A through H will each receive one of the strains of bacteria, with A receiving the wild type. Row 1 will contain only bacteria and medium, to serve as the cell control. Row 2 receives 0.05 ml of the test agent in solution. For most inorganic toxicants, a convenient initial concentration is 2 mg/ml in water, if possible (or dimethylsulfoxide if necessary). The test agent is added to row 2 with a 1-ml Tridak stepper. Serial dilutions are then performed from row 3 through row 11, using microtiter diluters (Cooke), which hold 0.05 ml and do automatic serial dilutions. The final dilutions of mutagen are shown in Figure 2 and range from 1:4 (row 2) to 1:2048 (row 11). Row 12 is left uninoculated as a sterility control for medium, mutagen, and solvent. Having been checked and stored as described above, each of the bacterial cultures is diluted in MH to a final viable count of ~ 1 x 10^6 colony forming units (CFU)/ml, for optimal sensitivity in microtiter. Because several of the repair-deficient strains (especially the multi-mutants; see Table 1) do not reach the same cell density overnight as does the wild type, the appropriate

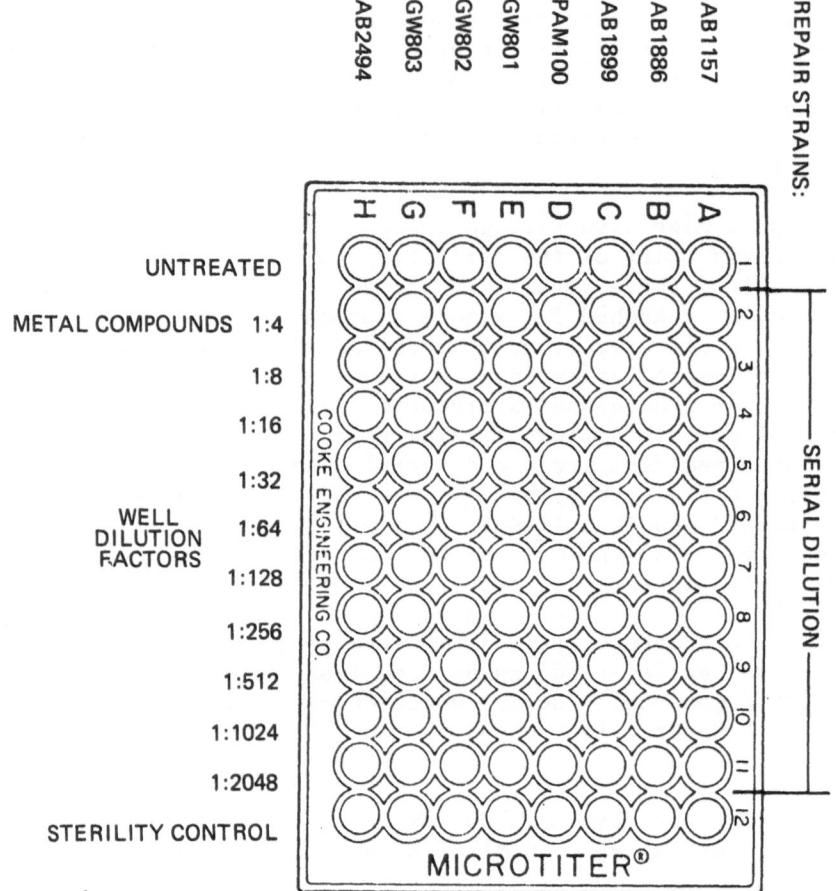

Figure 2.   Sample microtiter plate.

dilution must be established for each strain.   Bacterial culture
dilutions should be discarded daily; they may be used all day if
kept on ice.   With a 1-ml Tridak stepper, 0.05 ml of bacterial
culture dilution is added to each well of the appropriate row for
that strain.

Initial concentrations of test agents can be varied, within
limits of solubility, to test for effects throughout a very large
concentration range.   Plates are sealed with tape to retard
evaporation.   Sterile lids can be used for short-term treatments,
but incubation for over four hours results in significant fluid
loss by evaporation.   The standard incubation time is 24 h.   Some
tested compounds initially inhibit growth, but then either

resistant organisms overgrow the wells or the test agent is
metabolized and becomes ineffective. For such agents, a shorter
incubation time can be used.

Mammalian microsomal fractions can be used for activation in
the repair assay. Positive controls such as 2-aminoanthracene show
good responses with either of two methods of addition of rat liver
homogenate (S-9). The S-9 mix is prepared according to Ames et al.
(1975), and 0.05 ml is added with the mutagen to wells in row 2, in
place of the initial 0.05 ml of MH broth. Both substances are
simultaneously serially diluted through the wells. Alternatively,
one may add 0.05 ml of S-9 mix in place of MH broth to each well in
the plate and then dilute only the test mutagen. Throughout either
procedure, all plates are kept on ice. For all positive controls
we have studied, the former method results in more differential
lethality than the latter. After incubation, the plates are read
visually, using a microtiter mirror. For each of the lettered
rows, the totally inhibited well with the lowest test-agent
concentration is determined visually and recorded. From the
dilution factor for that well and the known concentration of test
agent, a minimal inhibitory concentration (MIC) can be determined
for each strain. Differential inhibition is given as the factor of
increase in MIC over that of the wild-type strain, in this case
AB1157 (i.e., MIC for repair-deficient strain/MIC for repair-
proficient strain).

RESULTS

Table 3 compares the results of the microtiter repair assay
and the disk assay, using eight strains of E. coli and ten rhodium
complexes recently synthesized by Abbott (Warren et al., in press
a).

In the optimal test, an inhibitory test-agent concentration
for the wild-type bacteria would be determined. In practice, this
is not possible for all agents, because the wild type often is
resistant to all concentrations within the range of solubility of
the test agent. In such cases, the highest concentration that
allows determination of an MIC for the most sensitive strain has
been used (less than 11 wells inhibited). The factor of increase
in MIC is therefore underestimated.

A series of substitutionally inert rhodium complexes was used
to demonstrate the utility of the method described here, not only
for detecting DNA damage but also as a screening test preceding the
Ames reversion assay (Warren et al., in press a). Dose
responsiveness was predictable, and as the dose was doubled, the
MIC also doubled, until the entire plate was killed. The factor of
increase in MIC for any one test agent did not change once a lethal

Table 3. Comparison of Microtiter and Disc Repair Assays, Using Rhodium Complexes

Inhibition of Strains Tested by the Disc (D)[a] and Microtiter (M)[b] Methods

Complex	AB1886 uvr A6		GW801 rec A56		AB2494 lex A1		GW802 uvr A6 rec A56		GW803 lex A1 rec A56		AB1899 lon-1	
	D	M	D	M	D	M	D	M	D	M	D	M
$[Rh(Pyr)_4Br_2]Br$	18.0	16	23.4	32	12.7	0	24.4	512	22.0	8	13.4	4
$[Rh(CH_3CN)_3Cl_3]$	9.3	0	7.3	4	2.0	0	12.0	12	8.7	4	2.7	0
$[Rh(Bipy)_2Cl_2]Cl$	7.5	0	0	0	0	0	16.5	16	0	0	0	0
$[Rh(3Pic)_4Cl_2]Cl$	15.3	16	23.0	2	12.7	0	28.3	128	20.0	0	14.7	16
$[Rh(Pyr)_4Cl_2]Cl$	23.0	8	22.6	0	7.5	0	27.0	256	235.0	0	123.0	0
$[Rh(Phen)_2Cl_2]Cl$	9.0	0	2.0	0	0	0	7.0	64	1.5	0	0	0
$[Rh(en)_2]Cl_2$	0	0	0	0	0	0	7.6	16	0	0	0	0
$RhCl_3(H_2O)_3$	6.6	0	3.3	0	0.5	0	8.6	8	4.3	0	2.0	0
$[Rh(NH_3)_4Cl_2]Cl$	0	0	0	0	0	0	12.0	16	1.3	0	0	0
$[Rh(Trien)Cl_2]Cl$	1.3	0	12.3	0	0.5	0	23.3	64	16.3	0	0.6	2

[a]For test agents at 20 mg/ml H_2O, 20 μl/disc; means of results from 6 discs; expressed as difference in diameter from wild type (mm).
[b]For test agents at 2 mg/ml H_2O, without S-9; means of results from duplicate plates; expressed as factor of increase in MIC over that of wild type.

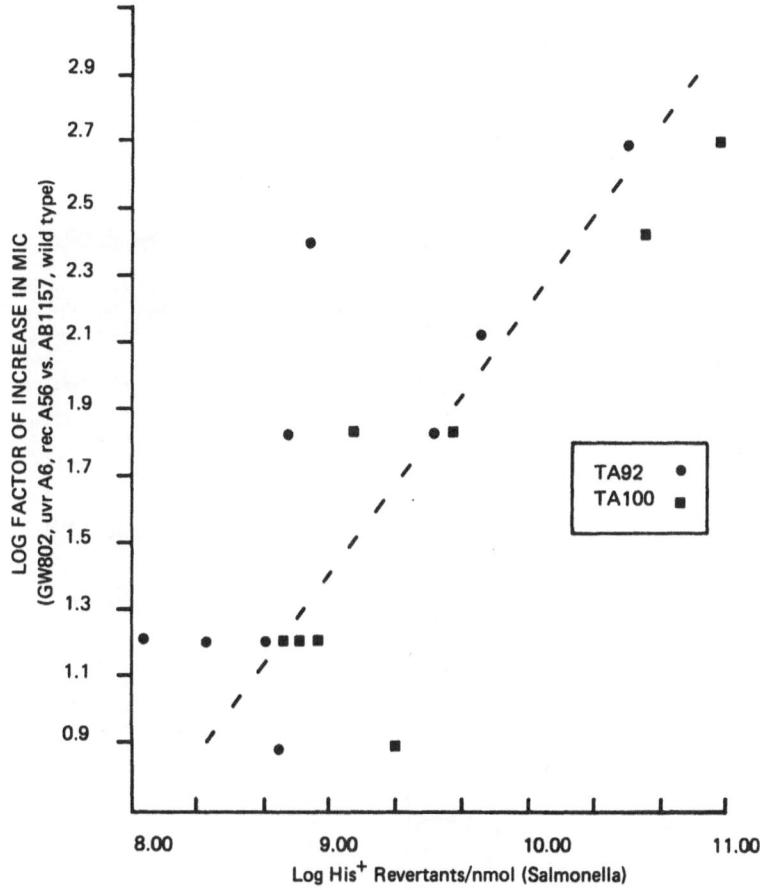

Figure 3. Correlation between sensitivity of the microtiter
repair assay (using strain GW802, uvr A6, rec A56) and
mutagenicity (strains TA92 and TA100) in the Ames
test for a group of rhodium complexes.

concentration was reached in the wild-type strain. A comparison
between the results of the microtiter repair assay and the Ames
test for these compounds is given in Figure 3. The coefficient of
correlation between the results of the two tests was 0.92,
indicating agreement. No Ames-positive test agents in a series of
23 rhodium complexes were missed by the repair assay, while one
repair-positive agent (very weak) was Ames negative.

An S-9 dose response in the microtiter repair assay with an
organic herbicide, diallate, is shown in Figure 4. Diallate, an
indirect-acting mutagen, was used because this compound is small

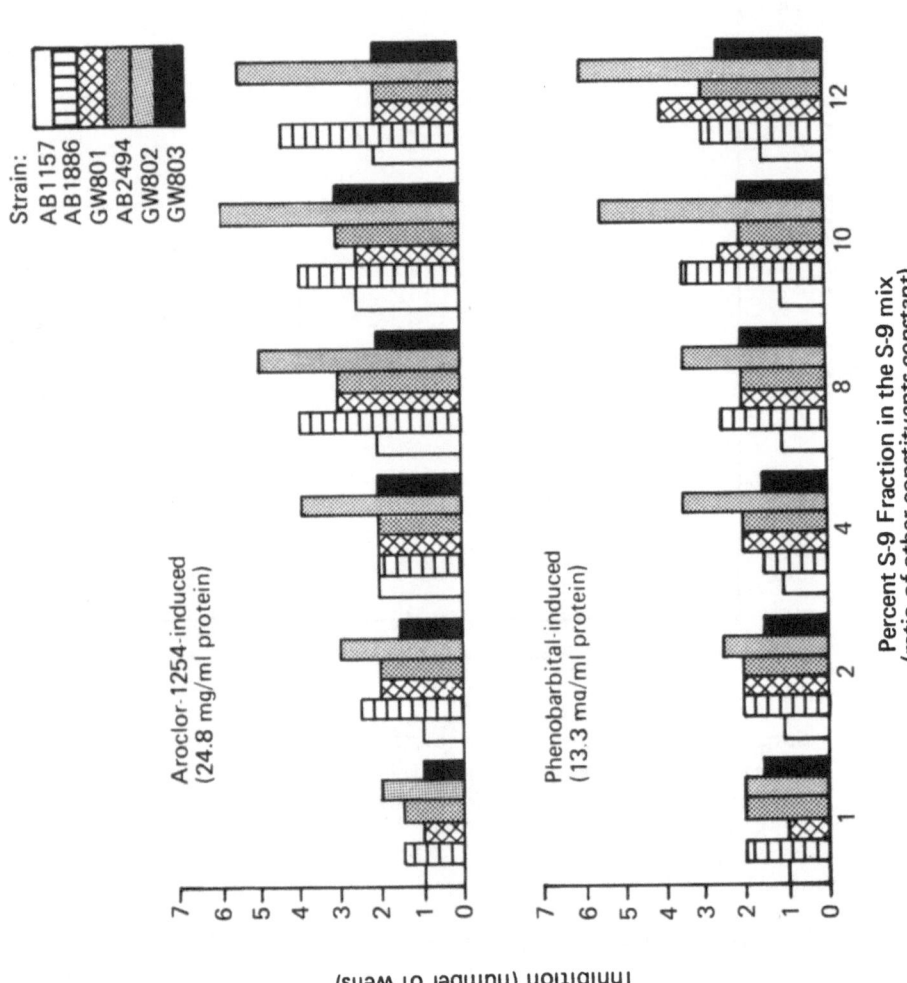

Figure 4.   Enzyme activation of diallate in a microtiter repair assay.

and should have no difficulty penetrating cell walls. The results
were related to the enzyme dose. Also, the type of inducer used
affected the response just as it does in a mutagenesis assay. We
have previously noted that a higher concentration of protein is
required of phenobarbital-induced enzyme than of Aroclor-1254-
induced S-9 for diallate mutagenicity in the Ames test. While the
inhibitory activity of several metal salts was reduced by S-9
(zinc, palladium, mercury, antimony, arsenic, selenium, nickel,
manganese, and cadmium), we have seen only one in which S-9
increases genetic toxicity (tellurium)(Rogers and Warren,
unpublished data).

This system is designed for testing complex mixtures. It has
been used to document the differential lethality of a sample of
smelter rafter dust, air filter particulate extracts, and urine
concentrates (Warren et al., 1979).

DISCUSSION

The microtiter repair assay techniques described here employ
standard microtiter equipment and can be automated. A scanning
device is available for reading optical density directly from the
plates for growth determinations. Various dyes can be used to
indicate growth (e.g., bromophenol blue)(Green and Muriel, 1976).
In cases of agents which kill cells but cause filamentation of
some strains, plate counting may be required, because filamentation
increases optical density without an increase in the number of
viable cells. This is particularly likely with strains AB1899 and
PAM100, both of which filament easily and in response to any
chemical insult.

Strains of E. coli have been constructed with mutations to
block incision (uvr A, B, C) and to partially block excision (pol
A$_1$, rec B, C), resynthesis and conditioned responses (rec A, lex
A), and recombination (rec A, rec F). Mutagenesis might be caused
by less-well-known repair mechanisms that could be unmasked if the
more rapid error-free repair were removed by mutation and post-
replicative repair were incapacitated as well. The strain GW802
has these defects and is easily the most sensitive strain in the
battery at this time. There is some evidence that an SOS-like
conditioned response system exists in mammals and may be
responsible for tumor induction by some mutagens (Radman, 1977).
Many of the metal mutagens and all of the substitutionally inert
transition metal complexes we have tested require the pKM101
plasmid in Salmonella for mutagenesis and do not require a
functional uvr system. Venturini and Monti-Bragadin (1978) found
that no mutation is caused by platinum without the lex function in
E. coli, but that lex$^+$ E. coli can be mutated without carrying
pKM101. Therefore, these strains are suitable for metal-ion

mutagenesis without the pKM101 repair system. Because many
mutagens can function as curing agents, a system that does not
require a plasmid would seem to be advantageous.

If an agent's lethal and mutagenic effects on each bacterial
strain are known, then the general type of DNA damage caused by
that agent can be predicted. For instance, uvr⁻ strains cannot
recognize DNA helix distortions; agents differentially lethal to
uvr⁻ strains but not mutagenic are most likely to be strand cross-
linkers (Murray, 1979). Green and Muriel (1976) give five classes
of damaging agents and note that some mutagens fit in more than
one category.

It has also been shown that repair assays correlate
better than do mutagenesis or prophage induction with anti-tumor
activity of alkylating agents (Tamaro et al., 1977), indicating
another useful screening capacity for the microtiter repair system.

CONCLUSIONS

The microtiter technique is rapid, cost effective, and
sensitive and requires very little sample. The bacterial strains
are easily grown and checked and provide an accurate screening
test for both organic and inorganic genetically toxic chemicals
and complex mixtures, regardless of their activity in mutagenesis
systems. This test is a very useful prescreen for mutagenesis
assays.

REFERENCES

Ames, B.N., J. McCann, and E. Yamasaki. 1975. Methods for
    detecting carcinogens and mutagens with the Salmonella/
    mammalian-microsome mutagenicity test. Mutation Res.
    31:347-363.

Cooke Engineering. 1972. Handbook of Microtiter Procedures.
    T.B. Conrath, ed. Dynatech Corp.: Cambridge, MA. 475 pp.

Flessel, C.P., A. Furst, and S.B. Radding. 1979. A comparison of
    carcinogenic metals. In: Metal Ions in Biological Systems,
    Volume 10, Chapter 2, Carcinogenicity and Metal Ions. H.
    Sigel, ed. Marcel Decker: New York.

Green, M.H.L., and W.J. Muriel. 1976. Mutagen testing using Trp⁺
    reversion in Escherichia coli. Mutation Res. 38:3-32.

Greenberg, J. 1967. Loci for radiation sensitivity in Escherichia
    coli strain B_{s-1}. Genetics 55:193-201.

Hanawalt, P.C., P.K. Cooper, A.K. Ganesan, and G.A. Smith. 1979. DNA repair in bacteria and mammalian cells. Ann. Rev. Biochem. 48:783-836.

Hsie, A.W., N.P. Johnson, D.B. Couch, J.R. San Sebastian, J.P. O'Neill, J.D. Hoeschele, R.O. Rahn, and N. Forbes. 1979. Quantitative mammalian cell mutagenesis and a preliminary study of the mutagenic potential of metallic compounds. In: Trace Elements in Health and Disease. N. Kharasch, ed. Raven Press: New York. pp. 55-69.

Hyman, J., Z. Leifer, and H.S. Rosenkranz. 1980. The E. coli polA$_1^-$ assay: a quantitative procedure for diffusible and non-diffusible chemicals. Mutation Res. 74:107-111.

Kanematsu, O., M. Hara, and T. Kada. 1980. Rec assay and mutagenicity studies on metal compounds. Mutation Res. 77:109-116.

Kimball, R.F. 1978. The relation of repair phenomena to mutation induction in bacteria. Mutation Res. 55:85-120.

McCarroll, N.E., B.H. Keech, and C.E. Piper. 1980a. A comparative evaluation of microsuspension microbial DNA repair systems. Environ. Mutagen. 2:270. (abstr.)

McCarroll, N.E., C.E. Piper, G.M. Fukin, B.H. Keech, and G. Gridley. 1979. A serial dilution multiwell suspension assay for DNA damage in E. coli. Environ. Mutagen. 1:123. (abstr.)

McCarroll, N.E., C.E. Piper, and B.H. Keech. 1980b. Bacterial microsuspension assays with benzene and other organic solvents. Environ. Mutagen. 2:281. (abstr.)

Murray, M.M. 1979. Substrate specificity of uvr excision repair. Environ. Mutagen. 1:347-352.

Nishioka, H. 1975. Mutagenic activities of metal compounds in bacteria. Mutation Res. 31:185-189.

Radman, M. 1977. Inducible pathways in deoxyribonucleic acid repair, mutagenesis and carcinogenesis. Biochem. Soc. Trans. 5:1194-1199.

Sunderman, F.W.,Jr. 1978. Carcinogenic effects of metals. Fed. Proc. 37:40-46.

Tamaro, M., S. Venturini, C. Eftimiadi, and C. Monti-Bragadin. 1977. Interaction of platinum compounds with bacterial DNA. Experientia 33:317-319.

Tindall, K.R. 1977. The mutagenicity of inorganic ions in
    microbial systems. Unpublished master's thesis, Montana
    State University. University Microfilm: Bozeman, MT.

Tindall, K.R., G.R. Warren, and P.D. Skaar. 1978. Metal ion
    effects in microbial screening systems. Mutation Res. 53:9091.
    (abstr.)

Tomizawa, J.-i., and H. Ogawa. 1968. Breakage of DNA in rec$^+$
    and rec$^-$ bacteria by disintegration of radiophosphorous atom
    in DNA and possible cause of pleiotropic effects of recA
    mutation. Cold Spring Harbor Symp. Quant. Biol. 33:243-251.

Venturini, S., and C. Monti-Bragadin. 1978. R plasmid-mediated
    enhancement of mutagenesis in strains of Escherichia coli
    deficient in known repair functions. Mutation Res. 50:1-8.

Warren, G., E. Abbott, P. Schultz, K. Bennett, and S. Rogers.
    (in press a). Mutagenicity of a series of octahedral rhodium
    III compounds. Mutation Res.

Warren, G.R., S.J. Rogers, and E.H. Abbott. (in press b). The
    genetic toxicology of substitutionally inert transition metal
    complexes. In: Inorganic Chemistry in Biology and Medicine,
    ACS Advances in Chemistry Series.

Warren, G., S. Rogers, S.G. Mevec, and M.D. Roach. 1979. Mutagen
    screening in an isolated high lung cancer mortality area of
    Montana. Montana Air Pollution Study, Air Quality Bureau,
    Dept. of Health and Environmental Sciences, Helena, MT.
    30 pp.

Warren, G., P. Schultz, D. Bancroft, K. Bennett, E.H. Abbott, and
    S. Rogers. (MS). Mutagenicity of a series of octahedral
    chromium III compounds.

Warren, G.R., P.D. Skaar, and S.J. Rogers. 1976. Genetic activity
    of dithiocarbamate and thiocarbamoyl disulfide fungicides in
    Saccharomyces cerevisiae, Salmonella typhimurium, and
    Escherichia coli. Mutation Res. 38:391-392 (abstr.).

Witkin, E.M. 1976. Ultraviolet mutagenesis and inducible DNA
    repair in Escherichia coli. Bacteriol. Rev. 40:869-907.

Yanofsky, C. 1963. Amino acid replacements associated with
    mutation and recombination in the A gene and their
    relationship to in vitro coding data. Cold Spring Harbor
    Symp. Quant. Biol. 28:581-588.

Zakour, R.A., L.A. Loeb, T.A. Kunkel, and R.M. Koplitz. 1979.
    Metals, DNA polymerization, and genetic miscoding. In:
    Trace Elements in Health and Disease. N. Kharasch, ed.
    Raven Press: New York. pp. 135-153.

A CULTURE SYSTEM FOR THE DIRECT EXPOSURE OF MAMMALIAN CELLS TO AIRBORNE POLLUTANTS

Ronald E. Rasmussen and T. Timothy Crocker
Department of Community and Environmental Medicine
College of Medicine
University of California
Irvine, California

INTRODUCTION

Most airborne pollutants first enter the body through the respiratory tract. In mammals and most other air-breathers, mechanisms have evolved to deal with these pollutants, especially those of a particulate nature. In recent times, however, air pollutant gases have been introduced into the environment at higher concentrations than before. Well-known examples are ozone (O_3) and oxides of nitrogen (NO_2, NO). Other gaseous pollutants whose effects are not so well known include short-lived, highly reactive species produced photochemically in the urban air mixture called smog. Peroxyacetyl nitrate (PAN) is one example (Stephens, 1969).

There has not been sufficient time for animals to evolve resistance to atmospheric oxidant gases. This lack of inherent resistance can be shown dramatically in rats exposed to 0.8 ppm of O_3 in air. At rest, the rats can tolerate this concentration for several hours, but if they are forced to exercise (i.e. on a treadmill) they quickly succumb to lung edema and hemorrhage (R.F. Phalen, University of California, Irvine, personal communication, 1980). Rats can be acclimatized to O_3 by repeated exposure at rest, so that they can resist the effects of O_3 when forced to exercise. Thus, substantial adaptational changes must have occurred in the lung. Studies of various enzymes in lungs of rats exposed to O_3 and NO_2 also have shown alterations (Chow et al., 1976). It can be supposed that similar effects occur in the human lung, since oxidant gas concentrations in some areas frequently approach, and even exceed, 1 ppm.

The cell culture and exposure system described in this report represents an in vitro approach to the study of the initial interactions between oxidant gases and respiratory cells. The goals of the present studies were to document some of the cytotoxic, biochemical, and cytogenetic effects on mammalian cells of exposure to low concentrations of O_3 and NO_2. Certain features of this system allow the exposure of cell cultures to ambient polluted atmospheres in a manner resembling exposure at the surface of the respiratory epithelium. Living cells are separated from the test atmosphere only by a thin layer of nutrient medium held over the cells by capillary attraction. As this situation allows close contact between cells and the ambient atmosphere, it may be especially useful in studying the primary interactions between cells and airborne pollutants.

MATERIALS AND METHODS

Exposure System

The following features are required in a cell system for testing exposure to gaseous pollutants:

1) accurate generation and monitoring of the pollutants to be studied,

2) maintenance of the cells in close contact with the test atmosphere without allowing them to dry out,

3) exposure under biologically sterile conditions, and

4) provision for recovery of the cells for further culturing and analysis.

The polluted atmospheres were generated by the measured addition of NO_2 or O_3 to a stream of clean air. The initial exposure system consisted of one control and one experimental chamber. Results with this early system have been reported elsewhere (Samuelsen et al., 1978; Crocker et al., 1979). The system was later expanded to a total of six chambers that could be arranged as desired for control or text exposures. Figure 1 is a diagram of the system, showing interconnections for using three different gases, either individually or in combination. Atmospheres containing NO_2 or sulfur dioxide (SO_2) were generated by diluting gas from stock cylinders in a stream of clean air. Clean air was provided by treating the building air supply sequentially with Purafil, activated charcoal, and filtration through a 2-μm bacteriological filter. This treatment reduced the level of NO_2 to < 10 ppb and the O_3 level to below detectable

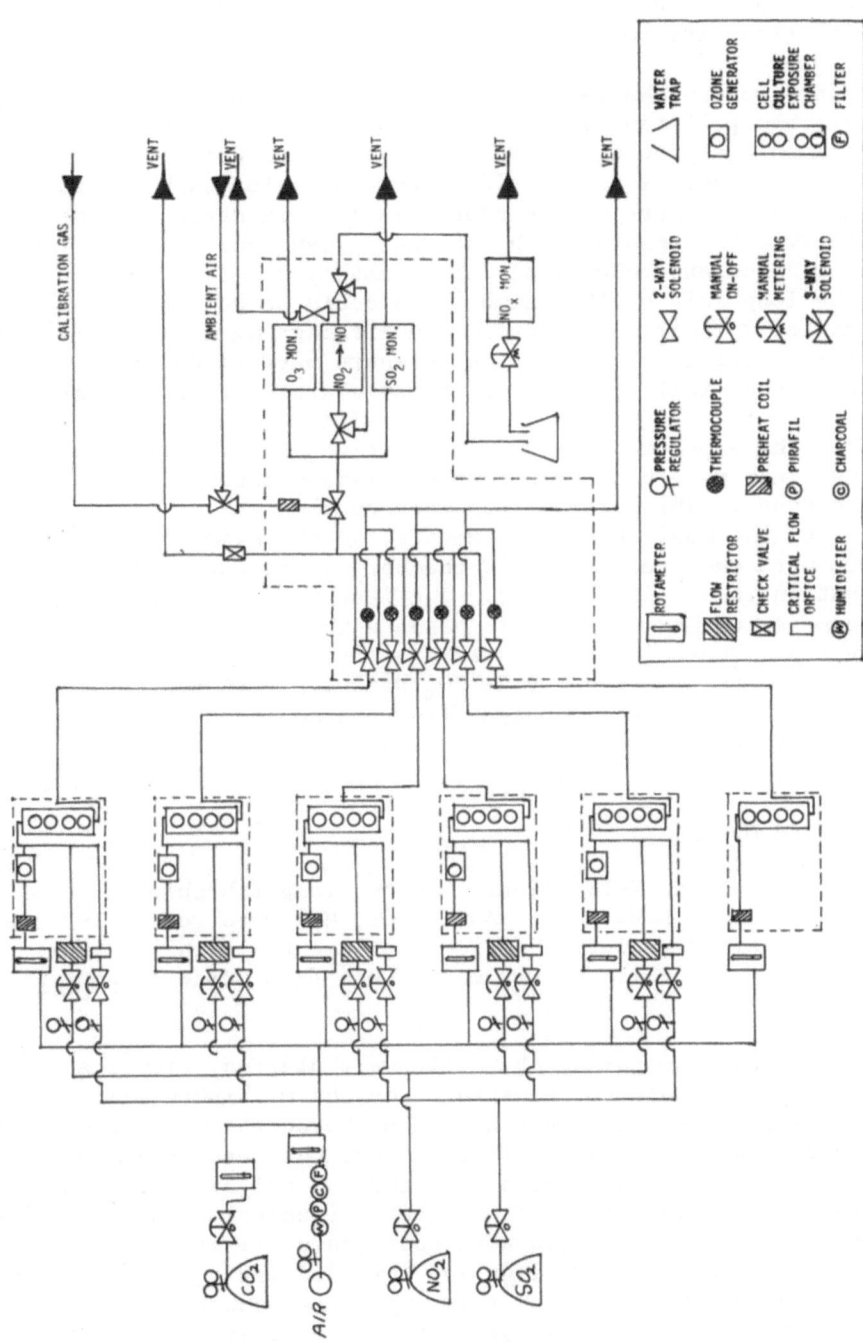

Figure 1. The exposure system.

amounts (i.e., < 1 ppb). Because the cell culture medium was
bicarbonate-buffered, carbon dioxide (CO_2) was added to the air
stream to raise its concentration to 5%.

Each exposure chamber was provided with flow controls for the
clean air and pollutant gases. The gases were mixed with the air
stream immediately before they entered the exposure chambers.
Ozone was provided by an individual generator for each exposure
chamber. The chambers were enclosed in 37°C incubators. The
chambers were rectangular with internal dimensions of approximately
10 x 10 x 35 cm and volumes of approximately 3.5 1. Gas flow was
along the long axis of the chamber and could be adjusted up to a
flow rate of 4 1/min. All tubing and fittings were either Teflon
or stainless steel. The exposure chambers were lined with
stainless steel foil; chambers of Lexan plastic are also available.

From the exposure chambers, the gas was carried via heated
sampling lines to a bank of solenoid valves controlled by a
microprocessor. The outflow from each chamber was sequentially
directed to the monitoring instruments, which (except for the NO_x
monitor) were enclosed in a 37°C chamber to prevent moisture
condensation in the instruments. The instruments were selected to
provide measurements of specific gases without interference from
other gases. These instruments were a Dasibi model 1003 AH O_3
monitor and a Beckman model 952A NO_x analyzer. (A monitor for SO_2
had yet to be installed.) Output from the monitors was fed to a
multichannel recorder.

Cell Culture Method

Cells of strain V-79 Chinese hamster lung fibroblasts
(obtained from Dr. E.H.Y. Chu, Ann Arbor, MI) were routinely grown
in Eagle's minimum essential medium supplemented with 10% fetal
bovine serum. All cell culture media were from Grand Island
Biological Co.

For exposure to gases, the cells were planted, either as
dispersed cells or as confluent cultures, on Millipore filters
(HAWP, 0.47-μm pore size) that had been thoroughly washed and
autoclaved. The cell-bearing filters were then assembled into the
specially designed holder shown in Figures 2 and 3. These holders
were fabricated from either Lexan plastic (General Electric Co.) or
stainless steel. The filters were positioned cell-side up and
sealed in place with the O-rings and threaded cap. Growth medium
was provided to the cells with a syringe pump connected to the
fitting in the base of the holder. Medium perfusing through the
filter was drawn off through a small tube in the membrane holder,
shown in Figure 2 as a diagonal line at the right-hand side of the
membrane holder. With this arrangement the cells were kept moist

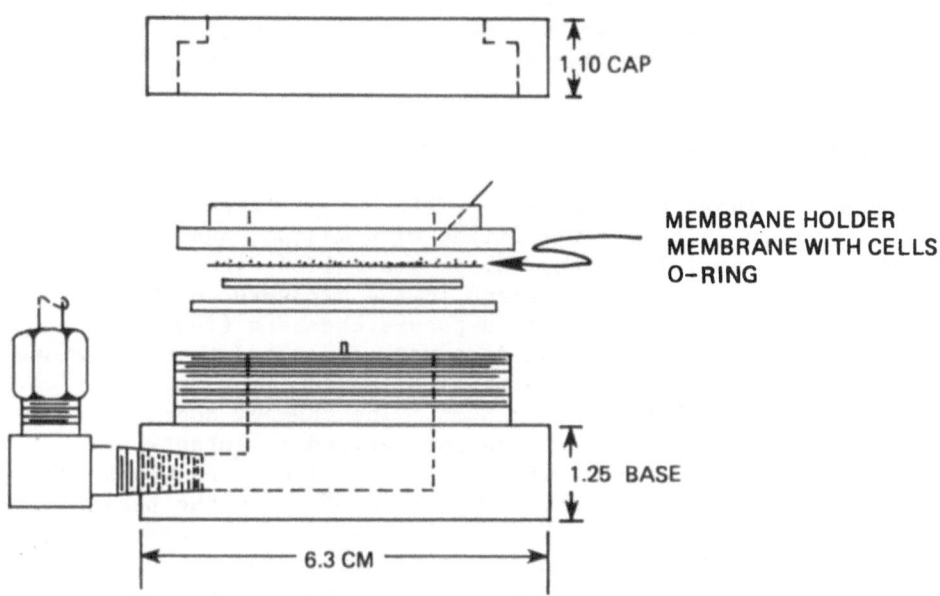

Figure 2.   Diagram of a filter holder.

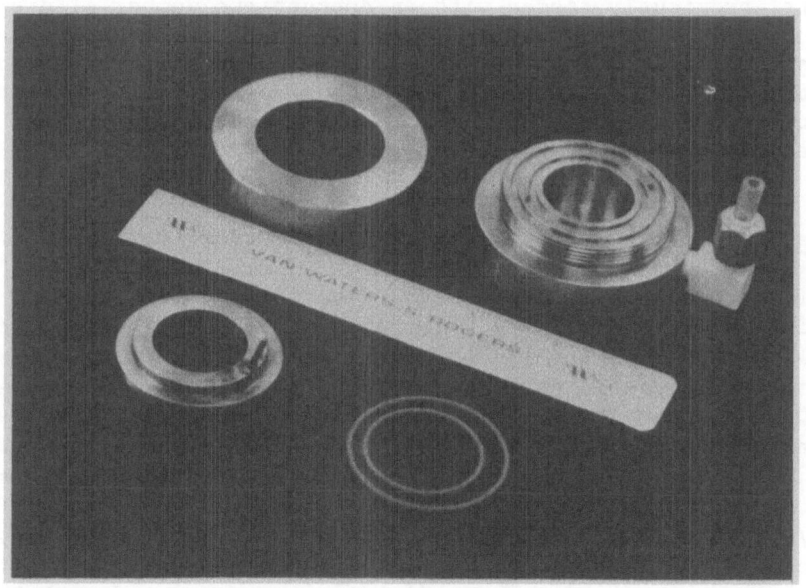

Figure 3.   Photograph of a disassembled filter holder.

and continuously provided with fresh nutrient, while in nearly
direct contact with the ambient atmosphere. In clean air, nearly
100% viability could be maintained for at least several hours.

Experimental Procedure

Figure 4 shows a filter with cells being placed in a holder.
Before assembly, the base was filled with medium, and after
assembly, the top well of the holder was filled with medium to
protect the cells before placement in the exposure chambers. The
holders were then placed in the exposure chambers (four per
chamber). The bases of the holders were connected to the syringe
pump, while the tubes for withdrawal of medium from the upper wells
were connected to a peristaltic pump. The chamber door was sealed,
and gas flow was initiated. When the desired pollutant
concentration within the chamber had been reached and was stable,
the medium overlaying the cells was drawn off with the peristaltic
pump, and cell exposure was begun. The syringe pump was turned on
at the same time to slowly perfuse nutrient medium through the
filter. To conclude the exposure, the exposure chambers were
opened, the holders removed, and the top well immediately filled
with nutrient medium. Because the gas exposure levels used were of
the same order of magnitude as in the ambient atmosphere (0.15 ppm
NO_2; 0.05 ppm O_3), no special precautions were necessary.

Subsequent procedures depended on the nature of the
experiment. To estimate colony formation by surviving cells, the
filters, previously seeded with an appropriate number of dispersed
single cells, were transferred directly to petri dishes containing
nutrient medium and incubated for 7 to 10 days to permit colony
development. The cells could then be harvested from the filters
with trypsin and subcultured for further study, such as for
mutagenesis or chromosomal effects.

Measurement of Effects of O_3 and NO_2 on DNA Replication

The procedures for this study were adapted from those
described by Painter (1977). Cultures of V-79 cells were grown for
48 h with carbon-14-labeled thymidine (^{14}C-TdR) at 0.01 µCi/ml (50
mCi/mmol). The cells were then planted on filters and exposed to
O_3, NO_2, or clean air at the concentrations and for the times
indicated in the Results section, below. After the gas exposure,
the filters were returned to nutrient medium; at intervals, sample
cultures were labeled for 10 min with tritiated thymidine (^{3}H-TdR;
5 µCi/ml, 60 Ci/mmol). After this labeling, the filters were fixed
in ice-cold 5% trichloroacetic acid, washed several times with 70%
ethanol, and air dried. The radioactivity associated with the
filters was measured by scintillation counting, and the ratio of

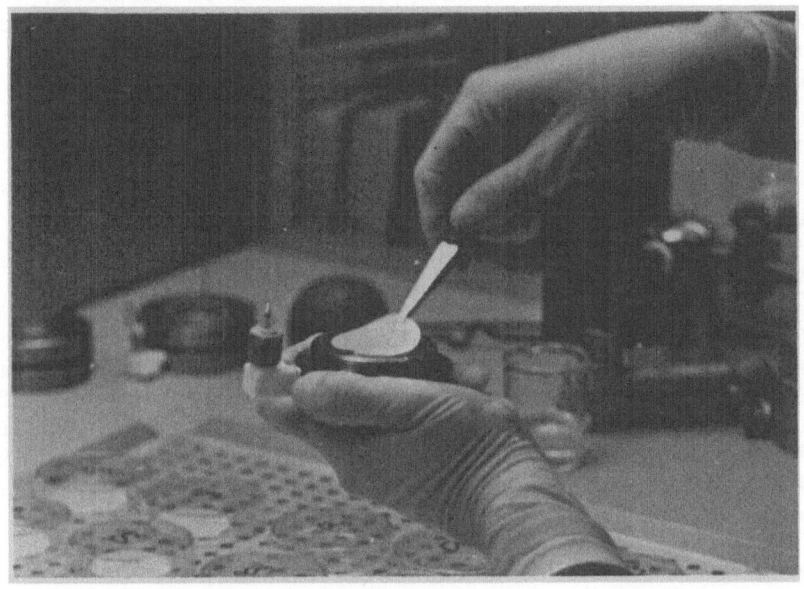

Figure 4. The filter with cells being placed in the holder.

^{3}H dpm:^{14}C dpm was calculated to give an index of the rate of DNA synthesis at the time of labeling with ^{3}H–TdR.

Measurement of Cytotoxic Effects of O_3 and NO_2

Strain V–79 cells were seeded into filters as dispersed single cells and allowed three to four hours for attachment. At that time, the filters were placed in holders, which were then put in the exposure chambers and exposed to O_3 or NO_2 at the concentration and for the times indicated below. After exposure, the filters were removed from the holders, transferred to petri dishes containing nutrient medium, and incubated for 7 to 10 days to permit colony development by survivors. Colonies were visualized by staining the filters with hematoxylin.

Measurement of Direct Effects of NO_2 on V–79 Cells

Growing cultures of V–79 cells were double-labeled with ^{14}C amino acids (0.1 µCi/ml) and ^{3}H–TdR (1.0 µCi/ml) for 24 h and then seeded onto filters, which were then placed into Lexan filter holders. After another 24 h, the cells were exposed for 2 h to NO_2

at 5 ppm. The filters were then removed from the holders, washed
gently with 0.9% sodium chloride, and air dried. The radioactivity
remaining with the filters was determined by scintillation
counting.

EXPERIMENTAL RESULTS

Cytotoxic Effects of NO_2 and O_3

It was recognized early in these studies that V-79 cells were
very sensitive to the effects of NO_2 when the gas was in direct
contact with the cells (Samuelsen et al., 1978). Exposure of cells
to 0.15 ppm of NO_2 inactivated their colony-forming ability in a
dose-dependent manner; after 6 h, fewer than 90% of the originally
exposed cells could form macroscopic colonies during subsequent
incubation of the filters in immersed culture.

Ozone was somewhat more effective than NO_2 in activating
colony formation by V-79 cells. Figure 5 shows the results of a
series of studies with O_3 at 0.05 ppm in air. Colony-forming
ability was always compared with that of cells exposed to clean air
in separate chambers. In the clean-air controls, very little loss
of colony-forming ability was seen (less than 10%) for exposures as
long as 6 h.

Effect of NO_2 and O_3 on DNA Replication

In mammalian cells, many chemical and physical mutagens induce
DNA damage that interferes with DNA replication (Painter, 1978).
This can be shown by measuring, at intervals, the rate of DNA
synthesis in cell cultures after a single treatment with the test
agent. With most mutagens, the rate of DNA synthesis declines with
time after treatment. Chemicals that inhibit DNA synthesis, but do
not damage DNA, do not produce such an effect, and DNA synthesis
returns to its normal rate when the inhibitors are removed.

To test for the effect of NO_2 and O_3 on DNA replication, V-79
cells were labeled with ^{14}C-TdR as described above and exposed to
either O_3 (0.03 ppm, 1 h), NO_2 (0.15 ppm, 1 h), or 254-nm
ultraviolet (UV) light (5 J/m^2). Immediately after treatment and
one and two hours later, sample cultures from each group were
labeled for 10 min with ^3H-TdR and immediately fixed with cold 5%
trichloroacetic acid. The results are shown in Figure 6. The data
are presented as percentages of the rates of DNA synthesis in
appropriate controls (cultures exposed to clean air as controls for
NO_2 and O_3 and cultures sham-exposed to UV light).

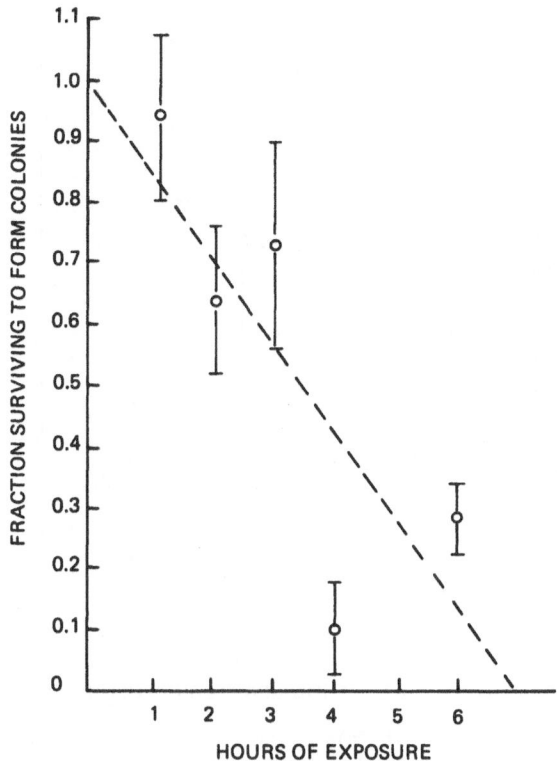

Figure 5.   Inactivation of V-79 cells by 0.05 ppm ozone.

Exposure to NO_2 produced a slight decrease in the rate of DNA synthesis, but the rate returned to normal when the cultures were placed in fresh culture medium. Both O_3 and UV light also produced a slight initial decrease in the rate of DNA synthesis; but in contrast to the results with NO_2-exposed cells, the rate of DNA synthesis continued to fall during subsequent incubation in fresh medium. This suggests that O_3 may have damaged the cellular DNA.

Mechanism of Action of O_3 and NO_2 on Directly-exposed Cells

The oxidizing effects of NO_2 and O_3 on membrane lipids are well known (Goldstein and Balchum, 1967; Thomas et al., 1968): the toxicity seen in the present exposure system may be due in part to such oxidations of the cell surface membranes. Exposure of thin films of peanut oil to 0.15 ppm of NO_2 in the present system produced detectable peroxidation (Goldstein et al., 1969).

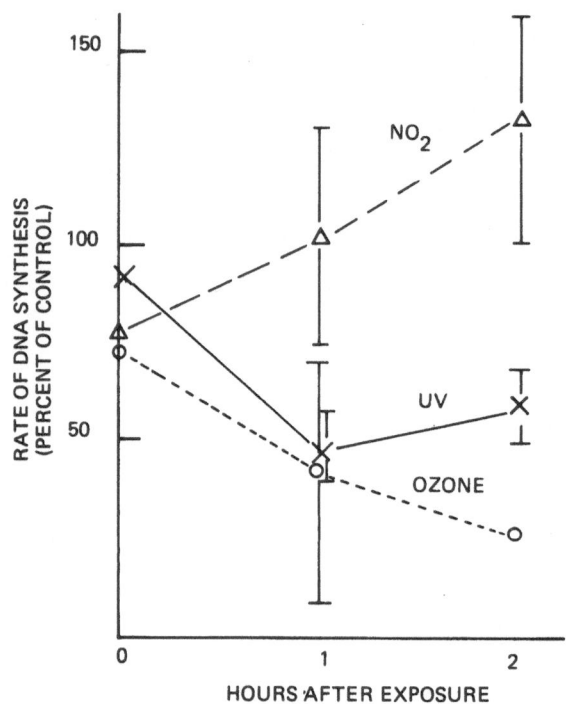

Figure 6.   The effect of NO_2, ozone, and UV-light on the rate of
DNA synthesis in V-79 cells.

Microscopic examination of cells after various periods of
exposure to NO_2 tended to support this hypothesis. The first
observable changes were that the cells seemed to become more
rounded and firmly attached to the filter substrate. With longer
times of exposure and higher NO_2 levels, the nuclei were distorted.
Finally, 5 ppm of NO_2 produced loss of cells from the filters. A
possible trivial explanation could be that NO_2 somehow affected the
Millipore membrane so that it could no longer support cell growth
and attachment. This was shown not to be the case through studies
in which filters were exposed to 5 ppm of NO_2 for 6 h. This
treatment did not alter cell attachment or the ability of the
filters to support cell growth.

To obtain some quantitative information on cell destruction
by NO_2, cultures of V-79 cells were labeled simultaneously with
[14]C-amino acids and ^{3}H-TdR as described above. The doubly-labeled
cells were then planted on filters so as to provide a nearly
confluent cell sheet at the time of exposure to NO_2.

Table 1 gives the results of several experiments in which exposure to NO_2 at 5 ppm for 2 h produced an average loss of cells from the filters of about 50%. Both ^{14}C and ^{3}H were lost from the filters. In every case, however, proportionately more tritium was lost from the NO_2-exposed filters; the ratio of $^{3}H:^{14}C$ remaining on the filters was always lower for the NO_2-exposed filters. Exposure to clean air did not produce any loss of label, as shown in experiments 18, 21, and 25.

Table 1. Results of Mammalian Cell Exposure to Nitrogen Dioxide

Experiment Number	Exposure	^{3}H DPM x 10^{-5} ± 1 SD	^{14}C DPM x 10^{-3} ± 1 SD	Ratio $^{3}H:^{14}C$
15	NO_2	1.72 ± 0.14	1.18 ± 0.05	14.6
	Air	7.58 ± 0.16	3.73 ± 0.10	20.3
17	NO_2	0.982 ± 0.074	3.78 ± 0.27	26.0
	Air	1.66 ± 0.19	5.88 ± 0.68	28.2
18	NO_2	0.850 ± 0.20	2.75 ± 0.57	30.9
	Air	2.50 ± 0.33	5.34 ± 0.82	46.8
	Not Exposed	2.31 ± 0.40	5.07 ± 0.91	45.6
21	NO_2	4.60 ± 0.82	25.00 ± 0.47	18.4
	Air	11.2 ± 0.79	44.50 ± 0.21	25.2
	Not Exposed	10.6 ± 0.40	44.10 ± 0.20	24.0
25	NO_2	3.62 ± 0.31	14.0 ± 1.3	25.9
	Air	5.33 ± 0.26	17.7 ± 1.3	30.1
	Not Exposed	5.55 ± 0.40	19.1 ± 1.4	29.1

DISCUSSION

This research project was begun with the goal of developing an in vitro analog of the respiratory epithelium. Cell culture methods were to be established for maintaining living cells in nearly direct contact with test atmospheres, while at the same time preventing the cells from drying out. In addition, these methods would allow recovery of the cells for further study. Previous systems for exposing cells to gaseous materials have relied on 1) solution of the pollutant in the medium bathing the cells or

2) periodic short exposures of the cell layer followed by immersion of the cells in liquid media.

A method for exposing cells for relatively long periods to almost any atmosphere has been developed; however, cells exposed in this system were extremely sensitive to very low levels of pollutant gases, which may not realistically indicate the effects of pollutants on experimental animals and humans. For example, O_3 and NO_2 concentrations of less than 1 ppm caused rapid destruction of cells in this test system. This great sensitivity may be related to the relatively large area of the cell surface that was exposed to the gases, since the cells were spread thinly on the membrane surface. In contrast, the cells of the respiratory tract in vivo are closely packed, as in the columnar epithelial areas, and only a small percentage of the total cell surface area is exposed to airflow in the airways. Also, in the normal lung the cells are covered by a mucous layer that shields them from exposure to oxidants and particulates. Simulation of this mucous layer in the present exposure system was not attempted.

The exposure system is not limited to the use of the cell cultures on membrane filters. The exposure chambers will accommodate conventional organ culture dishes, which are being used in current studies of the role of oxidant gases in neoplastic transformation. It is known that O_3 and NO_2 damage lung cells and induce hyperplasia (Hackett, 1979; Dungworth et al., 1975), but the role (if any) of this response in neoplasia is not clear. Using the present exposure system, we hope to provide evidence on this subject.

ACKNOWLEDGMENTS

This project has received support from the U.S. Environmental Protection Agency (contract no. 68-02-1204), the U.S. Air Force Office of Scientific Research (contract no. 77-3343), and the U.S. National Institute for Environmental Health Sciences (grant no. ES-01835). We thank Dr. G. Scott Samuelsen and John T. Taylor, of the School of Engineering of the University of California, Irvine, for invaluable aid and counsel. We also thank M.E. Witte and D.L. Swedberg for able technical assistance.

REFERENCES

Chow, C.K., M.Z. Hussain, C.E. Cross, D.L. Dungworth, and M.G. Mustafa. 1976. Effect of low levels of ozone on rat lungs. I. Biochemical responses during recovery and reexposure. Exper. Molec. Pathol. 25:182-188.

Crocker, T.T., R.E. Rasmussen, M.E. Witte, G.S. Samuelsen, and J.T. Taylor. 1979. A unique culture system for in vitro exposure of respiratory cells to pollutant gases. In: Nitrogenous Air Pollutants. Chemical and Biological Implications. D. Grosjean, ed. Ann Arbor Science Publishers: Ann Arbor, MI. pp. 179-188.

Dungworth, D.L., C.E. Cross, J.R. Gillespie, and C.G. Plopper. 1975. The effects of ozone on animals. In: Ozone Chemistry and Technology. J.S. Murphy and J.R. Orr, eds. The Franklin Institute Press: Philadelphia. pp. 29-54.

Goldstein, B.D. and O.J. Balchum. 1967. Effect of ozone on lipid peroxidation in the red blood cell. Proc. Soc. Exper. Biol. Med. 126:356-358.

Goldstein, B.D., C. Lodi, C. Collinson, and O.J. Balchum. 1969. Ozone and lipid peroxidation. Arch. Environ. Hlth. 18:631-635.

Hackett, N.A. 1979. Proliferation of lung and airway cells induced by nitrogen dioxide. J. Toxicol. Environ. Hlth. 5:917-928.

Painter, R.B. 1977. Rapid test to detect agents that damage human DNA. Nature 265:650-651.

Painter, R.B. 1978. DNA synthesis inhibition in HeLa cells as a simple test for agents that damage human DNA. J. Environ. Pathol. Toxicol. 2:65-78.

Samuelsen, G.S., R.E. Rasmussen, B.K. Nair, and T.T. Crocker. 1978. Novel culture and exposure system for measurement of effects of airborne pollutants on mammalian cells. Environ. Sci. Technol. 12:426-430.

Stephens, E.R. 1969. The formation, reactions, and properties of peroxy acyl nitrates (PANs) in photochemical air pollution. In: Advances in Environmental Sciences and Technology, Vol. 1. J.N. Pitts and R.L. Metcalf, eds. Wiley Interscience: New York. pp. 119-146.

Thomas, H.V., P.K. Mueller, and R.L. Lyman. 1969. Lipoperoxidation of lung lipids in rats exposed to nitrogen dioxide. Science 159:532-534.

SESSION 2

DRINKING WATER AND AQUEOUS EFFLUENTS

IS DRINKING WATER A SIGNIFICANT SOURCE OF HUMAN EXPOSURE TO CHEMICAL CARCINOGENS AND MUTAGENS?

Richard J. Bull
Health Effects Research Laboratory
U.S. Environmental Protection Agency
Cincinnati, Ohio

Drinking water has long been suspected as a medium through which chemical carcinogens and mutagens reach man. The first inquiries into this possibility were made by Heuper and Ruchhoft (1954), who tested carbon-chloroform extracts from the following samples: a gravity oil separator effluent from a petroleum refinery; raw water from a canal polluted by wastes from a petroleum refinery; raw water from Nitro, WV (the Kanawha River); and finished water from Cincinnati, OH (the Ohio River). These samples were applied topically to black male C57 mice in whole-animal carcinogenicity tests lasting one year. Samples from the refinery and the ship canal were carcinogenic; the raw water from Nitro and the finished water from Cincinnati gave negative results. While this study did not give direct evidence of carcinogens in drinking waters, it brought attention to the presence of such chemicals in surface waters used as sources for drinking water.

The development and general application of very sophisticated analytical tools in the past ten years have greatly altered our views on pollution of drinking waters. Coupled gas chromatographic and mass spectral analysis of drinking water has rapidly increased the numbers of chemicals known to occur in drinking water. Where previously the identification of a single chemical might take months or years, the general availability of mass spectrometry and the development of associated computer systems now allow almost instantaneous identification of hundreds of chemicals (if reference fragmentation patterns are available). To date, more than 1152 different chemicals have been identified in extracts of U.S. drinking waters (Melton, 1979); several of these chemicals are known to produce tumors in humans or experimental animals.

135

The most significant finding in recent years is that the organic chemicals usually found at the highest concentrations in drinking waters arise not from industrial pollution, but from disinfection of drinking water with chlorine (Rook, 1974; Bellar et al., 1974). The first such chemicals identified were the trihalomethanes (THMs), primarily bromo- and chloro-substituted methanes, with occasional traces of iodomethanes. The significance of these observations was increased by the National Cancer Institute Carcinogenesis Bioassay Program's finding that chloroform is carcinogenic in rats and mice. Results from animal studies were subsequently supported by a number of studies indicating a correlation between drinking water chlorination and gastrointestinal- and urinary-tract cancer mortality (see Wilkins et al., 1979, for a review). In addition, several chemicals found in drinking water have been identified as mutagens in the Ames test (Simmons et al., 1977).

As the above findings became known, workers began to investigate the biological properties of complex mixtures of chemicals in drinking water. A number of laboratories have found organic concentrates of drinking water to be mutagenic in the Ames test (Loper et al., 1978; Glatz et al., 1978). These studies have been followed up with demonstrations that such concentrates can transform BALB/3T3 cells (Lang et al., in press). When chemicals from water are sufficiently concentrated and fractionated, it is doubtful that any surface water supply will be found to be without any mutagenic activity.

Several investigators have shown that the THMs represent only a fraction of the products of chlorination. Many substances produced by chlorination of humic and fulvic acids and isolated from water remain to be identified; many of these are not available from commercial sources. Data recently reported by Symons (in press), of the Drinking Water Research Division of the Municipal Environmental Research Laboratory, in Cincinnati, indicate that more than 50% of the organic chlorine produced through chlorination of drinking water can be in products other than THMs. Although chloramination, an alternate disinfection method, suppresses THM formation, it does not suppress the production of non-THM organic chlorine to the same extent.

Mouse skin initiation/promotion studies with various disinfectants indicate that these non-THM products cannot readily be dismissed. In one such experiment (see Table 1), tumor-initiating activity was found only for drinking water disinfected with chlorine, chloramines, or ozone (Bull, 1980). Since these concentrates were prepared by reverse osmosis, they did not contain THMs.

Table 1.   Tumor Initiation by Reaction Products of Various
Disinfectants[a] After 20 Weeks of Promotion with PMA[b]

Sample[c]	Concentration Factor[c]	No. Animals with Tumors	Total Tumors	% Animals with Tumors
Non-disinfected	102	0/25	0	0
Chlorine	106	4/25	5	20
Chloramine	142	5/25	8	32
Chlorine dioxide	168	0/25	0	0
Ozone	186	7/25	9	36
Saline	–	1/25	1	4
7,12-dimethyl-benzo(a)anthracene	–	16/25	35	140

[a]Substrate was settled, coagulated, and filtered Ohio River water.
The total dose was 1.5 ml (6 x 0.25 ml) given subcutaneously at
the concentration factor indicated.
[b]PMA = Phorbol myristate acetate applied at a dose of 2.5 µg, three
times weekly.
[c]After treatment, water was subjected to reverse osmosis with
cellulose acetate to the level indicated by the concentration
factor (initial volume/final volume).

    With this brief background, we can summarize the major
categories of hazardous chemicals in drinking water:

1) Trace quantities of a wide variety of synthetic
   organic chemicals, usually at < 1 µg/l in surface
   waters.

2) Sporadic occurrence of high concentrations of
   individual industrial chemicals, typically involving
   groundwater contamination with bulk solvents.

3) Natural (or background) organic chemicals, such as
   humic and fulvic acids.

4) Products of the reaction of disinfectants with
   background chemicals (including THMs, the major
   class of organic compounds in drinking water).

5) Chemicals leached from distribution systems, such as
   lead, asbestos, and polycyclic aromatic hydrocarbons.

   6) Water-treatment chemicals (including polyelectrolytes,
      coagulants, and corrosion-control chemicals).

The third and fourth of these items are peculiar to and ubiquitous
in drinking water; they pose the greatest problems in assessing the
risks associated with drinking water.

   The overwhelming numbers of chemicals that appear in drinking
waters, their low individual concentrations, the complications
associated with drinking-water disinfection, and the need to
demonstrate actual reduction in carcinogenic risk through various
treatment options dictate a bioassay approach to drinking water
risk assessment. The presence of several hundred to several
thousand compounds in a drinking water at a fraction of a microgram
per liter is the rule rather than the exception. Individually,
most of these chemicals might contribute little to human disease
(with the possible exception of the THMs). However, they add up
to concentrations of from one to several milligrams of total
organic carbon per liter of drinking water and could account for
some of the hazards suggested by epidemiological data (Wilkins et
al., 1979).

REFERENCES

Bellar, T.A., J.J. Lichtenberg, and A.D. Kroner. 1974. The
     occurrence of organohalides in chlorinated drinking water.
     J. Amer. Water Works Assoc. 66:703-706.

Bull, R.J. 1980. Health effects of alternate disinfectants and
     their reaction products. J. Amer. Water Works Assoc.
     72:299-303.

Glatz, B.A., C.D. Chriswell, M.D. Arguello, H.J. Svec, J.S. Fritz,
     S.M. Grimm, and M.A. Thomson. 1978. Examination of drinking
     water for mutagenic activity. J. Amer. Water Works Assoc.
     70:465-468.

Heuper, W.C. and C.C. Ruchhoft. 1954. Carcinogenic studies on
     absorbates of industrially polluted raw and finished water
     supplies. Arch. Ind. Hyg. Occup. Med. 9:488-495.

Lang, D.R., H. Kurzepa, M.S. Cole, and J.C. Loper. (in press).
     Malignant transformation of BALB/3T3 cells by residue organic
     mixtures from drinking water. J. Environ. Pathol. Toxicol.

Loper, J.C., D.R. Lang, R.S. Schoeny, B.B. Richmond, P.M.
     Gallagher, and C.C. Smith. 1978. Residue organic mixtures
     from drinking water show in vitro mutagenic and transforming
     activity. J. Toxicol. Environ. Hlth. 4:919-938.

Melton, R.G. 1979. GC/MS Analysis of Organics in Drinking Water
    Concentrates and Advanced Waste Treatment Concentrates.
    Preliminary Report--Combined Results on Five Drinking Water
    Supplies. Prepared under contract no. 68-03-2458 for the U.S.
    Environmental Protection Agency, Cincinnati, OH.

Rook, J.J. 1974. Formation of haloforms during chlorination of
    natural waters. J. Water Treat. Exam. 22:234-243.

Simmons, V.F., K. Kauhanen, and R.G. Tardiff. 1977. Mutagenic
    activity of chemicals identified in drinking water. Presented
    at the Second International Conference on Environmental
    Mutagens, Edinburgh.

Symons, J.M. (in press). Utilization of various treatment unit
    processes and treatment modifications for trihalomethane
    control. In: Proceedings, Control of Organic Chemical
    Contaminants in Drinking Water. U.S. Environmental Protection
    Agency: Washington, DC.

Wilkins, J.R. III, N.A. Reiches, and C.W. Kruse. 1979. Organic
    chemical contaminants in drinking water and cancer. J.
    Epidemiol. 110:420-448.

ALTERNATIVE STRATEGIES AND METHODS FOR CONCENTRATING CHEMICALS FROM WATER

Frederick C. Kopfler
Health Effects Research Laboratory
U.S. Environmental Protection Agency
Cincinnati, Ohio

INTRODUCTION

The concentration of organic matter in water can range from several hundred micrograms per liter in groundwater to many milligrams per liter in industrial or sewage effluents. Generally, the biologically active materials will be present in concentrations too low to be detected by testing the small amount of aqueous sample that can be incorporated directly into a biological test system. The organic matter in drinking water and wastewater is a complex mixture and defies complete characterization by current technology. Consequently, these materials cannot be purchased or synthesized for biological testing, but must be obtained from the water to be evaluated. A paradoxical situation results, since the evaluation of methods for concentrating an organic substance depends on the existence of reliable analytical methods for the quantitative analysis of the substance. Therefore most of the data available concerning organic concentration techniques are based on performance with a few specific compounds or on general parameters such as total organic carbon (TOC) or total organic halogen.

Methods for producing samples of organics from water for biological testing can be divided into two types: concentration and isolation. In the former, water is removed, leaving the dissolved substances behind; in the latter, the organic substances are removed from the water. Combinations of methods have also been used to prepare samples for biological testing. This paper does not deal with the theoretical aspects of the methods, but rather reviews the advantages and disadvantages of representative examples of both approaches. This paper is intended as a guide to the interpretation of data obtained through biological testing of organic concentrates produced through methods currently in use.

The choice of approach to preparing samples for biological testing is determined by the biological test system to be used. The final sample must provide a sufficient quantity of material in a volume of solvent that is compatible with and can be accommodated by the test system. Generally, more sample is required for biological testing than for chemical analysis. Most of the methods included in this paper were developed and evaluated for preparing samples for chemical analysis and, in most cases, will have to be scaled up to prepare samples for anything other than small-scale in vitro tests.

CONCENTRATION TECHNIQUES

Some volume-reduction methods for preparing organic concentrates for biological testing include freeze concentration, freeze-drying, vacuum evaporation, and membrane processes (reverse osmosis and ultrafiltration). If aqueous concentrates are required, the degree to which the samples can be reduced in volume is limited by the concentration of inorganic substances in the sample and the aqueous solubility of the organic substances. If the final concentration required for testing is greater than these parameters will allow, then the volume-reduction method can be used as a first step in a combination method, as will be described later.

Freeze concentration is the process whereby a water sample is frozen into a shell of pure ice, leaving in the center the unfrozen water containing the dissolved substances. Shapiro (1961) proposed using this method for concentrating environmental water samples; it was subsequently evaluated by Baker (1969). The method allows effective recovery of all components tested, including volatile and ionized organic species. The components are equally but not totally recovered.

Freeze concentration works well in distilled water solutions, but recovery of organic solutes decreases with increasing salt content, due to alterations of the forming ice surface that result in the incorporation of solute-rich liquid into the ice (Baker, 1970). Baker demonstrated that initial concentrations of up to 310 mg/l of total dissolved solids have little effect on the recovery of the test substance m-cresol, with approximately 80% being recovered in the liquid after a 20-fold reduction in volume (Baker, 1969). Baker used a rotary evaporator with the flask immersed in a freezing bath to concentrate samples of a few hundred milliliters spiked with milligram quantities of test compounds. At one time, a freeze concentrator capable of concentrating 5-1 samples up to

30-fold was available, but it is no longer manufactured. If large samples are to be freeze concentrated, the equipment will have to be custom-made and evaluated.

Freeze-drying is the process of removing water vapor directly from the frozen sample by sublimation under vacuum. Large-scale equipment is available, but it is cumbersome and expensive, and it requires large amounts of energy. Freeze-drying is a slow process. In our laboratory, we use an apparatus that allows up to 40 l of water to be processed at one time; about 72 h are required to remove all of the water. The residue remaining is composed largely of inorganic salts and is hygroscopic, so that a solvent is required to recover it from the large stainless steel pans used as sample containers.

Vacuum evaporation (or vacuum distillation) is the boiling of the aqueous sample at reduced pressure at or near ambient temperature. This method has been used to concentrate water samples for chemical analysis (Jolley et al., 1975) and for biological testing (Johnston and Herron, 1979). The required apparatus can be assembled from commercially available components.

Freeze-drying and vacuum evaporation can achieve high degrees of concentration with little contamination, and only substances volatile at the temperature and pressure used will be lost. For most environmental water samples, large percentages of TOC can be recovered. The major drawback in both cases is the difficulty of recovering the organic substances from inorganic precipitates. This problem will be addressed at the end of this section.

In reverse osmosis, water is preferentially forced through a membrane by applying an external pressure that exceeds the osmotic pressure across the membrane. Reverse osmosis systems are commercially available in many sizes and generally contain either cellulose acetate or polyamide (nylon) membranes. The polyamide membranes are more stable at extremes of pH but are highly sensitive to chlorine. They cannot, therefore, be used to process chlorinated waters unless the residual is chemically reduced. One disadvantage of the commercial systems is that they contain plastic components and synthetic adhesives that could adsorb sample components or release contaminants into the concentrate.

A schematic of a reverse osmosis system used to concentrate water samples is shown in Figure 1. The water to be concentrated is circulated past the membrane under pressure; a fraction of the volume is removed, and the remainder is recirculated to the feed tank. Cellulose acetate membranes commonly used are rated to reject organic substances of molecular weight greater than 200 and 90 to 97% of the inorganic ions. Generally, this degree of inorganic rejection is achieved. However, while molecular weight

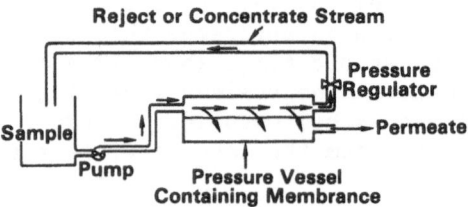

Figure 1.   Schematic of reverse osmosis concentrator.

influences the rejection of organics by the membrane, polar or
ionized species are rejected more effectively than hydrophobic
nonionized substances. In contrast to the other volume-reduction
methods described, reverse osmotic concentration works through an
exponential decay process in the recirculating system. Figure 2
illustrates the percent recoveries obtained for compounds with
rejections of 70, 80, and 90% at volume reductions between 10- and
1000-fold. Table 1 shows the actual concentration factors obtained
when sample volume is reduced 10-, 100-, and 1000-fold. While
reverse osmosis does not retain all of the substances equally, it
does retain much of the organic carbon in the drinking water
samples.

     Each of these concentration methods has its own advantages and
disadvantages, but the one disadvantage shared by all of these
methods is that inorganic species are concentrated along with the
organic substances of interest. The degree to which samples can be
concentrated by these methods before precipitation of inorganics
occurs varies with the types and concentrations of inorganics
originally present, but is generally 20- to 50-fold. If the
bioassay system is sensitive enough and will tolerate aqueous
samples, and the inorganic salts do not interfere, these
concentrates can be tested directly. Freeze concentration and
reverse osmosis produce only such aqueous concentrates; if more
concentrated samples are required, the samples must be processed

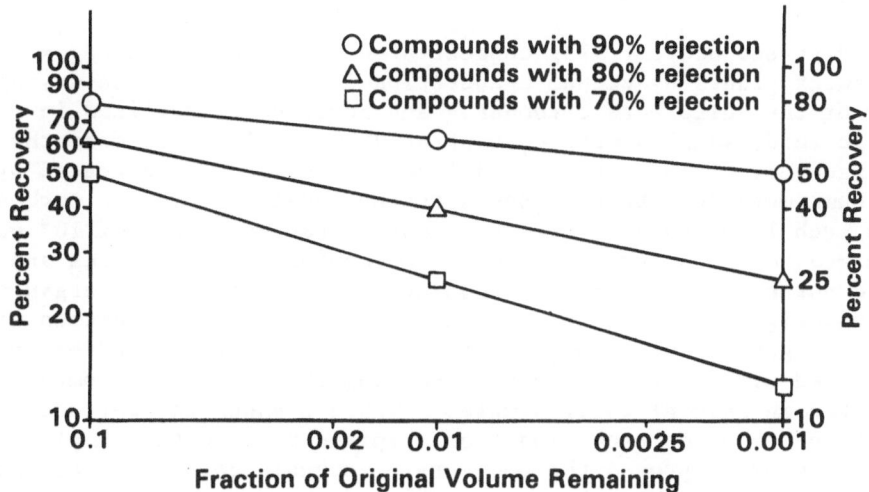

Figure 2.   Recovery of compounds by reverse osmosis at various
            stages of concentration.

Table 1.   Effect of Percent Rejection on the Actual Concentration
           of Constituents Obtained by Reverse Osmosis

Compound Percent Rejection	Actual Concentration Factor		
	Fraction of Original Volume Remaining		
	0.1	0.01	0.001
100	10	100	1000
90	8	60	500
80	5.2	40	250
70	5	25	125

further by other means. Freeze-drying and vacuum evaporation can
be used to produce either aqueous concentrates or a solid residue.
If inorganic constituents precipitate, recovery of the organics
from the residue is difficult.

      Pitt and Scott (1973) concentrated effluent from sewage-
treatment plants by vacuum evaporation followed by freeze-drying.
Many of the salts were carbonates and could be solubilized in
acetic acid; samples were concentrated 2000-fold, with more than
95% recovery of TOC. River and lake waters were concentrated by
the same methods. Because the starting levels of organic carbon
were much lower than in sewage, a concentration factor of 10^4 was
required. The inorganics were not predominantly carbonates and
could not be redissolved; consequently, about 75% of the organic
carbon originally present could not be recovered. Extraction with
methanol gave only slightly better yields. Recovery studies were
conducted with 13 compounds representing the classes of compounds
expected in natural water samples. Table 2 shows the percentages
of these compounds recovered from a spiked river water sample.
These workers improved the recovery of organic carbon by passing
the water sample through a weak cation exchange column prior to the
evaporative steps. Losses of organic carbon from four samples thus
treated still ranged from < 10 to 57%.

      Crathorne et al. (1979) studied the recovery of
5-chlorouracil, 5-chlorouridine, 4-chlororesorcinol, and
5-chlorosalicylic acid from residues obtained from spiked samples
of finished drinking water. The 14 solvents listed in Table 3 were
investigated for efficiency in recovering the compounds from the
residue. Only water and methanol yielded greater than 90%
recovery, but the mean recovery with methanol was about 60% (B.
Crathorne, Water Research Centre, Medmenham Laboratory, Manlaw,
Buckinghamshire, England, personal communication, 1980).

      Volume-reduction methods are most successful if precipitation
is prevented during concentration. This has been accomplished by
Pitt as described above, by Kopfler et al. (1977), using Donnan
dialysis to exchange sodium ions for calcium ions during reverse
osmosis, and by Johnston and Herron (1979), through a method to be
described below.

ISOLATION TECHNIQUES

      Isolation methods remove organics from water by concentrating
them in organic solvents. One method is to use immiscible organic
solvents to extract the water; another is to adsorb the organics
onto a solid medium and elute them with an organic solvent. Many
variations of these methods have been investigated, because they
have been used extensively to prepare samples for chemical analysis.

Table 2. Recovery of Individual Compounds in Aqueous
Solution after Evaporation-Lyophilization[a]

Compound	Anticipated Final Concentration (μg/ml)	Percent Recovered
Sucrose	10	>50[b]
Uracil	10	80
Guanosine	10	5
Xanthine	8	10
Uric Acid	25	<5
Hippuric acid	20	5
p-Cresol	13	0
p-Hydroxyphenylacetic acid	15	5
Syringic acid	15	15
p-Hydroxybenzoic acid	18	10
o-Chlorobenzoic acid	40	20
p-Chlorobenzoic acid	40	0
o-Chlorophenol	33	0

[a]From Pitt and Scott (1973).
[b]Quantification uncertain due to interference of naturally
occurring compound.

Table 3. Solvents Tested for Extracting Concentration Residues[a]

Solvent	Concentration Residue
Methanol	Diethyl ether
Water	Dimethylsulfoxide
Acetone	Dioxane
Acetonitrile	Ethyl acetate
Carbon tetrachloride	Isopropanol
Chloroform	Pyridine
Dichloromethane	Tetrahydrofuran

[a]B. Crathorne, Water Research Center, Medmenham Laboratory, Manlaw,
Buckinghamshire, England, personal communication, 1980.

Direct liquid-liquid extraction is suitable for the recovery
of organics from water samples of several liters. Large samples,
however, require continuous extractions using large volumes of
solvent or refluxing a smaller volume of solvent to provide pure
extractant.

As with any process employing solvents, impurities can be
concentrated along with sample components. Most highly purified
organic solvents contain preservatives--often an antioxidant--that
can react to add organic contaminants to samples. For example,
cyclohexene is an impurity present in the best grades of methylene
chloride. When methylene chloride is used to extract samples that
contain a chlorine residual (as do most drinking waters and
wastewaters), the cyclohexene produces mono-, tri-, and tetra-
chlorocyclohexenes and cyclohexanes by reacting with the chlorine
residual (Logsdon et al., 1977). Also, peroxides may contaminate
extracts prepared with ether or may react with sample components to
produce new substances not originally present in the sample or
solvent. Another critical area for investigation is whether
changes in the organic residues could occur in concentrates of
organics during storage between preparation and analysis or
biological testing.

The adsorption-elution methods require the least-complex
apparatus for isolating organics from water. The water sample is
passed through a column of the solid adsorbant, and the organics
are subsequently eluted with a smaller volume of suitable solvent.
The most common adsorption-elution methods used are activated
carbon, ion exchange resins, macroreticular resins, and reverse
phase high-performance liquid chromatography columns.

Activated carbon has been used to remove organics from aqueous
solutions (Buelow et al., 1973). However, recovery from carbon is
not as good as can be obtained with other media (Chriswell et al.,
1977). Formerly, organics were recovered from the carbon through
air-drying followed by prolonged Soxhlet extraction with chloroform
and ethanol. This method has been abandoned because of the air-
drying step and the continued boiling of the extract in the
extraction apparatus. A procedure using supercritical liquid
carbon dioxide as a solvent has recently been developed and is
being evaluated as a means of regenerating granular activated
carbon used in water treatment (Modell et al., 1978). The liquid
carbon dioxide is miscible with water, allowing the drying step to
be eliminated. The extraction takes place at about 30°C in a
closed system, eliminating the high temperature encountered with
organic solvents. The liquid carbon dioxide has also been shown to
be a good solvent for several classes of organics. This method has
not yet been evaluated for producing extracts for bioassay.

The XAD resins produced by Rhom and Haas have been used extensively to recover organic substances from water. Two types are available: a styrene-divinyl benzene copolymer and a methacrylate-based copolymer (Dressler, 1979; Gustafson and Paleos, 1971). Resins of both types are available in various pore sizes, giving different unit surface areas. These resins are produced for industrial use and contain many lower-molecular-weight contaminants. The resins must be prepared for laboratory use by serial extraction in a Soxhlet extractor with methanol, diethyl ether, and acetonitrile (Junk et al., 1974). Before use, the resins should be evaluated to insure that the extraction has, in fact, removed contaminants from the resin to the degree required. If resin beads are allowed to dry out, they can crack, exposing newly contaminated surfaces; thus, clean resin should be stored under methanol until used.

XAD resins have been used to isolate synthetic organic chemical contaminants from water for chemical analysis. They have a great affinity for hydrophobic substances and retain virtually all of these materials, even when the aqueous sample is passed through the resin column at high flow rates (Junk et al., 1974). Much of the organic matter in water is hydrophilic, however; Thurman and his co-workers (1978) have demonstrated that the capacity of the resins for these compounds is not great and that flow rates during the adsorption step must be in the range of 15 to 20 bed volumes/h for good recovery. Organic acids and bases are effectively adsorbed from water only after ionization has been suppressed by pH adjustment. Lowering the pH to protonate the organic acids generally presents no problem, but attempting to recover organic bases at a high pH can result in clogging of the column by inorganic hydroxides after only a small amount of water has passed through the column. It has been estimated that about 50% of the TOC in the average water sample can be concentrated onto a column of XAD-8 resin (Malcolm et al., 1977).

Ionizable organic substances can be recovered by elution with aqueous solutions of inorganic acids or bases. This will give an aqueous solution more concentrated than the original sample, but further concentration may still be required for biological testing. Elution with organic solvents is effective for recovering neutral organic substances. A variety of solvents have been used to elute the adsorbed organics from XAD columns; the most commonly used are diethyl ether, methanol, acetone, methylene chloride, and mixtures of these solvents. As discussed for liquid-liquid extraction, precautions must be taken to prevent impurities in the organic solvents from producing artifacts in the sample.

Workers from the U.S. Geological Survey have also found that humic and fulvic acids are adsorbed onto both types of XAD resins at pH 2 (Aikin et al., 1979). However, about 20% of the material

binds irreversibly to XAD-2 resin and cannot be eluted with alkali
or organic solvents. Much of the organic matter in surface waters
is composed of these naturally occurring acids or derivatives
produced during water disinfection. Because they can bind many
lower-molecular-weight organic substances, these acids should be
recovered to insure recovery of these bound materials. Owing to
its simplicity, it is tempting to use the resin technique for
producing samples for biological testing; however, it must be
remembered that not all of the organic substances are adsorbed and
those that are may not be fully recovered.

COMBINATION METHODS

Methods for isolating organics from water for biological
testing include combinations of several techniques, for convenience
or in attempts to approach 100% recovery of organics from the
water.

Kopfler et al. (1977) use reverse osmosis to reduce the volume
of water samples from thousands of liters to about forty liters.
To prevent precipitation of inorganic salts, sodium ions are
exchanged for calcium and magnesium ions in the concentrate through
a Nafion tubular membrane (Dupont) concurrently with the reverse
osmosis process. The concentrated aqueous sample is then
transported to the laboratory, where it is extracted with pentane
and methylene chloride and passed through a column of XAD-2 resin,
which is eluted with ethanol. Recovery of organics by this method
is estimated to be 35 to 40%. Residues have been tested for
mutagenicity and cellular transformation in vitro and
teratogenicity and carcinogenicity in vivo.

Johnston and Herron (1979) first pass the water through a
"parfait" column containing layers of silica gel, cation exchange
resin, and anion exchange resin. This step results in the
adsorption of some neutral hydrophobic organics as well as ionized
species including inorganic ions. The effluent from the column is
evaporated under vacuum to concentrate the hydrophilic nonvolatile
organic substances. Substances adsorbed on the column are
recovered by separating the layers in the column and eluting each
with 2 M triethylammonium carbonate buffer followed by acetone.
Samples prepared by this method have been tested for mutagenicity
in vitro.

Baird and co-workers (1980) use a series of stainless-steel
columns packed with microparticulate-sized weak ion exchange resins
and XAD resins. They report 85 to 90% removal of TOC from highly
treated wastewater passed through this series of columns. The
columns are eluted with acetonitrile and a 4.5-M sodium chloride
solution containing acetonitrile. Since the presence of

acetonitrile interferes with the TOC analysis, the actual recovery of organics cannot be determined. The saline eluates of the columns are concentrated further by extraction with acetonitrile at pH 7 and again at pH 1. After these extractions, the saline solution is still colored, indicating that organics have not been completely recovered. These extracts have been tested for mutagenicity in vitro.

Probably the most elaborate device for recovering organics from water for toxicity testing is that developed in France by Carbridenc and Sidka (1979). The apparatus is designed to extract 1000 liters of water with 100 liters of chloroform under an inert gas. The water is extracted first at pH 7, then at pH 2, and finally at pH 10. The aqueous sample is then neutralized and passed through small columns containing anion exchange resin (eluted with butanol), XAD-2 (eluted with a mixture of ethanol and methylene chloride), and activated carbon (washed with ethanol and extracted with chloroform). The recovery of 51 substances was determined using a 100-1 version of this apparatus. Forty-five of the compounds were detected in one or more of the fractions, and total recovery was calculated to be 88% by weight. Extracts of drinking water prepared by this method have been tested for cytotoxicity in vitro and for promotion in vivo.

CONCLUSIONS

The results from biological tests of organic concentrates in water can be used to estimate the hazards associated with the water, but only to the degree that the concentrate represents the organic materials actually present in the water. The concentrate should contain representative amounts of all the organic materials originally present, or at least a predetermined fraction of them. Also, the integrity of the chemicals must be maintained, with no contaminants present, or at least none that interfere with the biological tests. Until such methods or combinations are developed and validated, the information in this paper should serve as a guide to the representativeness of concentrates produced by various methods and should allow proper reservations to be made in the interpretation of biological test results.

REFERENCES

Aikin, G.R., E.M. Thurman, R.L. Malcolm, and H.F. Walton. 1979. Comparison of XAD macroporous resins for the concentration of fulvic acid from aqueous solution. Anal. Chem. 51:1799-1803.

Baird, R., J. Gute, C. Jacks, R. Jenkins, L. Neisess, B.
    Scheybeler, R. Van Sluis, and W. Yanko. (1980). Health
    effects of water reuse: a combination of toxicological and
    chemical methods for assessment. In: Water Chlorination,
    Environmental Impact and Health Effects, Vol. 3. R.L. Jolley,
    W.A. Brungs, and R.B. Cumming, eds. Ann Arbor Press: Ann
    Arbor, MI.

Baker, R.A. 1969. Trace organic contaminant concentration by
    freezing--III. ice washing. Water Res. 3:717-730.

Baker, R.A. 1970. Trace organic contaminant concentration by
    freezing--IV. ionic effects. Water Res. 4:559-573.

Buelow, R.W., J.K. Carswell, and J.M. Symons. 1973. An improved
    method for determining organics by activated carbon adsorption
    and solvent extraction. J. Am. Water Works Assoc. 65:57-72.

Carbridenc, R. and A. Sidka. 1979. Extraction des micropollutants
    organiques des eaux en vue de la realisation d'essais
    biologiques. Presented at the European Symposium on the
    Analysis of Organic Micropollutants in Water, Berlin, Federal
    Republic of Germany.

Chriswell, C.D., R.L. Ericson, G.A. Junk, K.W. Lee, J.S. Fritz, and
    H.J. Svec. 1977. Comparison of macroreticular resin and
    activated carbon as sorbents. J. Am. Water Works Assoc.
    69:669-674.

Crathorne, B., C.B. Watts, and M. Fielding. 1979. The analysis of
    non-volatile organic compounds in water by high-performance
    liquid chromatography. J. Chromatogr. 185:671-690.

Dressler, M. 1979. Extraction of trace amounts of organic
    compounds from water with porous organic polymers. J.
    Chromatogr. 165:167-206.

Gustafson, R.L., and J. Paleos. 1971. Interactions responsible
    for the selective adsorption of organics on organic surfaces.
    In: Organic Compounds in Aquatic Environments. S.J. Faust
    and J.V. Hunter, eds. Marcel Dekker, Inc.: New York. pp.
    213-237.

Hemon, D., P. Lazor, R. Cabridenc, A. Sidka, B. Festy, C.
    Gerinroze, and I. Chouroulinkov. 1978. I: Micropollution
    organique des eaux destinees a la consummation humaine. Rev.
    Epidem. et Sante Publ. 26:441-450.

Johnston, J.B., and J.N. Herron. 1979. A Routine Water Monitoring
    Test for Mutagenic Compounds. UIUC-WRC-79-0141. University
    of Illinois: Urbana, IL. 87 pp.

Jolley, R.L., S. Katz, J.E. Morchek, W.W. Pitt, and W.T. Rainey.
    1975. Analyzing organics in dilute aqueous solution. Chem.
    Tech. 5:312-318.

Junk, G.A., J.J. Richard, M.D. Grieser, D. Witiak, J.D. Witiak,
    M.D. Arguello, R. Vick, H.J. Svec, J.S. Fritz, and G.V.
    Calder. 1974. Use of macroreticular resins in the analysis
    of water for trace organic contaminants. J. Chromatogr.
    99:745-762.

Kopfler, F.C., E.W. Coleman, R.C. Melton, R.C. Tardiff, S.C. Lynch,
    and J.K. Smith. 1977. Extraction and identification of
    organic micropollutants: reverse osmosis method. Ann. N. Y.
    Acad. Sci. 298:20-30.

Logsdon, O.J., K.E. Nottingham, and T.O. Meiggs. 1977. Formation
    of nitrosamines and chlorocycloalkanes during analytical
    procedures. Presented at the 91st meeting of the Association
    of Official Analytical Chemists, Washington, DC.

Malcolm, R.L., E.M. Thurman, and G.R. Aiken. 1977. The
    concentration and fractionation of trace organic solutes from
    natural and polluted water using XAD-8, methylmethacrylate
    resin. In: Trace Substances in Environmental Health, Volume
    XI. D.D. Hemphill, ed. University of Missouri: Columbia,
    MO. pp. 307-314.

Modell, M., R.P. deFilippi, and V. Krukonis. 1978. Regeneration
    of activated carbon with supercritical carbon dioxide.
    Presented before the Division of Environmental Chemistry,
    American Chemical Society, Miami, FL.

Pitt, W.W., and C.D. Scott. 1973. Measurement of molecular
    organic contaminants in polluted water. In: Ecology and
    Analysis of Trace Contaminants. ORNL-NSF-EATC-1. Oak Ridge
    National Laboratory: Oak Ridge, TN. pp. 309-331.

Shapiro, J. 1961. Freezing out, a safe technique for
    concentration of dilute solutions. Science 133:2063-2064.

Thurman, E.M., R.L. Malcolm, and G.R. Aiken. 1978. Prediction of
    capacity factors for aqueous organic solutes adsorbed on a
    porous acrylic resin. Anal. Chem. 50:775-779.

DETECTION OF ORGANIC MUTAGENS IN WATER RESIDUES

John C. Loper and M. Wilson Tabor
Departments of Microbiology and Environmental Health
University of Cincinnati College of Medicine
Cincinnati, Ohio

INTRODUCTION

In previous studies (Loper and Lang, 1978; Loper et al., 1978; Lang et al., in press; Kurzepa et al., in press), we used short-term bioassays to demonstrate the mutagenicity, carcinogenicity, and toxicity of residues prepared from samples of drinking water from six U. S. cities. The samples were processed by Gulf South Research Institute (New Orleans, LA), using reverse osmosis plus XAD resin sorption-desorption as described by Kopfler et al. (1977). Using the Ames test, we found city-specific patterns of dose-dependent mutagenesis that were essentially independent of the microsomal activation system. One or more samples from each city showed reproducible transformation frequencies at least three times the spontaneous frequency. Focus formation induced by these samples was equivalent to malignant transformation as verified in nude mice. In these studies, quantitation of mutagenic and transformation responses were complicated by the toxicity and heterogeneity of the complex residue mixtures.

These findings justify further efforts at compound identification, and for that purpose, we have proposed the development of a coupled bioassay/chemical fractionation procedure (Loper and Lang, 1978). Such a method would be patterned after successful analyses of other complex environmental mixtures such as synthetic fuels (Guerin et al., 1978). Mutagenicity would be assayed using the Ames test. Initial partitioning of the sample by liquid/liquid extraction would be followed by high-performance liquid chromatography (HPLC) for separation into smaller subfractions. Active fractions sufficiently free of inactive and biocidal components would be analyzed by gas chromatography/mass spectrometry (GC/MS) for identification of peaks.

Drinking water residues are not usually generated in sufficient quantity for developing such methods. However, from Mr. Francis Middleton we obtained residue still on hand from previous use of the U. S. Public Health Service carbon-chloroform extractor. This mega sampler was used in the 1950's to early 1960's for the processing of 100,000-gal samples of drinking water and source water. The sample is a liquid solution in chloroform ($CHCl_3$) of 125 g of residue obtained about 1960 from 50,000 gal of drinking water (Middleton et al., 1962). When the $CHCl_3$ is removed under a stream of nitrogen, the residue has the sticky consistency of water residues recovered from XAD resins. We have termed this material the "carbon-chloroform extracted organics" (CCEO). In precise composition, it may or may not closely resemble residues from drinking water today, but for our work, it is invaluable as an abundant supply of a complex mixture of water residuals.

Using this CCEO, we have attempted to develop a general method of coupled bioassay/chemical fractionation for separating mixtures from current drinking water. With such a method we could test immediately whether the mutagenicity of such mixtures in the Ames test is due to summation of the low activity of many mutagens or to the effects of a few relatively active components. The method should allow isolation of active subfractions in yields suitable for compound identification. Some significant biohazardous properties of residue components may not be detected by our bioassay procedure of bacterial mutagenesis plating. For example, co-carcinogens, carcinogen promoters, and certain procarcinogens and teratological agents would not be recognized. So that other subfractions could be tested using short-term in vitro and in vivo mammalian assay systems, the method should permit maximum recovery of the total sample, distributed into multiple subfractions.

METHODS AND RESULTS

Preliminary Characterization and Primary Partitioning of CCEO

Aliquots of CCEO were shown to be reproducibly stripped of all toxic traces of $CHCl_3$ by streaming with dry nitrogen for 60 min at 60°C. Mutagenesis testing of the residue using Salmonella strains TA98 and TA100, as described elsewhere (Loper et al., 1978), revealed dose-related, microsomal-activation-dependent TA100 mutagenesis, with some toxicity at higher doses. Following experimentation with aqueous extraction of the residue from the original $CHCl_3$ solutions and from solutions in methylene chloride (CH_2Cl_2) (Tabor and Loper, 1980), a procedure was adopted for the semisolid/liquid extraction of neutrals, acids, and bases into

hexane. The resulting distributions of weight and mutagenic
activity for TA98 and TA100 with and without microsomal activation
have been detailed elsewhere (Tabor et al., 1980). The neutrals
are one third of the total sample by weight but contain nearly all
of the mutagenicity for TA100. All subsequent procedures were
conducted on portions of this CCEO neutral sample.

Method Development

Quantities of partitioned samples taken for mutagen isolation
had to be sufficient for repeated cycles of separation and
bioassay. To obtain useful data on both mutagenicity and toxicity
with minimum loss of material, tests were limited to single plates
of four dose levels chosen to induce colony counts of two to three
times those seen spontaneously (Loper, 1980). Reverse-phase HPLC,
employing mixtures and gradients of water/acetonitrile, was used
for chemical separation. The instrument was a Waters Associates
ALC/GPC 204 equipped with two 6000A pumps, UK6 injector, solvent
programmer, and 254-nm absorbance detector. A guard column (3.9 mm
x 2.5 cm) packed with pellicular particles bonded with
octadecylsilane (BONDAPAK--C_{18}/CORASIL; Waters Associates) was
followed by a radial compression module (RCM; Waters Associates)
containing a 8-mm x 10-cm column packed with 10-μm silica particles
bonded with a high load of octadecylsilane. To achieve adequate
separation levels, various gradient elution conditions were
investigated using the analytical column, as sample quantities were
increased from microgram to milligram levels. One milligram of the
CCEO neutral sample induced approximately 2000 TA100 colonies in
our assay, and for RCM chromatography, 20-mg samples were routinely
loaded in 200-μl volumes of acetonitrile.

A flow diagram of our procedure for partitioned samples is
given in Figure 1. Activity losses accompanying removal of the
bactericidal acetonitrile were minimized by solvent exchange into
CH_2Cl_2 using SEP-PAKS (Waters Associates) packed with μBONDAPAK-C_{18}
(Waters Associates) followed by evaporation of the CH_2Cl_2 at 40°C
under dry nitrogen in the presence of a small volume of
dimethylsulfoxide. Replicate separations reproduced the HPLC
fingerprint at the top of Figure 2, and with this added material,
we assayed fractions 1 through 6 using both TA98 and TA100 with and
without microsomal activation. As before, all the activity was
detected with TA100 and appeared in fraction 5B. Rechromatography
of this region gave the isolated major peaks at the bottom of
Figure 2. Subfractions 5B/5 and 5B/6 contained all the mutagenic
activity, and each of these was rechromatographed (Figure 3). Such
subfractions are currently undergoing further study; some
preliminary data are presented here. In one series of four dose-
level determinations, the summed mutagenesis from the active
subfractions (5B/5/2, 5B/6/2, 5B/6/3) was approximately half that

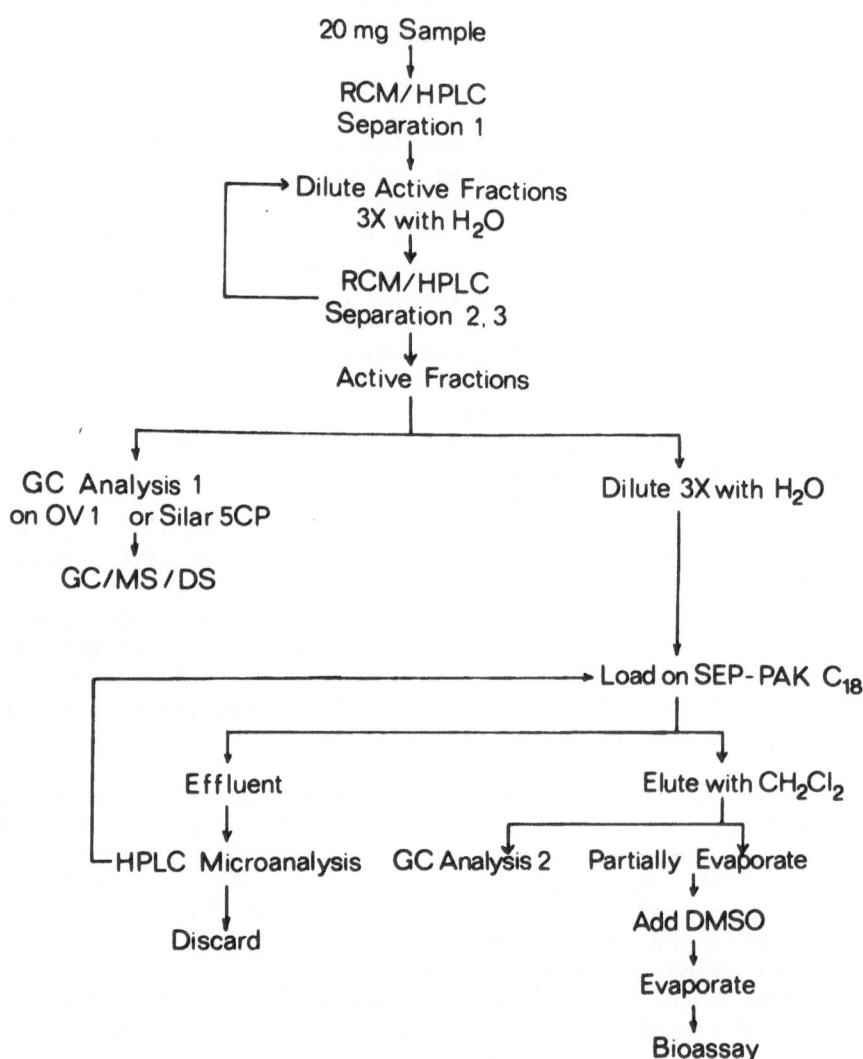

Figure 1.   Flow diagram for coupled bioassay/chemical fractionation
            of partitioned complex mixtures.

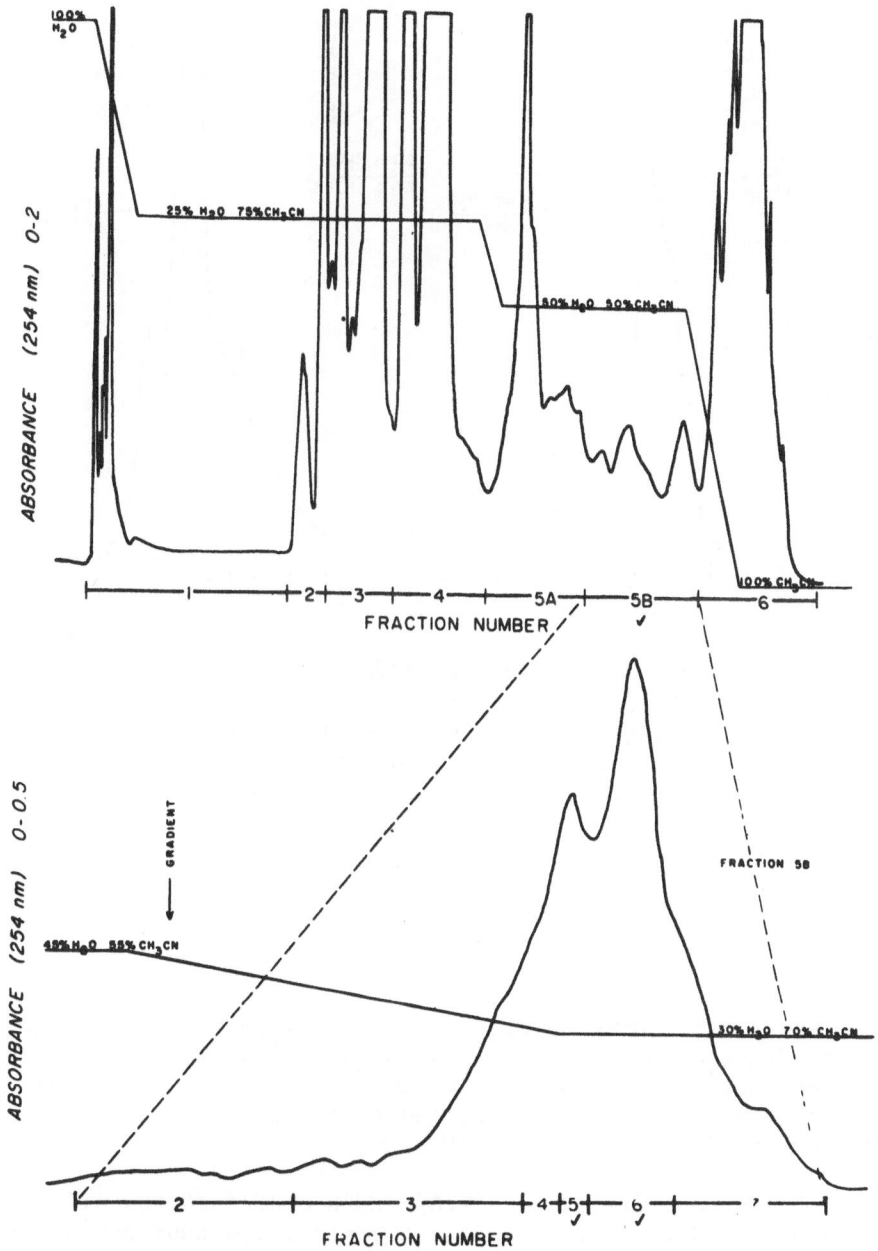

Figure 2.   RCM/HPLC reverse-phase separation of CCEO neutrals. A
20-mg sample was fractionated, using gradient elutions
as shown, and rechromatographed by the procedures given
in Figure 1.   Subfractions mutagenic to TA100 in the
presence of microsomal activation are indicated by a
check (√).

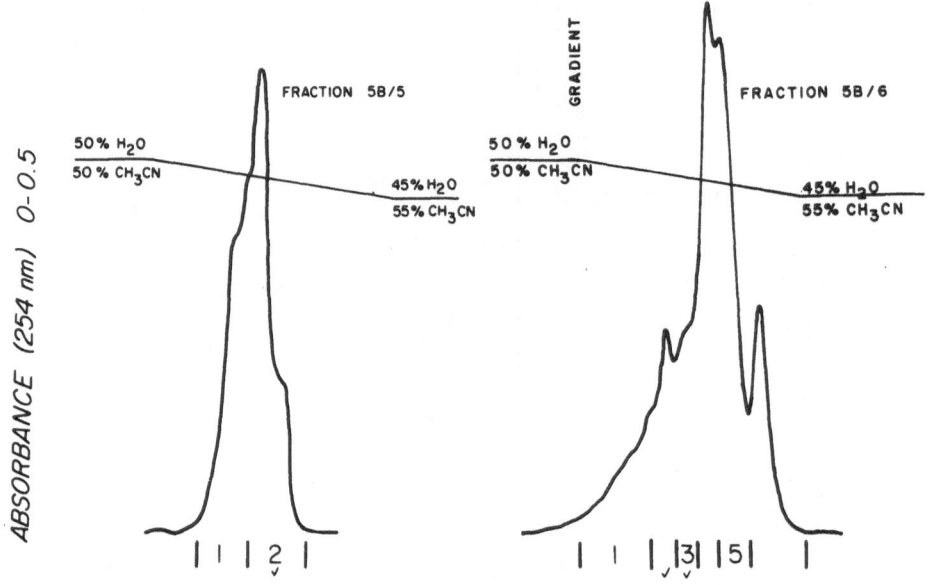

Figure 3.   Fractions 5B/5 and 5B/6, generated as shown in Figure
            2, were rechromatographed separately.   Mutagenic
            subfractions are indicated by a check (√).

from the initial 20-mg sample.   The relative purity of residue in
each subfraction was estimated by GC analysis.   Tracings of GC
chromatograms obtained using a column containing 10% SE30 on
Chromosorb WHP 80/100 mesh (Applied Science) as a stationary phase
are shown in Figure 4.

    Subfractions 5B/5/2 and 5B/6/2 each contained two different
components, and one other component constituted subfraction 5B/6/3.
To date, three 20-mg aliquots of CCEO neutrals have been carried
through the procedure as far as peak separation and GC analysis,
using both polar (10% Silar 5CP on Chromosorb Q 80/100 mesh:
Applied Science) and nonpolar stationary phases (10% SE30 on
Chromosorb WHP 80/100 mesh or 3% OV1 on Chromosorb W 80/100 mesh:
Applied Science):   the patterns shown in Figures 3 and 4 are
representative of the purity of all three aliquots.

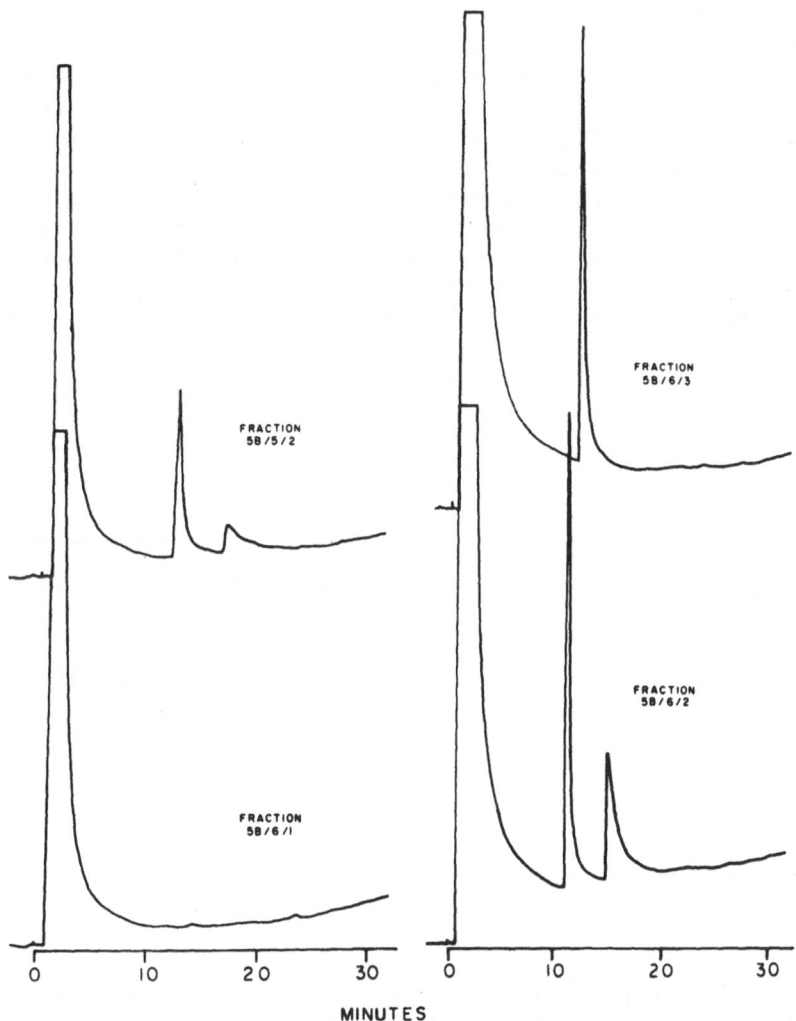

MINUTES

Figure 4.   Flame-ionization gas chromatograms of RCM/HPLC
            subfractions. Seven and one-half microliters of a
            water/acetonitrile solution (about 50:50 by volume) of
            each subfraction was slowly injected into a Perkin-
            Elmer Model 900 GC fitted with a 2-mm x 1-m stainless
            steel column containing 10% SE30 on Chromosorb WHP
            80/100 mesh. The nitrogen carrier gas flow was at 18
            ml/min, and the temperatures of injector and detector
            were 280°C and 350°C respectively. A linear temperature
            program (120°C to 220°C at 4°/min) was initiated at the
            time of injection. Data were collected continously and
            analyzed using a Spectra Physics Autolab System I
            Computing Integrator, and chromatograms were displayed
            on a 10-mV recorder at an attenuation of 160.

Weight values for a sample constituents were calculated based on peak areas compared with peak areas obtained from chromatography of 1 μl of a 1 mg/ml solution in chloroform of American Oil Chemists Society Reference Mixture No. 6, run under conditions identical to those for the experimental samples. These weights have been used in estimating specific activity, in net revertant colonies per milligram, shown in Table 1 (see Figure 4). Thus for 5B/5/2 and for 5B/6/2, the weights of their two GC components have been combined, so that results are expressed as net revertant colonies per total weight of each subfraction. Preliminary GC/MS analyses indicate that some of these compounds are polyhalogenated: this is indicated in Table 1 by an X with the number of components for each subfraction.

Table 1. Response of TA100 to Mutagens in the CCEO
Neutrals Mixture

	Net Revertant Colonies (+S-9)	
RCM/HPLC Subfractions[a]	Total	Per mg
5B/5/2 (1X + 1)	6000	3×10^5
5B/6/1 (?)	?	--
5B/6/2 (1X + 1)	4600	10^5
5B/6/3 (1X)	13000	4×10^5

[a]Number of major peaks in parentheses; X indicates the presence of halogen(s), based on preliminary MS data.

DISCUSSION AND CONCLUSIONS

We have demonstrated that this coupled bioassay/chemical fractionation procedure is reproducible for this complex residue mixture, and we feel it will serve as a general method. For the present, we assume that concentration methods yield residues representative of the non-volatile organics in drinking water. Based on our observations in the six-city study and our results from CCEO neutral sample, drinking water residues appear to contain a vast number of non-mutagenic compounds, possibly some of low mutagenicity, and a few highly mutagenic compounds. Evidence for

this is summarized in Table 2. Of the RCM/HPLC subfractions of the
CCEO neutrals, only three were mutagenic. Peak 5B/6/3, containing
a single component by GC, showed a specific activity comparable to
those of the carcinogens β-naphthylamine or 3-methylcholanthrene
(McCann et al., 1975). The other two subfractions contained two GC
components each, and their mutagens must have been comparably
potent (see Table 1). Of course, these subfractions were derived
from the old CCEO sample. For more recently isolated residues, the
presence of highly active compounds was suggested by our data for
the XAD eluate of Seattle sample 1. This fraction showed direct-
acting mutagenesis, which was increased 24-fold in specific
activity (revertant colonies per milligram of sample material) by
extraction into hexane (Loper et al., 1978: see Table 2). The
specific activity of the active compound(s) might be considerably
higher, based on our observation that this hexane-extracted
subfraction yielded 20 major 254-nm absorption peaks through HPLC
(Tabor et al., 1980).

Table 2. Mutagenesis of Drinking Water Residue Fractions
and Known Carcinogens in the Ames Test (TA100)[a]

Test Substance	Net Revertant Colonies/Plate/mg
XAD eluate of Seattle drinking water (sample 1)	4×10^2
Hexane extract of XAD eluate of Seattle drinking water (sample 1)	10^4
CCEO neutrals fraction 5B/6/3	4×10^5
β-propiolactone	5×10^4
β-naphthylamine	6×10^4
3-methylcholanthrene	2×10^5

[a]Data for XAD eluate and subfraction is from Loper et al., 1978:
that for known compounds is from McCann et al., 1975.

Final identification of the mutagenic components in the CCEO
may be of some direct benefit, even though the sample is now nearly
20 years old, and some of the halogen content may be due to storage

in CHCl$_3$. Identification by GC/MS is in progress. Should
previously unidentified mutagens be detected, their
characterization by MS would permit data searches, by peak
recognition, among MS profiles of residues of current drinking
water samples.

We propose to apply our procedure of coupled bioassay/chemical
fractionation to the identification of mutagens in recently derived
drinking water residues, obtained by desorption from XAD. Should
major mutagens in these samples be identified, a number of
questions concerning water quality could be investigated. These
topics include the reduction or avoidance of mutagens by alternate
disinfection procedures; seasonal changes in the types and amount
of mutagenic activity; and the mutagenic potential of discharges of
diverse industrial processes into surface or ground water destined
for human use. Knowledge of specific mutagens in a water sample
would provide an important criterion for evaluating alternate
residue-isolation procedures and could lead to simplification of
analytical chemical detection methods. Identification of
significant non-volatile bacterial mutagens in water would be a
step toward toxicological assessment of their risk to man.

ACKNOWLEDGMENTS

We are grateful to Debbie Spector for technical assistance;
to our colleagues Dr. Carl C. Smith, for advice and encouragement
in this work, and Dr. Joseph MacGee, for analytical contributions;
and particularly to Mr. Francis Middleton for providing us the
CCEO sample. This research was supported by a grant from the U.S.
Environmental Protection Agency.

REFERENCES

Guerin, M.R., B.R. Clark, C.-h. Ho, J.L. Epler, and T.K. Rao.
    1978. Short-term bioassay of complex organic mixtures: part
    I, chemistry. In: Application of Short-term Bioassays in the
    Fractionation and Analysis of Complex Environmental Mixtures.
    M.D. Waters, S. Nesnow, J.L. Huisingh, S.S. Sandhu, and L.
    Claxton, eds. Plenum Press: New York. pp. 247-268.

Kopfler, F.C., W.E. Coleman, R.G. Melton, R.G. Tardiff, S.C. Lynch,
    and J.K. Smith. 1977. Extraction and identification of
    organic micropollutants: reverse osmosis method. Ann. N. Y.
    Acad. Sci. 298:20-30.

Kurzepa, H., A.P. Kyriazis, and D.R. Lang. (in press). Growth characteristics of tumors induced by tranplantation into athymic mice of BALB/3T3 cells transformed in vitro by residue organics from drinking water. J. Environ. Pathol. Toxicol.

Lang, D.R., H. Kurzepa, M.S. Cole, and J.C. Loper. (in press). Malignant transformation of BALB/3T3 cells by residue organic mixtures from drinking water. J. Environ. Pathol. Toxicol.

Loper, J.C. 1980. Overview of the use of short-term biological tests in the assessment of the health effects of water chlorination. In: Water Chlorination: Environmental Impact and Health Effects, Vol. 3. R.L. Jolley, W.A. Brungs, and R.B. Cumming, eds. Ann Arbor Science: Ann Arbor, MI. pp. 937-945

Loper, J.C., and D.R. Lang. 1978. Mutagenic, carcinogenic, and toxic effects of residual organics in drinking water. In: Application of Short-term Bioassays in the Fractionation and Analysis of Complex Environmental Mixtures. M.D. Waters, S. Nesnow, J.L. Huisingh, S.S. Sandhu, and L. Claxton, eds. Plenum Press: New York. pp. 513-528.

Loper, J.C., D.R. Lang, R.S. Schoeny, B.B. Richmond, P.M. Gallagher, and C.C. Smith. 1978. Residue organic mixtures from drinking water show in vitro mutagenic and transforming activity. J. Toxicol. Environ. Hlth. 4:919-938.

McCann, J., E. Choi, E. Yamasaki, and B.N. Ames. 1975. Detection of carcinogens as mutagens in the Salmonella/microsome test: assay of 300 chemicals. Proc. Natl. Acad. Sci. USA. 72:5135-5139.

Middleton, F.M., H.H. Pettit, and A.A. Rosen. 1962. The mega sampler for extensive investigation of organic pollutants in water. In: Proceedings of the 17th Industrial Waste Conference. Engineering Ext. Ser. 112:454-460. Purdue University: Lafayette, IN.

Tabor, M.W., and J.C. Loper. 1980. Separation of mutagens from drinking water using coupled bioassay/analytical fractionation Int. J. Environ. Anal. Chem. 8:1-19.

Tabor, M.W., J.C. Loper, and K. Barone. 1980. Analytical procedures for fractionating non-volatile mutagenic components from drinking water concentrates. In: Water Chlorination: Environmental Impact and Health Effects, Vol. 3. R.L. Jolley, W.A. Brungs, and R.B. Cumming, eds. Ann Arbor Science: Ann Arbor, MI. pp. 899-912.

SHORT-TERM METHODS FOR ASSESSING *IN VIVO* CARCINOGENIC ACTIVITY OF COMPLEX MIXTURES

Michael A. Pereira and Richard J. Bull
Health Effects Research Laboratory
U.S. Environmental Protection Agency
Cincinnati, Ohio

INTRODUCTION

The carcinogenic activity of a chemical or a complex mixture derived from an environmental sample is assessed most efficiently through a three-tier decision scheme (Bridges, 1973; Weisburger and Williams, 1977; Bull and Pereira, in press). In tier 1, the samples are screened for evidence of carcinogenic and mutagenic activity. The Ames <u>Salmonella</u> mutation bioassay and <u>in vitro</u> and <u>in vivo</u> cytogenetic assays appear to sucessfully identify most chemicals and samples with carcinogenic activity. These two types of assays detect the two major classes of genotoxic agents, mutagens and clastogens, and would therefore form the backbone of tier 1. Other possible assays for tier 1 include mammalian cell mutation, sister-chromatid exchange, micronuclei, and unscheduled DNA synthesis. The nature of tier 1 bioassays--especially their lack of correlation to carcinogenic potency, the absence of a direct demonstration of cancer or neoplasia, and the number of false positives--requires that carcinogenic activity be confirmed in tier 2.

Tier 2 is the level in the decision tree where false positives are eliminated, insuring that the time-consuming and expensive tier 3 bioassay is used only for chemical and environmental samples that are virtually certain to contain carcinogenic activity. The function of the tier 3 bioassay is to provide data for assessing the quantitative risk associated with the carcinogenic activity of chemicals and samples. To be effective, tier 2 must confirm carcinogenic activity with a minimum of false positives, while maintaining an acceptably low level of false negatives.

Quantitation of carcinogenic activity in environmental samples
of complex mixtures, and thus ranking of mixtures according to
carcinogenic activity, poses a unique problem. Lifetime feeding or
exposure studies with rodents in a bioassay following the National
Cancer Institute (NCI) protocol is presently the only acceptable
procedure for obtaining data on carcinogenic potency that may be
extrapolated to humans (IRLG, 1979). This type of bioassay is not
feasible with most complex mixtures derived from drinking water and
other environmental samples. For example, the expense of obtaining
a single drinking-water sample for an NCI bioassay greatly exceeds
that of the bioassay itself, potentially costing millions of
dollars. Furthermore, the uniqueness of each drinking-water sample
(containing thousands of unknown chemicals) prevents the
generalization of the bioassay results to other drinking waters
(Bull et al., in press). The composition of drinking waters varies
depending on treatment (especially disinfectant), source (ground or
surface), seasons, and the effects of industrial use and municipal
waste disposal. Therefore, each drinking water would have to be
considered a unique test substance, and each would require a
separate bioassay, costing millions of dollars, at least until
enough assays were performed to determine whether generalizations
among drinking waters were possible.

CARCINOGENESIS TESTING MATRIX

The lack of a tier 3 bioassay for complex mixtures derived
from environmental samples means that any ranking or quantitation
of carcinogenic activity must be obtained at tier 2. To accomplish
this, tier 2 bioassay results would have to relate quantitatively
to carcinogenic potency. Also, since no short-term bioassay
appears to be sensitive to all chemical carcinogens, a
Carcinogenesis Testing Matrix (CTM)(Bull and Pereira, in press)
has been proposed for tier 2. It includes mouse lung adenoma
(Shimkin and Stoner, 1975; Stoner and Shimkin, in press), mouse
skin initiation/promotion (Slaga et al., in press), rat liver foci
(Pereira, in press), in vitro cell transformation (DiPaolo, 1979;
Styles, 1979), and in vivo sister-chromatid exchange (Latt et al.,
1979). These bioassays were chosen because evidence, or some
reasonable rationale, indicates that their results will relate to
carcinogenic potency, at least when taken together. The in vivo
bioassays appear most attractive, because their results should be
influenced by the same pharmacokinetic and metabolic factors as in
lifetime-exposure carcinogenicity bioassays. Where possible,
bioassays that directly measure the acquisition of neoplastic
properties, such as benign tumors, preneoplastic lesions, and cell
transformation, are included in the CTM.

The increased sensitivity of the CTM bioassays (compared with
lifetime exposure bioassays) greatly reduces the sample requirement.

These bioassays require a few applications of sample, at most. This increased sensitivity can be illustrated with the rat liver foci bioassay (Pereira, in press). Briefly, the bioassay protocol is as follows:

1) the sample is administered by any convenient route in single or multiple doses;

2) the rats are given 500 ppm sodium phenobarbital in their drinking water, starting four to seven days after the sample and continuing for seven weeks; and

3) the rats are then sacrificed, and their livers excised and examined histochemically for foci of gamma glutamyl transpeptidase.

A two-thirds hepatectomy can be performed either 18 to 24 h before or 14 days after administering the sample. Performing the partial hepatectomy 18 to 24 h before exposure increases the spectrum of chemical classes of carcinogens to which the assay is sensitive and increases the response. Performing it 14 days after exposure also increases the response and keeps the initiation step distinctly separated from the promotion.

Table 1 shows the results for diethylnitrosamine (DENA) in the rat liver foci bioassay. A single dose of DENA (30 mg/kg) gave positive results when administered either 24 h after or 14 days prior to partial hepatectomy. When the partial hepatectomy was performed 14 days after DENA exposure, a single dose as low as 300 µg/kg was detected. Since partial hepatectomy 24 h prior to the DENA increases sensitivity to DENA (Scherer and Emmelot, 1976), one would expect an even lower single dose to be detected. Ten low daily doses of 300 µg/kg, for a total dose of 3 mg/kg, appear to be additive in the bioassay (Table 1). The additivity of multiple low doses can also result in increased sensitivity (Ford and Pereira, in press). The use of tumor promoters, preneoplastic lesions, and benign tumors in short-term bioassays greatly decreases the amount of sample required so that acquiring sufficient amounts of sample becomes feasible.

DIFFICULTIES IN RANKING POTENCY FROM SINGLE BIOASSAYS

The use of two or more bioassys to rank the carcinogenic activities of complex mixtures could result in different rankings in the different bioassays, or in some mixtures giving a positive result in one bioassay while other mixtures were positive only in another bioassay. Table 2 outlines a hypothetical situation where four environmental samples of complex mixtures (1, 2, 3, and 4) were tested in bioassays A and B. The possible results include

Table 1. Rat Liver Foci Bioassay of Diethylnitrosamine (DENA)

DENA[a] (mg/kg)	No. of Animals	Partial Hepatectomy	GGTase-Positive Foci/cm^2 (mean ± std. error)
Experiment 1[b]			
30	10	-24 h	18.8 ± 4.6
30	10	+14 days	3.92 ± 0.91
0	10	-24 h	0.40 ± 0.14
Experiment 2[c]			
0.3	10	+14 days	1.02 ± 0.17
0.3 x 10	9	+14 days	2.16 ± 1.02
3	52	+14 days	1.51 ± 0.25
0 x 10	23	+14 days	0.59 ± 0.20

[a]Rats were administered DENA in 0.3 ml distilled water by gastric intubation. One week later the rats received 500 ppm phenobarbital in their drinking water for one week. After partial hepatectomy on day 14, the phenobarbital was decreased to 250 ppm. The rats were maintained on this concentration of phenobarbital for four to five weeks. Cyrostatic sections of liver were stained for GGTase activity and at least 2 cm^2 examined for foci.
[b]Pereira, in press.
[c]Ford and Pereira, in press.

Table 2. Possible Outcomes of a Two-Bioassay Matrix Testing of Drinking Water

	Bioassay	
Drinking Water Sample	A	B
1	+	−
2	+	−
3	−	+
4	−	−

samples 1 and 2 positive in bioassay A, and 1 and 3 positive in bioassay B. Of the three environmental samples possessing carcinogenic activity, the question then is, which sample is the most hazardous? To rank samples 2 and 3, which are positive in different bioassays, one must critically compare the results from the two bioassays. That sample 4 is negative in both bioassays does not mean it is not a potent carcinogen, unless it is demonstrated that the two bioassays as a set can detect all the carcinogens conceivably present in the environmental samples. Since most bioassays appear at least somewhat selective in their responses to well-established chemical carcinogens, it seems inappropriate to judge relative hazard on the basis of how many tests are positive.

The relative hazard of the environmental samples could be derived from the relative responses in the bioassays of the CTM, if the responses in the various bioassays were calibrated to the same standardized estimate of carcinogenic potency. A test substance would have a standardized estimate of carcinogenic potency derived from each bioassay of the CTM. A decision would have to be made on a procedure for using these individual estimates to arrive at a single estimate. This might be accomplished by averaging the individual estimates or by accepting the estimate of highest carcinogenic potency.

The following examples more explicitly illustrate the problem:

1) The Ames Salmonella mutagenicity bioassay has been proposed as correlating with carcinogenic potency (Meselson and Russell, 1977). This correlation requires the exclusion of nitrosamines, since their Ames test response is very weak compared with their carcinogenicity. When environmental samples of complex mixtures of unknown chemical composition were assayed, the Ames test would predict an erroneously low carcinogenic activity if highly active nitrosamines were present. When a liquid preincubation is used with the Ames test, nitrosamines can be detected, though still not at a level reflecting their potency. The use of liquid preincubation could also change the ranking of the other carcinogens. The results of the two bioassays (standard and preincubation Ames tests) would have to be compared as if they were two separate tests.

2) As another example, when drinking water concentrates from five cities were assayed in the Ames Salmonella mutagenicity and mouse skin initiation/promotion tests, the results of these two tests were not correlated (Loper et al., 1979; Robinson et al., 1980). The mouse skin initiation/promotion assay is very sensitive to polycyclic aromatic hydrocarbons (PAH) and nitrosamines (Pereira, in press). Comparing the responses of PAH and nitrosamines in the mouse skin and Salmonella mutagenicity tests

reveals many discrepancies (Andrews et al., 1978a, b). Among 25 polycylic aromatics, mutagenicity and carcinogenicity were positively correlated for 58% and negatively correlated for 41% (Andrews et al., 1978a). Since the two assays rank these chemicals differently, the responses of the two assays to drinking water samples are not expected to be correlated if either PAH or nitrosamines are present. The critical question is how to rank the carcinogenic activity of environmental samples based on results from more than one assay.

VALIDATION OF THE CARCINOGENESIS TESTING MATRIX

Results from two or more bioassays can be compared by calibrating each bioassay with respect to the carcinogenic potency of the chemicals to which each bioassay is sensitive. Calibration curves similar to the one used by Meselson and Russell (1977) can be determined; these investigators used the reciprocal of the log response in the short-term bioassay versus the reciprocal of the log carcinogenic potency. The reciprocal of the carcinogenic potency of each chemical can be calculated from lifetime exposure bioassay results as the dose (Dose 1/2) required to produce tumors in 50% of the animals (rats and mice) in two years. The highest estimate of carcinogenic potency is to be used in the calibration curves, since this is the value employed for extrapolation to man by the Carcinogen Asessment Group of the U.S. Environmental Protection Agency, Office of Health and Environmental Assessment.

For certain chemicals or chemical classes, individual bioassays of the CTM will respond poorly relative to the carcinogenic potencies of the chemical classes. As long as other bioassays of the CTM correctly indicate the potencies of such compounds, the responses for these compounds will be dropped from the calibration curve of any bioassay that responds poorly. For the matrix, it is the overall correlation with carcinogenic activity that is important. This method of selecting chemicals to be incorporated into the calibration curve of a bioassay results in a nonrandom grouping of chemicals, making further testing necessary for validation of the CTM.

If a chemical is extremely potent in a particular bioassay, compared with its carcinogenic potency derived from lifetime exposure, it may be necessary to delete that bioassay from the CTM. The CTM is being proposed for use with complex mixtures of undefined chemical composition and where a bioassay is not feasible. Therefore, such a response in a bioassay would tend to result in consistent overestimates of carcinogenic risk. Before a decision on keeping the bioassay in the CTM is made, the data for the chemical in the long-term carcinogenesis bioassay and the short-term bioassays will be carefully reviewed. Additional

chemicals of the same general class will be tested in the bioassay
to determine whether this is a general problem with the bioassay or
whether it is confined to a single chemical.

Validating the CTM to rank the carcinogenic hazard of
chemicals and environmental samples will involve three rounds of
testing the ability of each component bioassay and of the CTM as a
whole to predict carcinogenic potency. Archetypal chemicals
representing the various chemical classes of carcinogens will be
tested in round one to determine calibration curves. In round two,
carcinogenic and noncarcinogenic (or weakly carcinogenic) analogues
will be tested to determine the ability of the CTM and its
component bioassay to predict relative carcinogenic potency. At
this time, it should be possible to decide whether the CTM is valid
and practicable and what each bioassay contributes to the CTM. The
third round will involve testing additional carcinogenic and
noncarcinogenic analogues, surrogate mixtures containing two or
more carcinogens, and complex mixtures spiked with known
carcinogens. The resulting CTM will then be ready for use in
determining the relative carcinogenic hazard associated with
drinking water and other environmental samples of complex mixtures.

REFERENCES

Andrews, A.W., L.H. Thibault, and W. Lijinsky. 1978a. The
     relationship between carcinogenicity and mutagenicity of some
     polynuclear hydrocarbons. Mutation Res. 51:311-318.

Andrews, A.W., L.H. Thibault, and W. Lijinsky. 1978b. The
     relationship between mutagnicity and carcinogenicity of some
     nitrosamines. Mutation Res. 51:319-326.

Bridges, B.A. 1973. Some general principles of mutagenicity
     screening and possible framework for testing procedures.
     Environ. Hlth. Perspect. 6:221-227.

Bull, R.J., and M.A. Pereira. (in press). Development of a
     short-term testing matrix for estimating relative carcinogenic
     risk. J. Environ. Pathol. Toxicol.

Bull, R.J., M.A. Pereira, and K.L. Blackburn. (in press).
     Bioassay techniques for evaluating the possible
     carcinogenicity of absorber effluents. In: Conference on
     Practical Application of Adsorption Techniques.

DiPaolo, J.A. 1979. Quantitative transformation by carcinogens
     of cells in early passage. In: Environmental Carcinogenesis.
     P. Emmelot and E. Kriek, eds. Elsevier Press: Amsterdam.
     pp. 365-380.

Druckrey, H., A. Schildback, D. Schmahl, R. Preusmann, and S. Ivankovic. 1963. Quantitative Analyse der carcinogenen Wirkung von Diathynitrosamin. Arzneimittel-Forsch. 13:841-851.

Ford, J.O., and M.A. Pereira. (in press). Short-term in vivo initiation/promotion bioassay for hepatocarcinogens. J. Environ. Pathol. Toxicol.

IRLG, Interagency Regulatory Liaison Group, Work Group on Risk Assessment. 1979. Scientific basis for identification of potential carcinogens and estimation of risks. J. Natl. Cancer Inst. 63:241-248.

Latt, S.A., R.R. Schreck, K.S. Loveday, and C.R. Shuler. 1979. In vitro and in vivo analysis of sister chromatid exchange. Pharmacol. Rev. 30:501-535.

Loper, J.C., D.R. Lang, R.S. Schoeny, B.B. Richmond, P.M. Gallagher, and C.C. Smith. 1978. Residue organic mixtures from drinking water show in vitro mutagenic and transforming activity. J. Toxicol. Environ. Hlth. 4:919-938.

Meselson, M., and K. Russell. 1977. Comparisons of carcinogenic and mutagenic potency. In: Origins of Human Cancer, Book C. H.H. Hiatt, J.D. Watson, and J.A. Winsten, eds. Cold Spring Harbor Laboratory: Cold Spring Harbor, NY. pp. 604-628.

Pereira, M.A. (in press). Rat liver foci bioassay. J. Environ. Pathol. Toxicol.

Robinson, M., J.W. Glass, D. Cmehil, R.J. Bull, and J.G. Orthoefer. 1980. Initiating and promoting activity of chemicals isolated from drinking waters in the SENCAR mouse: a five-city survey. Presented at the U.S. Environmental Protection Agency Second Symposium on the Application of Short-term Bioassays in the Fractionation and Analysis of Complex Environmental Mixtures, Williamsburg, VA.

Scherer, E., and P. Emmelot. 1976. Kinetics of induction and growth of enzyme-deficient islands involved in hepatocarcinogenesis. Cancer Res. 36:2544-2554.

Shimkin, M.B., and G.D. Stoner. 1975. Lung tumors in mice: Application to carcinogenesis bioassay. Adv. Cancer Res. 21:1-58.

Slaga, T.J., S.M. Fischer, L.L. Triplett, and S. Nesnow. (in press). Comparison of complete carcinogenesis and tumor initiation in mouse skin: Tumor initiation-promotion, a reliable short-term assay. J. Environ. Pathol. Toxicol.

Stoner, G.D., and M.B. Shimkin. (in press). Strain A mouse lung tumor bioassay. J. Environ. Pathol. Toxicol.

Syles, J.A. 1979. Cell transformation assays. In: Mutagenesis in Sub-mammalian Systems. G.E. Paget, ed. Baltimore University Press: Baltimore. pp. 53-71.

Weisburger, J.H., and G.M. Williams. 1977. Decision point approach to carcinogen testing. In: Structural Correlates of Carcinogenesis and Mutagenesis. HEW Publication No. (FDA) 78-1046. Rockville, MD. pp. 45-52.

THE INITIATING AND PROMOTING ACTIVITY OF CHEMICALS ISOLATED
FROM DRINKING WATERS IN THE SENCAR MOUSE: A FIVE-CITY
SURVEY

Merrel Robinson, John W. Glass, David Cmehil,
Richard J. Bull, and John G. Orthoefor
Health Effects Research Laboratory
U.S. Environmental Protection Agency
Cincinnati, Ohio

INTRODUCTION

Means of properly evaluating the carcinogenic risk posed by
organics in drinking water are of utmost concern to the U.S.
Environmental Protection Agency (EPA). While rodent lifetime
exposure and human epidemiological studies serve as the only
generally accepted means, the cost and time involved are highly
prohibitive. In addition, formulation of human epidemiological
data requires human exposure of sufficient magnitude to allow
separation of confounding factors from the relationship in
question. Since a major part of EPA's regulatory activities is
directed towards preventing significant increases in the
carcinogenic risk to the population, quick and reliable
investigative methodology is a necessity. Short-term bioassays can
be used to identify most potential problems and to provide an
initial risk assessment.

Loper et al. (1978) showed that all the organic material
isolated from the drinking water of six cities contained measurable
mutagenic activity in Salmonella tester strains. These cities were
selected to represent the most common types of drinking water
sources. Since a large number of chemicals that are known
carcinogens react positively in the Ames test, while noncarcinogens
do not (McCann et al., 1975), these results suggest that chemical
carcinogens are to be found in most drinking water.

Mouse skin initiation/promotion studies offer one means of
confirming the presence of chemical carcinogens in complex
mixtures. A positive response in this assay system may be
classified as a true carcinogenic response, on the basis of

evidence that strongly indicates a quantitative association of
benign papillomas with malignant tumors (Boutwell, 1974: Burns et
al., 1976; Shubik et al., 1953; Van Duuren et al., 1973). The
system may therefore be used as a short-term in vivo method of
assessing carcinogenic activity. Through the use of appropriate
experimental designs, the system allows the activity of tumor
initiators and tumor promoters to be clearly differentiated
(Barenblum, 1941; Hennings and Boutwell, 1970: Mufson et al., 1977;
Sivak and Van Duuren, 1971; Van Duuren, 1969).

The mouse skin bioassay was applied in our study to test both
the tumor-initiating and tumor-promoting potential of a complex
mixture of organic chemicals concentrated by reverse osmosis (RO)
from drinking water of five cities. These samples were obtained
from the same cities and processed by the same methods as employed
for Ames testing by Loper et al. (1978).

METHODS

The test samples for this study were concentrated from
drinking water supplies of Miami, Seattle, Philadelphia, Ottumwa
(IA), and New Orleans. These cities (Table 1) were selected to
represent both surface and ground types of water supply; sources
potentially contaminated by agricultural runoff, industrial wastes,
or municipal wastes; and uncontaminated sources (Tardiff and
Denizer, 1973).

Table 1. Cities Selected for Extraction and Bioassay
of Residual Organics in Drinking Water

City	Origin of Water Supply	Type of Water Supply
Miami, FL	Ground	Uncontaminated[a]
New Orleans, LA	Surface	Industrial wastes
Ottumwa, IA	Surface	Agricultural runoff
Philadelphia, PA	Surface	Municipal wastes
Seattle, WA	Surface	Uncontaminated[a]

[a]No known contamination from municipal, agricultural, or industrial
wastes; however, contamination from decomposition products of
natural origin is possible.

The concentrated organics were prepared for EPA by Gulf South Research Institute using the procedure described by Kopfler et al. (1977). The samples were concentrated from multiple 200-1 quantities of tap water, which received sufficient concentrated hydrogen chlorine (HCl) to maintain a pH of 5.5, by RO using a cellulose acetate (CA) membrane. The reject from the CA membrane was passed through a heat exchanger to maintain the water temperature at < 15°C. Part of the reject stream was diverted through a Donnan softening unit to avoid salt precipitation. The CA permeate was adjusted to pH 10 and then concentrated by RO using a nylon membrane, and the nylon permeate was discarded. Both the CA and the nylon concentrates were then adjusted to neutral pH and extracted with pentane and methylene chloride. The aqueous phases were adjusted to pH < 2 with HCl and again extracted with methylene chloride. For the purposes of this study, these fractions were combined and the solvent removed to produce the reverse osmosis extract (ROE) sample. The residual aqueous concentrate was passed through a column of purified XAD-2 resin. After removing metallic oxides and other inorganic agents by elution with 1 $\underline{M}$ HCl and distilled water, the organics were then eluted from the column with 95% ethanol. The ethanol was removed from the eluate by vacuum distillation, and the eluates from both columns were combined to produce the XAD sample.

Once the samples were concentrated, they were administered to mice subcutaneously (into the back). The following studies were conducted using the mouse skin bioassay.

Drinking Water Concentrates as Initiators

Male SENCAR mice (sensitive to carcinogens) were obtained from Dr. T.J. Slaga, of Oak Ridge National Laboratories, Oak Ridge, TN. The mice were 8 to 10 weeks old when the study began. The ROE and the XAD samples were administered over a two-week period in six injections of 0.1 ml of a 10% Emulphor (a polyoxyetheylated vegetable oil), for a total dose of 4.5 mg/mouse (total dose = 150 mg/kg body weight). To maintain control over dosage and to allow comparison of results with prior studies (Bull, 1980), the subcutaneous route of administration was chosen. The 7,12-dimethyl benz(a)anthracene (DMBA-positive control) was given in six injections of 0.1 ml in 10% Emulphor, for a 25 µg/mouse total dose. The DMBA was obtained from Eastman Kodak Company (Rochester, NY) and was purified by thin-layer chromatography by Dr. F. Bernard Daniel of the EPA Health Effects Research Laboratory (Cincinnati, OH). There were 60 animals in each exposure group.

Two weeks after the last initiating dose, the promoting phase was begun. Forty mice of each group received 1.0 µg phorbol myristate acetate (PMA) in 0.1 ml acetone applied to the shaved

back three days a week for 20 weeks. The remaining 20 animals in
each group received only acetone. The PMA was obtained from Dr.
Peter Borchert (University of Minnesota) and required no further
purification. The animals were weighed weekly and observed for
tumor incidence. The incidence of both papillomas and carcinomas
was charted weekly. Any of these that persisted for three weeks or
more were included in the cumulative count.

Following completion of the promotion period, the animals were
held for a total of one year for study and then were sacrificed.
Moribund animals were sacrificed as needed. Major organs and all
macroscopically evident lesions were sectioned and fixed in 10%
buffered formaldehyde solution for subsequent histopathological
evaluation.

Drinking Water Concentrates as Promoters

The tumor-promoting potentials of the ROE and XAD samples were
also tested in the SENCAR mouse. Groups of 20 mice (for ROE) and
30 mice (for XAD) received an initiating dose of 2.56 µg DMBA in
0.1 acetone topically to the shaved area of the back. Two weeks
later, the promoting schedule with water concentrate samples was
begun. The ROE from each city was applied at a dose of 100 µg per
mouse per application in 0.1 ml acetone, three times a week for 18
weeks. The XAD dose was 500 µg/mouse in 0.1 ml acetone, three
times a week for 18 weeks. The only reason for the differing doses
was the availability of sample, which was much more limited for the
ROE. A positive control group received 1 µg PMA in 0.1 ml acetone
per application, following the same initiating dose of DMBA. After
completing the treatment, surviving animals were held for
observation of tumor incidence until they were one year old and
then sacrificed. Histological evaluation was done in the same
manner as in the tumor-initiating-potential study.

Drinking Water Concentrates as Complete Carcinogens

A third study was done with the concentrate samples to test
their potential as complete carcinogens. In groups the same sizes
as above, the ROE and XAD samples were administered topically at
dose levels of 100 µg/mouse and 500 µg/mouse, respectively, in 0.1
acetone, three times a week for 20 weeks. Thereafter the same
protocol was followed as in the other two studies.

RESULTS

Initiating Activity

Table 2 presents the results at the end of 50 weeks of the study testing the initiating activity of the water concentrate samples. Positive results are apparent with several of the samples. The data indicate significantly greater numbers of papillomas per mouse in the animals treated with Ottumwa ROE (0.40) and New Orleans XAD (0.33). Marginal responses occurred in Ottumwa XAD (0.28) and Philadelphia XAD (0.25). Smaller response rates were observed with Miami XAD (0.23), New Orleans ROE (0.18), and Seattle XAD (0.20), compared with the vehicle control (0.10). Essentially negative results were seen in Miami ROE (0.15), Philadelphia ROE (0.10), and Seattle ROE (0.13). Using the increase over the control response and applying a normalization factor based on the amount of water processed to obtain samples, the cities were ranked according to relative activity per unit volume. The ranking was as follows: Miami (48), Ottumwa (34), New Orleans (28), Philadelphia (26), and Seattle (7).

The time course of tumor development is presented in Figures 1 and 2. In the groups receiving ROE, no persistent papillomas occurred after week 25, except in the Ottumwa group, where the cumulative count continued to rise throughout the remainder of the 50-week period. More variation was seen in the groups receiving the XAD; in all of these groups, tumor incidence tended to increase with time relative to the control group. The New Orleans group attained the highest level of tumors per animal within the last five weeks of the study, and this change accounted for its statistically significant difference.

All lesions that were observed grossly at the time of necropsy were histologically examined; Table 3 gives the final distribution of skin tumors, but not the total count, since some lesions meeting the criteria later regressed or coalesced. The skin lesions were predominantly papillomas. As their incidence was low, the low incidence of squamous cell carcinomas observed in the experimental group was not surprising. Interestingly, the group giving rise to the most carcinomas (Seattle XAD) was negative by papilloma count. However, due to limited numbers, the incidence of these tumors did not correlate with papilloma incidence. This fact illustrates the types of problems encountered when testing relatively small quantities of complex mixtures where it is not possible to test at levels approaching a maximally tolerated dose because of sample preparation expense. Another problem was that a few fibrosarcomas were observed that could have been injection-site related.

Table 2. Cumulative Tumor Count at 50 Weeks[a]

Sample	Mice with Tumors[b]	Total Number of Tumors	Tumors Per Mouse	Normalization Factor[c]	Normalized Activity[d]
Miami					
ROE	6	6	0.15	23	1.2
XAD	8	9	0.23	360	46.8
					48.0
ROE	11	16	0.40	12	3.6
XAD	10	11	0.28	170	30.6
					34.2
ROE	4	5	0.13	14	0.4
XAD	8	10	0.25	170	25.5
					25.9
ROE	5	5	0.13	1.3	0.0
XAD	8	8	0.20	66	6.6
					6.6
ROE	5	7	0.18	10	0.8
XAD	8	13	0.33	120	27.6
					28.4
DMBA (25 µg)	38	286	7.15	-	-
Emulphor	4	4	0.10	-	-

[a] Total dose of 150 mg/kg applied subcutaneously to each animal, followed by 20 weeks of promotion with 1.0 µg PMA three times weekly.

[b] Out of a total of 40 mice.

[c] Adjustment for amount of water processed to obtain concentrates =

$$\text{normalization factor} = \frac{\text{total recovered} \times \text{volume processed}}{\text{fraction wt} \times 10^6}$$

[d] Obtained by multiplying tumors per mouse minus control incidence by the normalization factor.

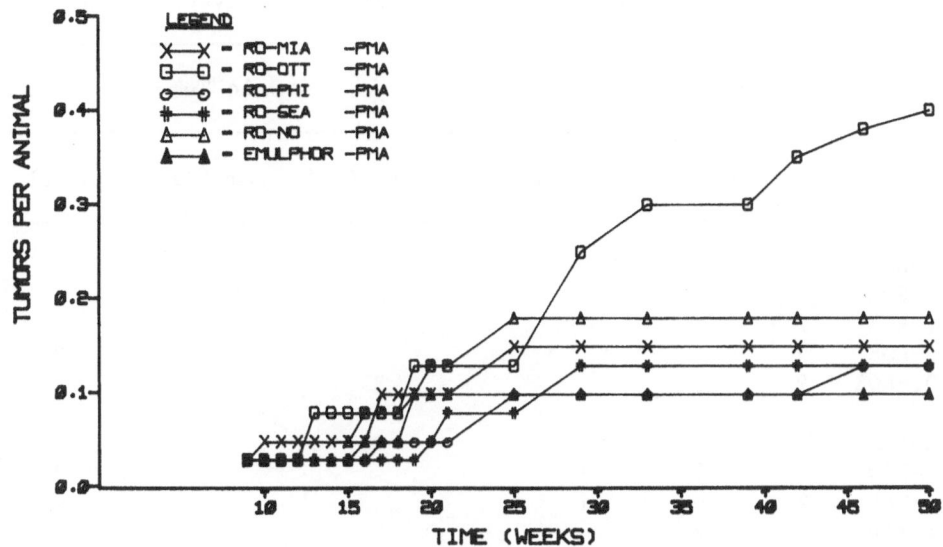

Figure 1.   Tumor incidence through week 50 after animals received
an initiating dose of 150 mg/kg ROE sample s.c. and PMA
as a tumor promoter three times a week from week 0 to 20.

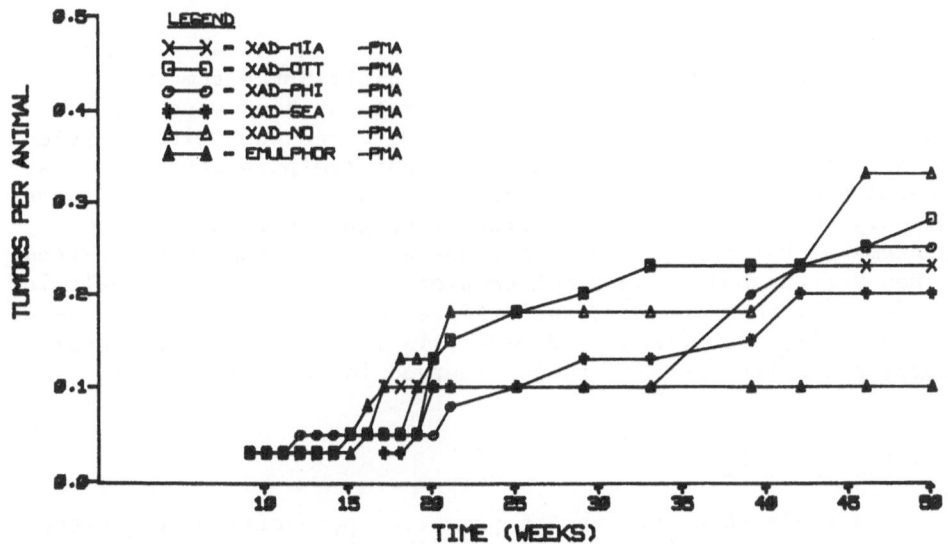

Figure 2.   Tumor incidence through week 50 after animals received
an initiating dose of 150 mg/kg XAD sample s.c. and PMA
as a tumor promoter three times a week from week 0 to 20.

Table 3. Summary of the Macroscopically Observed Lesions

Sample	Skin Tumors			Systemic Tumors		
	Pap.[a]	Car.[b]	Fib.Sa.[c]	Pul.Ad.Ca.[d]	Hem.(li.)[e]	Hepat.[f]
Mia-ROE	4					
Ott-ROE	8			3		
Phi-ROE	2					
Sea-ROE	2					3
N.O.-ROE	2	1		3		
Mia-XAD	5					
Ott-XAD	2	1	1			2
Phi-XAD	2		1			2
Sea-XAD		2		1		3
N.O.-XAD	8	1		1	1	4

[a]Papilloma.
[b]Carcinoma.
[c]Fibrosarcoma.
[d]Pulmonary adenocarcinomas.
[e]Hemangioma (liver).
[f]Hepatoma.

The systemic tumors observed were distributed somewhat unevenly among the groups. The most frequent kind were hepatomas. Four were observed with New Orleans XAD, three each with Seattle ROE and XAD, and two with Ottumwa and Philadelphia XAD. Animals treated with the other four samples gave no evidence of hepatomas, and only one animal in the control group gave evidence of hepatomas. Pulmonary adenocarcinomas were also somewhat elevated in New Orleans and Ottumwa ROE samples. Due to the relatively low incidence of these tumors, the differences could not be considered statistically significant. However, incidences in experimental groups generally exceeded those observed in control animals.

Promoting Activity

The study to determine the promoting potential of the water concentrate samples is continuing, and Table 4 shows the results through week 38. The groups receiving 500 µg XAD sample per application from Miami, New Orleans, and Ottumwa have yielded one papilloma each. No papillomas have occurred in groups treated with

Table 4. Drinking Water Concentrates as Promoters: Number of
Papillomas Per Number of Mice[a]

Sample	Vehicle	PMA	City				
			Mia.	N.O.	Ott.	Phil.	Sea.
Controls	0/20	319/20					
ROE (100 µg)			.0/20	0/20	0/20	0/20	0/20
XAD (500 µg)			1/20	1/30	1/30	0/30	1/30

[a]Initiator equals DMBA (2.56 µg/mouse applied topically). Drinking
water concentrate or PMA (1.0 µg in 0.1 ml acetone) applied
topically three times a week for 18 weeks. Results at 38 weeks.

Philadelphia and Seattle XAD, nor have papillomas occurred in any
of the ROE samples. In the positive control group, using 1 µg PMA
three times weekly, 19 of 20 mice had tumors, with a total of 319
papillomas. At this point, it appears that drinking water samples
at the doses applied do not promote DMBA tumorigenesis.

Complete Carcinogenic Activity

Table 5 gives the results after 38 weeks of studying the
potential of the ROE and XAD samples as complete carcinogens. Only
one papilloma has been observed, and that is in the New Orleans ROE
group. To this point in time, no evidence of complete
carcinogenesis of organic chemicals from drinking water has been
demonstrable. Again, the chemicals isolated from drinking waters
do not seem to be complete carcinogens at the doses applied.

DISCUSSION

Within the limits of possible dosage in the present work, it
appears that tumorigenic and/or carcinogenic substances were
present in the drinking waters. These chemicals were primarily
initiators in the mouse skin rather than promoters or complete
carcinogens. However, this conclusion must be clearly couched in
terms of the doses that were actually administered. Compared with
PMA, organic chemicals present in the ROE and XAD fractions were
less than 1/100 and 1/500 as potent as promoters, respectively.

Table 5. Drinking Water Concentrates as Complete Carcinogens:
Number of Papillomas Per Number of Mice[a]

	City				
Sample	Miami	N.O.	Ottum.	Phila.	Seattle
ROE (100 mg)	0/20	1/20	0/20	0/20	0/20
XAD (500 mg)	0/30	0/30	0/30	0/30	0/30

[a]Drinking water concentrate applied topically three times a week
for 20 weeks. Results at 38 weeks.

In view of the extreme potency of PMA, this was not altogether a
satisfying result. A further reservation is that no evidence
exists to indicate that mouse skin is a universal target tissue for
tumor promoters.

In the case of the Ottumwa ROE sample and all of the XAD
samples, tumor development was late. This contrast with the time
course of tumor development for the positive control DMBA (Figure
3) suggests that the chemicals in drinking water responsible for
initiating tumors may differ from DMBA with respect to underlying
mechanism(s). In terms of initiating activity, individual samples
produced significant increases in the number of tumors. At equal
doses of organic material, however, there was little to distinguish
positive from negative responses in the different samples. On the
other hand, if the data were adjusted to the amount of water
processed, the total units of activity present could vary among
water samples by a factor of seven. Although this calculation was
based on somewhat nonsignificant data, it does suggest that the
total risk observed might parallel the level of organic material
present. Undoubtedly, a wide variety of variables underlie this
parallel. For example, previous work has shown that disinfecting
drinking water can increase the levels of carcinogens isolated from
drinking water (Bull, 1980). However, the observation argues that
a prudent course of action in drinking water treatment might
involve reducing the total organic carbon present in the finished
drinking water. Although the data obtained in the present study
cannot be used to estimate risks to populations consuming these
drinking waters, it certainly justifies further research into
carcinogenic risks associated with drinking water.

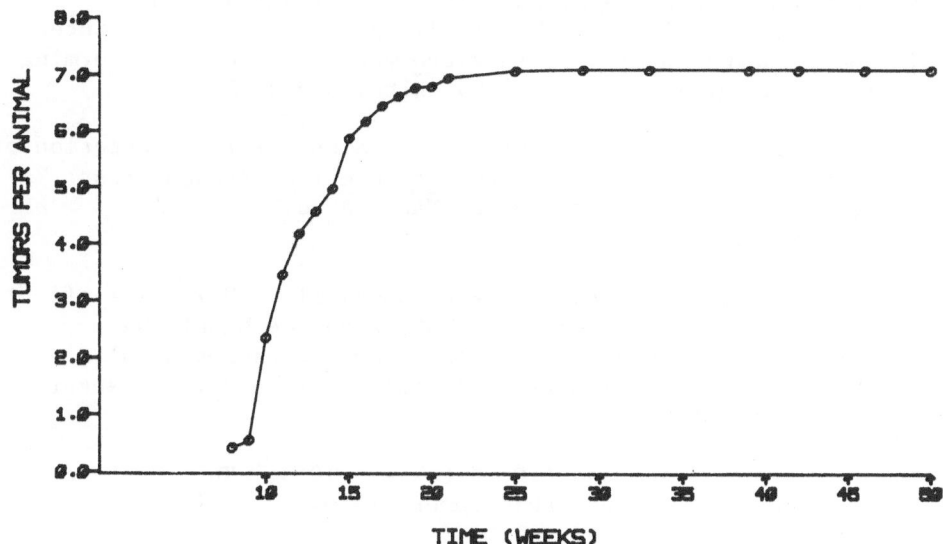

Figure 3.   Tumor incidence through week 50 after animals received
an initiating dose of 25 μg DMBA s.c. and PMA as a tumor
promoter three times a week from weeks 0 to 20.

REFERENCES

Barenblum, I. 1941. The cocarcinogenic actions of croton resin.
    Cancer Res. 1:44-48.

Boutwell, R.K. 1974. The function and mechanism of promoters of
    carcinogenesis. Critical Rev. Toxicol. 2:419-443.

Bull, R.J. 1980. Health effects of alternate disinfectants and
    reaction products. J. Am. Water Works Assoc. 72:299-303.

Burns, F.J., M. Vanderlaan, A. Sivak, and R.E. Albert. 1976.
    Regression kinetics of mouse skin papillomas. Cancer Res.
    36:1422-1427.

Hennings, H., and R.K. Boutwell. 1970. Studies on the mechanism
    of skin tumor promotion. Cancer Res. 30:312-320.

Kopfler, F.C., W.E. Coleman, R.G. Melton, R.G. Tardiff, S.C. Lynch,
    and J.K. Smith. 1977. Extraction and identification of
    organic micropollutants: Reverse osmosis method. Ann. N.Y.
    Acad. Sci. 298:20-30.

Loper, J.C., D.R. Lang, R.J. Schoeny, B.B. Richmond, P.M. Gallagher, and C.C. Smith. 1978. Residue organic mixtures from drinking water show in vitro mutagenic and transforming activity. J. Toxicol. Environ. Hlth. 4:919-938.

McCann, J., E. Choi, E. Yamasaki, and B. Ames. 1975. Detection of carcinogens as mutagens in the Salmonella/microsome test: assay of 300 chemicals. Proc. Natl. Acad. Sci. USA 72:5135-5139.

Mufson, R.A., R.C. Simsiman, and R.K. Boutwell. 1977. The effect of the phorbol ester tumor promoters on the basal and catecholamine-stimulated levels of cyclic adenosine 3':5'-monophosphate in mouse skin and epidermis in vivo. Cancer Res. 37:665-669.

Shubik, P., R. Baserga, and A.C. Ritchie. 1953. The life and progression of induced skin tumors in mice. Brit. J. Cancer 7:342-351.

Sivak, A., and B.L. Van Duuren. 1971. Cellular interactions of phorbol myristate acetate in tumor promotion. Chem.-Biol. Interact. 3:401-411.

Slaga, T.J., G.T. Bowden, and R.K. Boutwell. 1975. Acetic acid, a potent stimulator of mouse epidermal macromolecular synthesis and hyperplasia but with weak tumor-promoting ability. J. Nat. Cancer Inst. Vol. 55. 4:983-987.

Tardiff, R.G., and M. Denizer. 1973. Toxicity of organic compounds in drinking water. Water Quality Conference, University of Illinois, Urbana-Champaign. U.S. Government Printing Office 1974-657-053/1084. pp. 23-37.

Van Duuren, B.L., A. Sivak, A. Segal, I. Seidman, and C. Katz. 1973. Dose-response studies with a pure tumor-promoting agent, phorbol myristate acetate. Cancer Res. 33:2166-2172.

Van Duuren, B.L. 1969. Tumor-promoting agents in two-stage carcinogenesis. Progress Exp. Tumor Res. 11:31-68.

AQUEOUS EFFLUENT CONCENTRATION FOR APPLICATION TO BIOTEST SYSTEMS

William D. Ross, William J. Hillan, Mark T. Wininger,
JoAnne Gridley, Lan Fong Lee, and Richard J. Hare
Monsanto Research Corporation
Dayton, Ohio

Shahbeg S. Sandhu
Health Effects Research Laboratory
U.S. Environmental Protection Agency
Research Triangle Park, North Carolina

INTRODUCTION

Potential chemical mutagens in industrial effluents may be present at concentrations below the detection limits of biotests such as the Ames mutagenicity test. These chemicals may accumulate in biological food chains. Many insecticides and other chemicals are known to accumulate in living organisms where tissues act as effective storage depots for toxic compounds (Loomis, 1978). This effect is especially significant for human health when dilute toxicants enter the human food chain, such as through seafoods. Mollusks such as the oyster tend to accumulate toxicants, because they filter-feed, which concentrates and magnifies the effects of toxic materials. Because of this potential for bioaccumulation, methods a-e needed to determine the bioactivity of low concentrations of potential toxicants in industrial effluents.

The objective of the research program discussed in this paper was to evaluate and compare three methodologies for concentrating potential chemical mutagens in typical industrial effluents for application to in vitro biotest systems. For this study, the Ames Salmonella mutagenicity assay (Ames et al., 1975) was used. The optimum concentration methodology would ideally meet the following criteria:

1) concentration of relatively large quantities (> 3 l) of aqueous sample;

2) concentration factors of > 200 times;

3) little or no loss of volatile compounds;

4) efficient concentration and extraction of potential
   toxicants;

5) maintenance of high integrity of chemicals by preventing
   artifact formation;

6) maintenance of the relative concentrations of all
   compounds;

7) use of methods and reagents that are compatible with the
   biotest system; and

8) maintenance of microbial sterility of the resulting
   sample.

This paper describes the experimental approach and resulting
data for three concentration methods: adsorption using
macroreticular resins (XAD), freeze-drying (lyophilization), and
reverse osmosis (ultrafiltration). These methods were used to
concentrate added standard chemical compounds (i.e., potential
toxicants) in "typical" aqueous effluents for application to the
Ames mutagenicity test.

METHODOLOGY

Three methods were used to concentrate aqueous effluents for
application to in vitro biotest systems: sorbent extraction,
lyophilization, and reverse osmosis. A schematic is presented in
Figure 1.

Five-gallon samples of raw wastewater obtained from an
industrial plant served as typical standard effluent samples and
were used to evaluate each of the concentration methodologies.

To check for microbial contamination, aliquots of the neat
effluent samples were streaked with a sterile applicator onto both
Difco Bacto nutrient agar and Ames histidine-free bottom agar. The
plates were highly contaminated, indicating the need for filter
sterilization. The samples were sterilized by drawing them through
a series of Millipore filters of 1.3-, 0.45-, and 0.2-μm pore size.
The Ames Salmonella mutagenicity assay (Ames et al., 1975) was used
to test the neat filtrate for mutagenicity, using two histidine-
requiring strains, TA98 and TA100. The specific procedure for
analyzing the neat effluents used five concentrations in
triplicate: 0.01, 0.1, 0.5, 0.75, and 1.0 ml/plate. Each
concentration was tested with and without rat liver S-9 fractions
in the plate-incorporation test. Spot and toxicity tests were also

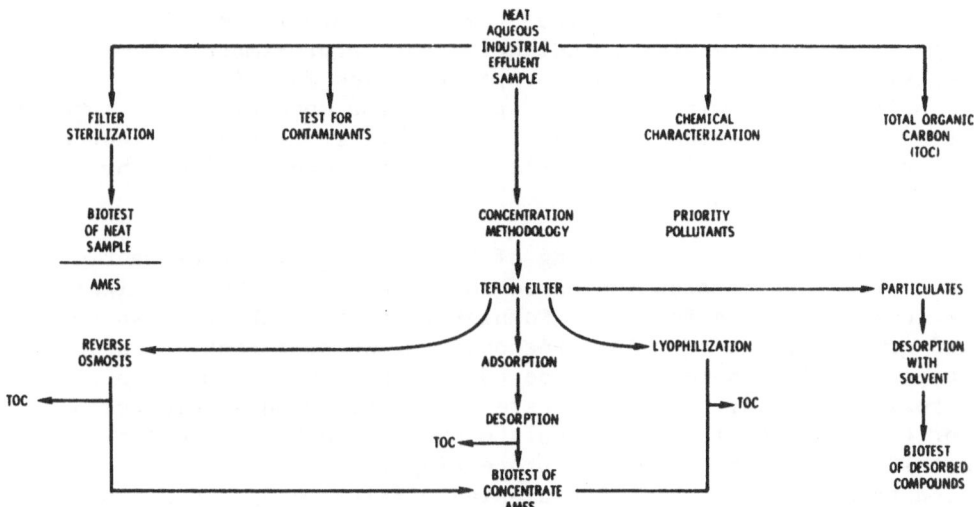

Figure 1.   Schematic of concentration and biotesting of textile-
            industry effluents.

performed. The positive controls used were 2-nitrofluorene (2NF),
sodium nitrite, 9-amino-acridine, benzo(a)pyrene (B[a]P), and
2-amino-anthracene, while tap water served as a negative control.

     The concentrates from sorbent extraction and lyophilization
were tested using six concentrations in which the samples were not
toxic: 0.1, 0.4, 2.0, 10, 30, and 100 µl/plate. If the samples
were found to be toxic at a designated concentration, a lower,
nontoxic concentration was used as the highest concentration.

     Many effluent samples contain large amounts of particulate
material that impede flow through XAD resins, reverse osmosis
membranes, and sterilizing filters. The large particles (> 5 µm)
in the effluent used in this study were removed by filtration.
Potential toxic and mutagenic compounds adsorbed to particles were
also removed. These compounds were assayed for bioactivity. The
industrial effluent was filtered under vacuum through polyester
drain discs placed in series with 5-µm Teflon type LS Millipore
filters. Twenty-four-hour Soxhlet extractions were carried out on
filter blanks with four fresh polyester drain discs and six 5-µm
Teflon type LS Millipore filters, using 200 ml of methylene
chloride. A cellulose extraction thimble retained the filters.
The methylene chloride was reduced to 5 ml with a Kuderna-Danish

concentrator over a steam bath. Five milliliters of dimethylsulfoxide (DMSO) was added, and attempts were made to remove the
rest of the methylene chloride; however, approximately 6 ml of
organic extract remained, indicating that approximately 1 ml of
methylene chloride would not evaporate. The material was
transferred to a micro-Snyder apparatus for more efficient removal
of methylene chloride; however, the volume still was not reduced.
The presence of residual methylene chloride was confirmed by
infrared spectrophotometry.

Ames mutagenicity testing of the Soxhlet filter extract of
recovered particulates indicated no mutagenicity, but a slight
toxic response was found. A Soxhlet reagent blank was also tested
and gave a slight mutagenic response in the spot test and a toxic
response. A methylene chloride control gave both a mutagenic and
a toxic response. Although there was no clear mutagenic response
for the Soxhlet filter particulate extract, methylene chloride
should not have been used in processing samples for biotest systems
because of its potential for causing a false mutagenic response.
Acetone or other nonbioactive solvents are recommended for future
Soxhlet extractions.

Three known mutagenic materials were used in this evaluation:
acridine orange (AO), B(a)P, and 2NF. All are positive mutagens in
the Ames microbial test system. AO, a precursor to some dyes which
might be found in textile effluents, is a highly colored compound.
AO was used first in all of the experiments because it could be
easily traced by visual methods. B(a)P is a chemical mutagen
commonly found in environmental samples. This compound requires
S-9 activation and also represents less-stable compounds. It is
light sensitive and presents some concentration problems. A
commonly used positive standard in the Ames test is 2NF, which does
not require activation. The selection of nonmutagenic
concentrations of standard compounds for spiking the neat effluent
sample was based on previous experimental data (unpublished). All
of the concentrations of positive mutagens added to the neat
effluent theoretically should be mutagenic when concentrated by at
least a factor of 200 times. The spiked concentrations of mutagens
prior to concentration are listed in Table 1.

The standard EPA method (EPA, 1977) was used for chemical
characterization of these selected priority organic compounds.
This procedure involves solvent extraction and gas chromatography/
mass spectrometric (GC/MS) analysis of the effluent. The GC/MS
system used was a Hewlett-Packard 5985A GC/MS/Data System. Total
organic carbon (TOC) analyses were performed on the neat standard
effluent samples and on the concentrates to determine efficiencies
of recovery of organic compounds. The TOC analysis was performed
on a Technicon II Autoanalyzer using the TOC cartridge at 550 nm;
the recorder read full scale at 200 ppm.

Table 1. Standard Mutagenic Compounds and Concentrations
Added to Effluent Samples

Compound	Concentration (mg/l)
Benzo(a)pyrene	0.17
Acridine orange	7.60
2-Nitrofluorene	0.17

Sorbent Extraction

Sorbent materials have been used primarily to remove organic compounds from potable water whose matrix is relatively clean (Junk et al., 1974; Loper et al., 1978). More recent investigations (Rappaport et al., 1979) have used macroreticular resins such as Amberlite (Rohm and Haas) to remove potential mutagens from effluents and wastewaters. Compared with drinking water, industrial effluent samples are usually higher in particulates, contain larger numbers of much higher concentrations of organics, and have more extreme pH values. Processing waste effluents with XAD resins presents additional problems not encountered in the treatment of drinking water. Considerably more research is required with the various sorbent materials to determine the optima of parameters such as depth of bed, flow rates, desorption methods, and solvents, as well as breakthrough limits. Such studies would be highly complex because of the different interactions of each chemical compound with the sorbent materials.

The adsorbent used in this study was Rohm and Haas Amberlite XAD-2, a low polarity styrene-divinylbenzene copolymer possessing the macroreticular characteristics necessary for high sorptive capacity. Recovery efficiencies of about 80 organic compounds in water have been reported (Junk et al., 1974). The efficiencies vary from 35 to 100%; on average, however, recovery efficiencies are above 78%.

The initial XAD-2 (Applied Science) column was prepared by washing the resin with a 50:50 mixture of methanol and water and pouring off the excess solvent to leave a slurry of XAD-2. The system consisted of a 300-ml burette with a plug of silanized glass wool placed in the bottom to retain the particulate XAD resin. The XAD-2 slurry was added to the column, and a glass wool plug was placed on top of the slurry to prevent disruption of the XAD particulates forming a column of XAD-2 resin 2 cm long by 1 cm in diameter. The resin bed was washed with 3- to 30-ml aliquots of deionized water and maintained wet at all times.

Three liters of raw wastewater effluent was processed by filtering the sample through a 5-μm type LS Millipore filter to remove coarse particulates. This filter became plugged after 200 ml of sample was processed, and a pre-filter consisting of a Unipore polyester drain disc (Bio-Rad Laboratories) was placed ahead of the 5-μm Millipore filter. This modification allowed filtration of 600 to 700 ml of sample before the system became plugged with particulates. The 3-l sample was subsequently processed in 600-to 700-ml aliquots. Filters were replaced whenever plugging occurred. The filters with particulates were retained for solvent extraction of organics. After about half of the effluent had been processed, the flow rate slowed considerably, and highly purified nitrogen under pressure was applied to facilitate the filtering process. The average flow rate was about 1.2 l/h.

Direct extraction with DMSO was evaluated as a means of reducing experimental time by eliminating desorption with one solvent followed by exchange with DMSO. Three milliliters of DMSO were added to the XAD column, saturating the XAD resin. The solvent was permitted to stand for 30 min. Then the DMSO solution was drained into a sterile test tube.

Unconcentrated raw wastewater was filter-sterilized by passage through a series of Millipore filters, as described earlier. Ames mutagenicity and toxicity tests were also performed. The sample was found to be nonmutagenic and nontoxic. The DMSO-extracted XAD-2 concentrate solution was then tested for mutagenicity. The concentrated effluent processed through the methanol/water-washed XAD-2 columns indicated some Ames mutagenicity bioactivity, but no dose response. The XAD-2-processed control tap water gave a similar response.

A new XAD-2 column was prepared by the method of Junk et al. (1974), whereby three solvents (methanol, acetonitrile, and diethyl ether) are used to wash the resin in a Soxhlet extraction apparatus. The XAD resin was refluxed with each solvent for 8 h and then stored in methanol. A 3-l tap water blank was processed with washed XAD-2 and then tested for mutagenicity. No mutagenicity was found. Three liters of effluent were concentrated on the washed XAD-2; again, no mutagenicity was found.

We concluded from this study that 1) the neat effluent sample and the 600-fold concentrate were nonmutagenic; 2) DMSO extracts were marginally mutagenic if the XAD-2 was washed only with methanol/water; and 3) DMSO could be used directly as a desorbent solvent if the XAD-2 were washed properly (i.e., by the Junk method).

Standard mutagens were added to the standard effluent in order to determine the efficiency of recovery and to evaluate the mutagenicity of the concentrate using the Ames test system. The three standard mutagens were AO, B(a)P, and 2NF.

The zinc chloride salt of AO was made up by adding 22.8 mg AO to 3 1 of particulate-free raw wastewater. This solution, containing 170 ppb AO, was processed (concentrated by adsorption) by the procedure described previously. The flow rate through the XAD averaged 18 ml/min. The AO content of the effluent was measured by a colorimetric technique with an Aminco DW-2 dual wavelength UV-Vis spectrophotometer scanning the range of 400 to 650 nm. The wavelength monitored was 433 nm. The adsorption efficiency for AO was determined by comparing the concentration prior to processing with that of the XAD-2 filtrate. The starting material (neat effluent) contained 7.6 µg/ml, and the filtrate contained 4.5 µg/ml, indicating a collection efficiency of 40.8%. Desorption was achieved by adding 5 ml of DMSO for a residence time of 30 min. A comparison of the amount of AO collected (9.3 mg) with the desorbed amount in the DMSO (7.5 mg) indicated a desorption efficiency of 80.6%. Mutagenicity was tested by applying the unconcentrated effluent containing 7.6 mg/l (7.6 ppm) AO to the Ames plate-incorporation test. No mutagenicity was found. The final XAD-2/AO extract, concentrated from 3 1 of starting sample to 5 ml of DMSO (with a combined 32.9% recovery and desorption efficiency), contained 1.5 mg/ml (1,500 ppm). The total concentration factor was 196. A dose response was found using TA98 with S-9.

Five hundred micrograms of 2NF was added to 3 1 of particulate filtered sample effluent, making a concentration of 167 µg/1 (167 ppb). The XAD-2 was prepared, as described previously, with three solvents in a Soxhlet system. Three liters of sample effluent containing the 167 ppb 2NF was processed. The recovered 2NF was desorbed with 5 ml of DMSO. Mutagenicity testing indicated no response in the Ames test to the blank or to the neat, unconcentrated sample. A positive dose response with TA98 and S-9 and a response to TA100 (no dose response) was found for the XAD-2 concentrate.

Addition of 500 µg of B(a)P to 3 1 of neat effluent gave a concentration of 167 µg/1 (167 ppb). The 3 1 of neat effluent was concentrated by passing it through the XAD-2 column. Five milliliters of DMSO was used to desorb the recovered B(a)P. The blank sample and the neat sample containing 167 ppb B(a)P gave negative responses in the Ames test. The concentrated B(a)P sample indicated bioactivity in tests performed on two different dates: on 7/13 at 30 µl/plate with TA98 and S-9 (with no dose response) and on 7/19 at 30 and 40 µl/plate (with no consistent dose response). Recovery experiments were performed to determine extraction efficiency of B(a)P by XAD-2 from spiked effluent with subsequent desorption with DMSO. Much care was taken to isolate all samples containing B(a)P from light in order to eliminate potential light-degradation problems. Five milligrams of B(a)P was added to 3 l of industrial effluent, and the sample was processed through the XAD-2 column. Recovery efficiencies were determined by measuring the B(a)P solutions with an Aminco DW-2 dual wavelength UV-Vis spectrophotometer. The aqueous permeate was analyzed at 270 nm by scanning the range of 230 to 410 nm. The UV analysis showed that no B(a)P passed through the XAD-2, indicating a possible 100% recovery of B(a)P. The column was desorbed with 5 ml of DMSO. The UV analysis indicated a desorption efficiency of 45% of the B(a)P. This low desorption efficiency may indicate a permanent bonding of the B(a)P to the XAD-2 resin. The UV data gave no evidence of degradation of the B(a)P.

Lyophilization

     This freeze-drying approach to concentration is best applied to effluents that contain water-soluble, nonvolatile, heat-labile pollutants. Inorganic salts and biological compounds of large molecular weight are retained by this method. Bieri et al. (1979) have reported successful use of freeze-drying to remove water from Chesapeake Bay samples for application to chemical characterization tests. These researchers reported the potential loss of volatile compounds below C_{12} hydrocarbons and problems with chemical contamination from vacuum-pump oils. Van De Meent et al. (1977) suggested that freeze-drying led to catalytic conversion of alcohols to olefins. However, Bieri found no evidence of this problem (Bieri et al., 1979). In the present study, contamination in lyophilized samples was demonstrated. Consequently, all samples had to be filter-sterilized before treatment or after concentration prior to adding them to the Ames test system.

     In this experiment, the lyophilization system was built around a stainless steel drum manifold (Virtis Model 10-MR-ST). The other components were a vacuum pump, a backup trap, a vacuum gauge, and the sample-containing filter-seal flask. A vacuum of 0.133 mbar was maintained over processing times of 48 to 50 h. The system could run unattended during much of this time, including overnight. The backup trap retained much of the volatile material and could be analyzed for volatile compounds.

     Three liters of neat effluent sample was processed to dryness, leaving 0.8 g of a dry white residue. An inorganic chemical characterization analysis (EDAX) indicated the presence of sulfur, silicon, potassium, calcium, and iron. TOC analysis of the residue using the Technicon II Autoanalyzer, was 6.3%. This dry residue was dissolved in 10 ml of 50:50 DMSO:water solution. A

concentration factor of 300 times was achieved. Problems were
encountered in completely dissolving the residue. A "dissolved"
sample was applied to the Ames mutagenicity assay using the plate-
incorporation test and indicated no mutagenicity in the lyophilized
neat effluent. Microbial contamination was encountered, but not
enough to prevent a valid test. Filter sterilization is
recommended prior to lyophilization.

AO was added at a concentration of 0.0076 mg/ml to 3 l of
effluent sample. The sample was lyophilized in approximately 55 h.
Using the colorimetric method, 96.3% of the AO was retained in the
lyophilized sample. The freeze-dried powder was dissolved in a
50:50 mixture of sterile distilled water and DMSO and applied to
the Ames mutagenicity assay. A two-point dose response was
obtained at 10 µl and 2 µl with TA98 and S-9 activation. Microbial
contamination was so high that the plates had to be hand counted;
lower concentrations could not be counted because of the
contamination.

Reverse Osmosis

Reverse osmosis (RO) has become a versatile separation and
purification method. This process physically separates contaminant
from water by circulating the aqueous solution at high pressures
over the surface of a semipermeable membrane. Two factors
influence the concentration of contaminants: the physical and
chemical properties of the contaminants and the properties of the
membrane. Recent RO technology has improved rapidly as membrane
technology has advanced. Much of the development and applied
research has been directed toward the purification of aqueous
effluents, whereas this study is concerned with the concentrate.

The RO system used in this investigation was manufactured by
Abcor, Inc. (Cambridge, MA). The system is a bench-type static RO
and ultrafiltration test cell designed specifically for studying
membrane selectivity in a laboratory situation. The RO unit has a
200-ml capacity, maximum operating pressure of 105 kg/cm (gauge),
and a membrane diameter of 3 in. (7.6 cm). It is 6.25 in. high by
3.75 in. in diameter (15.9 x 9.5 cm) and weighs 6 lbs (2.7 kg).
The stainless steel unit has a Teflon-coated magnetic stirring bar
for constant agitation at the membrane surface and uses a SEPA-97
(Osmonics) cellulose acetate membrane of nominal 5-Å (0.5 nm) pore
size.

Three liters of raw wastewater was filtered to remove
particulates to prevent clogging of the RO membrane. A Unipore
polyester drain disc (Bio-Rad Cat. No. 334-0659) was placed over a
type LS Millipore filter with a 5.0-µm pore size. The filtered
effluent was spiked with AO to obtain a concentration of

0.0076 mg/ml. The spiked effluent was processed by RO. A
processing time of about 48 h was required to reduce the sample to
168 ml. Problems were encountered with a much-reduced flow rate
toward the end of the processing experiment. The RO unit was
dismantled and a large buildup of what appeared to be inorganic
materials highly colored with AO was found on the membrane,
impeding flow through the membrane. The retained solution
apparently became saturated with both dissolved salts and AO as the
concentration increased; this phenomenon will limit the
concentration factor attainable by RO. The efficiency of
concentration was tested colorimetrically: the concentration of AO
in the starting unprocessed spiked sample was 7.6 mg/ml and in the
processed concentrate 7.1 mg/ml. A conductivity measurement of the
filtrate (EP Meter, Myron L Co.), reduced from 3 1 to 200 ml,
indicated that 77% of the conductivity was removed by the process.
This result is an obvious limitation of the RO system.

CONCLUSIONS

    This investigation has demonstrated that an industrial
effluent spiked with subtoxic amounts of standard mutagens (AO,
2NF, and B[a]P) can be concentrated by XAD-2 resin. The work also
showed that AO could be concentrated by lyophilization and RO. The
other two mutagens were not tested with the latter two techniques.
The concentrated spiked effluent exhibited positive dose responses
in the Ames Salmonella biotest. The RO technique was found to
retain 84% of the TOC. However, only a tenfold concentration was
achieved, because of clogging of the membrane by precipitated
compounds.

    This evaluation also showed that the adsorbent concentration
procedure has the following advantages over lyophilization and RO
techniques: low cost of equipment, short processing time, ease of
portability, and potential for selectivity for specific chemical
classes by resins. A disadvantage of this approach is the need for
a desorption solvent, which increases the chances of sample
alteration, chemical contamination, loss of sample, and possible
lack of compatibility with biotest systems.

    The lyophilization procedure has the advantage of good
recovery of inorganic components and relatively high organic
compound recovery. However, it requires much processing time, and
the equipment costs are high.

    Reverse osmosis also concentrates inorganic components. It is
portable and has the potential advantage of recovering specific
chemical classes of compounds by use of selective membranes.

ACKNOWLEDGMENTS

This report was supported by the U.S. Environmental Protection
Agency, contract no. 68-02-1874.

REFERENCES

Ames, B., J. McCann, and E. Yamasaki. 1975. Methods for detecting
carcinogens and mutagens with the Salmonella/mammalian-
microsome mutagenicity test. Mutation Res. 31:347-364.

Bieri, R.N., M.K. Cueman, R.J. Huggett, W. MacIntyre, P. Shou, C.W.
Su, and G. Ho. 1979. Investigation of organic pollutants in
the Chesapeake Bay: Report #1, Grant R806012010. U.S.
Environmental Protection Agency: Annapolis.

EPA, U.S. Environmental Protection Agency. 1977. Sampling and
analysis procedures for the screening of industrial effluents
for priority pollutants. U.S. Environmental Protection
Agency: Cincinnati, OH.

Junk, G.A., J.J. Richard, M.D. Grieser, D. Witiak, J.L. Witiak,
M.D. Arguello, R. Vick, H.J. Svec, J.S. Fritz, and G.V.
Calder. 1974. Use of macroreticular resins in the analysis
of water for trace organic contaminants. J. Chromatogr.
99:745-762.

Loomis, T.A. 1978. Essentials of Toxicology. Lea and Febiges:
Philadelphia.

Loper, J.C., and D.R. Lang. 1978. Mutagenic, carcinogenic, and
toxic effects of residual organics in drinking water. In:
Application of Short-Term Bioassays in the Fractionation and
Analysis of Complex Environmental Mixtures. M.D. Waters, S.
Nesnow, J.L. Huisingh, S.S. Sandhu, and L. Claxton, eds.
Plenum Press: New York. pp. 513-528.

Rappaport, S.M., M.G. Richard, M.C. Hollstein, and R.E. Talcott.
1979. Mutagenic activity in organic wastewater concentrates.
Environ. Sci. Technol. 13:957-961.

Van De Meent, D., W.L. Maters, J.W. Peheew, and P.A. Schenck.
1977. Formation of artifacts in sediments upon freeze-
drying. Org. Geochem. 1:7-9.

SESSION 3

TERRESTRIAL SYSTEMS

POTENTIAL UTILITY OF PLANT TEST SYSTEMS FOR ENVIRONMENTAL MONITORING: AN OVERVIEW

Shahbeg Sandhu
Health Effects Research Laboratory
U.S. Environmental Protection Agency
Research Triangle Park, North Carolina

Research over the past decade has shown a significant proportion of genetic diseases in man to be caused by natural and man-made chemical mutagens. Human cancer is one of the diseases for which direct associations with certain environmental factors have been established (such as the association between cigarette smoking and lung carcinoma). The recently published "Atlas of Cancer Mortality" (Mason et al., 1975) provides further evidence for the association between human cancer and environmental factors. In this study, certain specific types of cancer appear to be associated with certain industrial activities. Chemical mutagens are also believed to contribute to birth defects, behavioral abnormalities, and aging. It has been suggested that environmental chemicals play a role in causing atherosclerosis (Benditt, 1973) and, most damaging of all, in deteriorating the human gene pool.

Awareness of the role of environmental chemicals as human health hazards has led to the development and use of various techniques for identifying potentially toxic chemicals and, if possible, eliminating exposure to them. These methodologies include several nontraditional types of short-term bioassays. A variety of test systems, ranging from the use of viruses to that of circulating human lymphocytes and sperm cells, have been used to identify potentially harmful chemicals or mixtures of chemicals. Over 100 different assays have been developed for detecting toxic chemicals (Hollstein et al., 1980).

Despite their historical role in the formulation of principles of genetics and genetic toxicology, plant test systems have not been employed in the recent rapid advances in the development of genotoxin-detection technology. This general neglect may perhaps

be attributed to the apparently great phylogenetic distance of
plants from animals. However, the use of plants as test organisms
for detecting the genetic effects of individual chemicals or as
monitors of environmental quality offers the following advantages:

1) Plants, like animals, are eukaryotes; thus, their
    organization and cell structure are similar.

2) Plants undergo mitosis and meiosis, thus making it
    possible to evaluate somatic and germ cell mutations and
    their transmission to future generations. This capacity
    is especially important when we consider that most of the
    short-term bioassays use bacteria or mammalian cells in
    culture, which do not undergo meiotic cell division.

3) Plants can easily be propagated from vegetative tissue or
    even from a single cell. Thus any variant line can be
    genetically characterized.

4) Plants show a wide array of genetic endpoints, including
    gene mutation, DNA repair, primary DNA damage, and
    chromosomal aberrations. In certain plant species (e.g.,
    barley and Arabidopsis), multiple-locus forward-mutation
    test systems have been developed that may be particularly
    relevant to human genetic systems. Using Arabidopsis, it
    is already feasible to monitor one hundred loci at once.
    This attribute is significant when we consider that
    different chemicals elicit qualitatively and
    quantitatively different responses at different loci in
    the same genome and that most of the commonly used
    short-term bioassays monitor genetic alterations at only
    one or two loci.

5) Plants are relatively easy and inexpensive to work with.
    Furthermore, mutational events are very easily scored by
    technicians.

6) Perhaps the most significant attribute of plants as test
    systems is their suitability for in situ monitoring.

     For the last three years, the U.S. Environmental Protection
Agency (EPA) and the National Institute of Environmental Health
Sciences (NIEHS) have made concerted efforts to develop and use
plant bioassays for environmental monitoring. There are two areas
in which plant systems show promise for immediate use: 1) as part
of a short-term first- or second-level laboratory test battery for
evaluating the mutagenicity of specific environmental chemicals or
chemical mixtures, and 2) in field monitoring studies, as
indicators of the mutagenicity of the total environment.

Possible Role of Plants for Mutagenicity Evaluation in the
Laboratory

A number of articles have been published emphasizing the need
for short-term bioassays (see Hollstein et al., 1979, for a review)
and for their integrated use with in vivo animal bioassays in
identifying environmental chemicals and estimating risk from
exposure to them (Waters et al., 1980). An extensive data base is
not yet available for comparing the genetic responses to chemicals
of plant test systems with those of other in vitro and in vivo
animal test systems. However, preliminary comparisons based on
review of the literature show a fairly good concordance. The
genetic potencies of eight chemicals in various in vitro and in
vivo test systems were compiled by Clive and Spector (1978). The
data in Table 1 show that mutation responses in plants correlate
well with those in mammalian systems.

Table 1. Comparative Mutational Potencies of Eight Chemicals in
Bacteria, Plants, Insects, and Mammals[a]

| Chemical | Bacteria | Plants | Insects | Mammals | |
				in vivo	in vitro
Trenimon	2	1	4	1	4
Mitomycin C	4	2	1	1	2
MNNG	5	3	7	5	2
Triethylene- melamine	8	4	8	3	5
Ethylenemelamine	3	5	3	4	8
Ethylmethane sulfonate	7	6	6	7	3
Methylmethane sulfonate	1	7	2	8	7
Dimethyl- nitrosamine	6	8	5	6	6

[a]Data from Clive and Spector (1978).

Rédei et al. (1980) have compiled data from the literature on
the mutagenic response of the multilocus Arabidopsis test system to
a number of known animal carcinogens. The list includes several
compounds that require metabolic activation to produce genetic

effects. This report shows an 84% correlation of genetic responses
in Arabidopsis with those in animal tests. It also shows that
Arabidopsis can provide the enzymes needed to transform promutagens
into reactive metabolites. Because it is a multiple locus system
and because Arabidopsis has a short generation time and can grow on
a variety of media, this test seems especially well suited for
further development for screening environmental chemicals in the
laboratory. This is the only existing short-term bioassay
for analyzing the genetic effects in progeny of environmental
exposure of the parents. The EPA, in a cooperative effort with the
University of Missouri, is currently validating this assay.

Nearly every biology student first visually encounters
chromosomes in onion or broad bean root tips--the chromosomes from
these materials are large and easy to manipulate. A few chemical
pesticides have been tested for their ability to induce chromosomal
aberrations in plant root tips; Table 2 compares the clastogenic
response to these pesticides in plant root tips with that in
mammalian cells in culture.

Table 2. A Comparison of Responses by Plant Root Tip Cells and
          Mammalian Cells in Culture to Pesticides[a]

| | Chromosome Aberrations | |
Compound	Plant Root Tips	Mammalian Cells in Culture
Apholate	+	+
Atrazine	+	+
2,4-D	+	+
DDT	+	+
Dichlorvos	+	+
Dieldrin	+	+
Ethylene dibromide	–	–
Griseofulvin	+	+
Hempa	–	–
Heptachlor	+	+
Maleic hydrazide	+	–
Mercury compounds	+	+
Phosphamidon	+	+
2,4,5-T	+	+
Tepa	+	+

[a]Data from W. F. Grant (1978).

In general, the positive or negative responses to these
pesticides are very similar in plant root tips and mammalian cells
in culture. The only exception seems to be maleic hydrazide. It
has recently been observed by Dr. Michael Plewa (University of
Illinois, personal communication, 1980) that although animals
cannot transform this compound to genetically reactive metabolites,
plants (at least Zea mays) do convert it into biologically active
forms.

The Health Effects Research Laboratory of EPA at Research
Triangle Park, NC, is in the process of evaluating the Vicia faba
root tip assay for possible inclusion in the level one test
battery. Several pesticides for which we have fairly extensive
data will be tested for chromosomal effects in this assay.

The presentation by Dr. Constantin in this symposium
(Constantin et al., 1980) further illustrates the utility of plant
cytogenetics assays in concert with microbial mutagenicity assays
for evaluating the potential health effects of complex
environmental mixtures. With few exceptions, plant bioassays do
not have as well-defined gene markers or genetically engineered
tester strains as are found in microbial bioassays. On the other
hand, very few microbial assays can be used to evaluate the
chromosomal effects of exposure to environmental chemicals.

The point of this discussion is that plant bioassays could be
profitably used for toxicological evaluation of environmental
chemicals. These assays will not be able to replace microbial test
systems in the foreseeable future, but will be useful in furnishing
complementary information.

The Role of Plants for In Situ Environmental Monitoring

Perhaps the most useful testing application for plants in the
future will be in monitoring the mutagenicity of the total
environment. Environmental chemicals exert their effects not in
isolation but in concert with other chemicals and environmental
factors. This environmental milieu is impossible to reproduce in
the laboratory. By growing experimental plants in the ground at
the site to be tested, one can evaluate the multimedia exposure
effects. Several plant systems have shown a great deal of promise
for in situ environmental monitoring. The Tradescantia stamen hair
assay (Schairer et al., 1978) has been used to monitor ambient air
quality at several industrial sites in the U.S. The waxy pollen
maize assay (Plewa, 1978) has been developed and applied to detect
the mutagenicity of agricultural chemicals. Klekowski (1978) has
developed a very useful bioassay for monitoring mutagens in
effluent streams, rivers, and lakes. An excellent review of these
test systems has been edited by de Serres (1978).

In the present symposium, Dr. Ma describes the potential use of a newly developed bioassay for monitoring complex environmental mixtures (Ma et al., 1980). The <u>Tradescantia</u> micronucleus assay appears to be even more sensitive than the <u>Tradescantia</u> stamen hair assay. This assay is under further validation with financial support provided by EPA.

Another assay, developed by Dr. Vig, of the University of Nevada (Vig, 1980), uses a soybean system for measuring point mutations and chromosomal aberrations in somatic cells. By using a series of promutagens (compounds that require mammalian metabolic activation enzymes for biotransformation to express their genetic potential), Dr. Vig has shown that plants have the ability to activate these compounds to mutagenic levels.

In the past, the lack of concurrent controls has caused some difficulties in interpreting data from <u>in situ</u> bioassays. Data from historical controls cannot be used as a substitute for on-site controls. <u>In situ</u> monitoring with plant test systems is not intended to take the place of more rigorous testing to evaluate the health hazards of exposure to a particular environment. None of the <u>in situ</u> plant bioassays have reached a stage of development where they could be used to identify specific mutagenic compounds from the environment. Their main utility so far appears to be in raising a "red flag," so that priorities can be set for applying more specific bioassays and chemical analysis to track down the sources of toxic chemicals.

REFERENCES

Benditt, E., and F. Benditt. 1973. Evidence for a monoclonal origin of human atherosclerotic plaques. Proc. Nat. Acad. Sci. US 70:1753.

Constantin, M.J., K. Lowe, T.K. Rao, F.W. Larimer, and J.L. Epler. 1980. The detection of potential genetic hazards in complex environmental mixtures using plant cytogenetics and microbial mutagenesis assays. Presented at the U.S. Environmental Protection Agency Second Symposium on the Application of Short-term Bioassays in the Fractionation and Analysis of Complex Environmental Mixtures, Williamsburg, VA.

Clive, D., and J.F.S. Spector. 1978. Comparative chemical mutagenesis: an overview. In: Proceedings Comparative Chemical Mutagenesis Workshop. F.J. de Serres, ed. National Institute of Environmental Health Sciences: Research Triangle Park, NC.

de Serres, F.J. 1978. Introduction: utilization of higher plant systems as monitors of environmental mutagens. Environ. Hlth. Perspect. 27:3-6.

Grant, W.F. 1978. Chromosome aberrations in plants as a monitoring system. Environ. Hlth. Perspect. 27:37-43.

Hollstein, M., J. McCann, F. Angelosanto, and W. Nichols. 1979. Short-term tests for carcinogens and mutagens. Mutation Res. 65:133-226.

Klekowski, E. 1978. Screening aquatic ecosystems for mutagens with fern bioassays. Environ. Hlth. Perspect. 27:99-102.

Ma, T.-H., V. Anderson, and S. Sandhu. 1980. A preliminary study of the clastogenic effects of diesel exhaust fumes using the Tradescantia micronucleus assay. Presented at the U.S. Environmental Protection Agency Second Symposium on the Application of Short-term Bioassays in the Fractionation and Analysis of Complex Environmental Mixtures, Williamsburg, VA.

Mason, T.J., F.W. McKay, R. Hoover, W.J. Blot, and J.F. Fraumeni, Jr. 1975. Atlas of Cancer Mortality for U.S. Counties: 1950-1969. Dept. of Health, Education, and Welfare publication no. 75-780. National Institutes of Health: Washington, DC.

Plewa, M.J. 1978. Activation of chemicals into mutagens by green plants: a preliminary discussion. Environ. Hlth. Perspect. 27:45-50.

Rédei, G.P., M.M. Rédei, W.R. Lower, and S.S. Sandhu. 1980. Idendification of carcinogens by mutagenicity for Arabidopsis. Mutation Res. 74:469-475.

Schairer, L.A., J. Van't Hof, C.G. Hayes, R.M. Burton, and F.J. de Serres. 1978. Measurements of biological activity of ambient air mixtures using a mobile laboratory for in situ exposures: preliminary results from the Tradescantia plant test system. Application of Short-term Bioassays in the Fractionation and Analysis of Complex Environmental Mixtures. M.D. Waters, S. Nesnow, J.L. Huisingh, S.S. Sandhu, and L. Claxton, eds. Plenum Press: New York. pp. 419-440.

Vig, B.K. 1980. Soybean system for testing the genetic effects of industrial emissions and liquid effluents. Presented at the U.S. Environmental Protection Agency Second Symposium on the Application of Short-term Bioassays in the Fractionation and Analysis of Complex Environmental Mixtures, Williamsburg, VA.

Waters, M.D., V. Simmon, A.D. Mitchell, T.A. Jorgenson, and R.
    Valencia. 1980. An overview of short-term tests for
    the mutagenic and carcinogenic potential of pesticides.
    J. Environ. Sci. Hlth. B.Pesticid. Food Contam. Agric.
    Wastes B15:867-906.

ARABIDOPSIS ASSAY OF ENVIRONMENTAL MUTAGENS

G.P. Rédei
Department of Agronomy
University of Missouri
Columbia, Missouri

INTRODUCTION

In the past, <u>Arabidopsis</u> assays have been employed for testing
the mutagenic effects of a variety of chemicals. Although it is
potentially useful for determining mutagenic hazards of complex
mixtures, this species has not been much used for this purpose.

An <u>Arabidopsis</u> assay system was initiated in the early 1960's
by Andreas Müller (1963, 1965b). The suggested procedure was based
on the principles first exploited by Gregor Mendel in his famous
pea experiments. In the autogamous species of pea, within
individual flowers, segregation of the alleles at meiosis and the
random combination of the gametes at fertilization resulted in the
reappearance of both dominant and recessive phenotypes of the
parents among the embryos developed on the F_1 plants. Therefore
when Mendel crossed two different varieties of peas (Yellow vs.
green and Smooth vs. wrinkled cotyledons), a study of only the F_1
plants, bearing the F_2 seeds, was frequently satisfactory for his
genetic analyses. Thus, he saved considerable time and labor,
enabling him to make more comprehensive studies.

But heterozygosity within the nucleus of a cell can also arise
by mutation. Mutation from a dominant to a recessive allele in the
diploid cells is concealed. If the mutation takes place early in
the diploid germline, and the heterozygous cell gives rise to a
sector encompassing both the pollen-producing (androecium) and the
egg-producing (gynoecium) lineages, segregation may become evident
among the embryos of the same individual. Since the germline is
commonly multicellular, the F_1 plant may be chimeric after a

211

mutational event. This is more of an advantage than a disadvantage
because it permits the screening of a large population at a low
cost.

     Such an analysis is impractical, however, in monoecious
plants, such as maize, because the silks and the tassel may not
differentiate concordantly (from the same cell lineage).
Obviously, dioecious plants and the majority of bisexual animals
are not amenable to such an analysis.

PURPOSE

     Though a large number of assay systems are available for
mutagens, none of them suits entirely the needs for testing all
hazardous compounds. The most efficient and the most widely used
microbial assays involve the detection of revertants at a specific
locus. The mutability of individual loci may vary by one or more
orders of magnitude (Table 1). Furthermore, the reversion assay
uses special alleles, one or another at a time. The mutability of
these special sites depends to a great extent on the nature of the
inducing agent (Table 2). Therefore it is not easy to draw firm
conclusions as to how complex mixtures of mutagens or even pure
compounds would affect other genes. Even worse, the mutability of
many important human genes cannot be directly measured, and the
inferences based on indirect methods are also somewhat tenuous.
Approximations based on mammalian assays, such as the mouse
specific locus assay (which is probably the best), are not much
better either, because we do not know for sure whether the six or
seven loci used represent accurately all the loci of mice (or those
of man).

Table 1. Rates of Spontaneous Mutation[a]

Organism	Mutant Phenotype	Rate
E. coli	Streptomycin resistance	4×10^{-4}
	Histidine auxotrophy	2×10^{-6}
Neurospora	Inositol independence	2×10^{-5}
	Adenine independence	4×10^{-6}
Drosophila	Yellow body	1×10^{-4}
	White eye	3×10^{-5}

[a]Rédei, 1980, p. 576.

Table 2. Site Specificity of Induced Mutations in Yeast[a]

	Revertants per 10^7 Cells			
Sites	EMS[b]	NMG[c]	UV[d]	γ-rays
cyl-131	1226	1775	41	18
cyl-133	4	1	83	8
cyl-9	8	8	2430	19

[a]Abridged from Prakash and Sherman, 1973.
[b]Ethylmethane sulfonate.
[c]Nitrosomethylguanidine.
[d]Ultraviolet.

There is obviously a need for improvement in the testing systems; we must search for approaches capable of detecting a variety of genetic alterations (gene mutations and chromosomal effects) at as large a number of positions in the genome as possible. We must also strive to learn much more about the capabilities of metabolism in activating or detoxifying the potentially hazardous chemical compounds present in the human environment. The tests must be relatively fast, reliable, reproducible, and inexpensive, and the information obtained should be applicable for predicting human hazards. The Arabidopsis assay outlined here appears quite valuable for meeting many of these goals.

THE ARABIDOPSIS ASSAY

Culture of the Plants

Arabidopsis thaliana (L.) Heynh. is a plant of the family Cruciferae (mustards), with a haploid chromosome number of five. The early genotypes may produce eight generations a year in the laboratory, where the life cycle can be hastened by continuous illumination (long-day plant). Under short daily light regimes, the vegetative period is quite long, and only a few generations can be grown in a year. Under such conditions, the plants grow much larger and may produce 50,000 seeds each.

The plants can be grown in pots on any good soil or other media. In the greenhouse, we culture them on soil in 5-in (12.7-cm) pots, each with several plants. In growth chambers, we have successfully raised 200 or more plants on Pro-Mix medium (Premier Brands) to maturity in petri plates 10 cm in diameter.

Before planting, the medium was moistened with a nutrient solution
of the following composition:

ammonium nitrate (NH_3)	200 mg/l
magnesium sulfate ($MgSO_4 \cdot 7H_2O$)	100 mg/l
calcium phosphate, monobasic ($CaH_4(PO_4)_2 \cdot H_2O$)	100 mg/l
potassium phosphate, monobasic (KH_2PO_4)	100 mg/l
potassium phosphate, dibasic (K_2HPO_4)	50 mg/l
ferricitrate	2.5 mg/l

After germination, the lid was removed, and the lost moisture was
replaced by distilled water as needed (at least two times a day).
In soil culture, supplemental nutrients are not needed. In any
case, the seeds should not be covered after planting, because
Arabidopsis requires light for germination.

When large quantities of seeds need to be planted, time may be
saved by suspending the seeds in a fluid yet viscous agar solution
(ca. 0.15%) and spreading them dropwise with the aid of a
separatory funnel or a pipette. After planting, the seeds must not
be allowed to dry for any length of time. To avoid washing them to
the rim of the container, watering is best done initially with a
fine mist. Later, after germination, the pots can be placed in a
tray containing distilled water. In the growth chamber, the
intensity of the light should not exceed 800 foot-candles (8608
lux), and a constant temperature of about 24°C is satisfactory.

The seeds of Arabidopsis loose their germination ability
within two to three years at higher temperatures and humidity.
Therefore, it is advisable to store the seeds under a regime where
the sum of the temperature in degrees Fahrenheit and the relative
humidity is below 100; the lower this figure, the longer the
viability.

Mutagen Assay

Generally, for laboratory assays, seeds are exposed to
mutagens. Though the mature embryo contains about 6000 to 7000
cells, only two of the cells represent the diploid germline at this
stage. As the seeds germinate and the seedlings develop, the
number of cells in the germline increases. The consequences of
mutation at two different stages of the growth of the germline are
diagrammed in Figure 1.

Either the mutants can be detected at the embryo stage (Figure
2) or the M_2 generation can be planted and classified for seedling
and plant traits. The most convenient method of assessing the
mutagenic effects is to determine the frequency of mutational
events expressed at the embryo stage. For this purpose the fruits

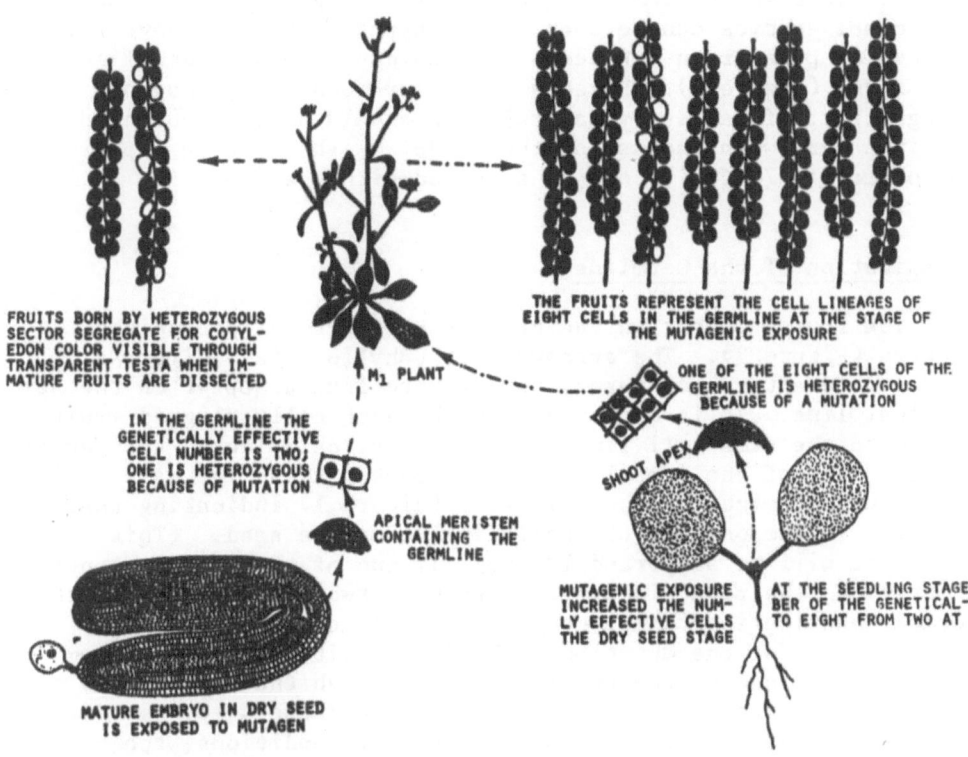

Figure 1. Consequences of mutagenic treatment at stages when the germline consists of two diploid cells (left) and when it has increased to eight (right). Explanations are on the diagram.

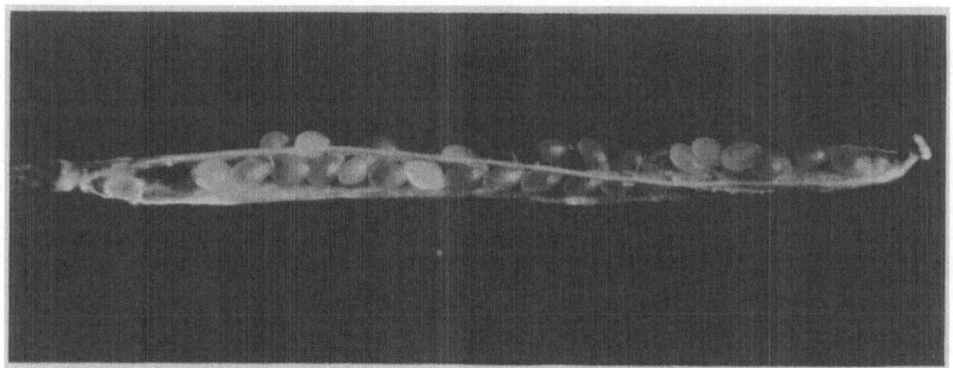

Figure 2. An <u>Arabidopsis</u> fruit, showing segregation for the color of the embryos.

are opened with sharp forceps before the seed coat turns brown and
opaque. Through the immature seed coat, the green cotyledons of
the normal embryos can be seen. Many types of mutants have white,
yellow, or pale green cotyledons, which can also be classified at
this stage (Figure 2). Persons with very good vision may not need
a magnifier; if necessary, an enlarger should be used which does
not interfere with the use of the hands. With a little experience,
approximately 70 fruits can be screened within an hour.

Organization of the Germline

The fruits appear on the stem in a rather well-organized
pattern (Figure 3). The arrangements (phyllotaxis) follows a
spiral, and after some turns, two or more fruits appear on the same
vertical line. It is expected that the vertically aligned fruits
belong to the same cell lineage. In chimerical plants, the genetic
constitution of the same lineage is expected to be the same, while
other cell lineages may be different (Figure 1) indicating that the
germline is composed of two cells in the mature seed. (This
statement will be supported later.) If one of these cells contains
a mutation, the plant is expected to have two sectors, one of the
normal constitution (homozygous wild type) and another that is
heterozygous for the mutation. The fruits situated on the stem
may represent one or the other sector. Though the periodicity of
the fruit arrangement on the stem may be influenced to some extent
by developmental factors or by the external conditions, the
experimenter must be familiar with the phyllotaxis in order to make
the work efficient. As a rule of thumb, it is advisable to scan
fruits which sit on opposite sides of the stem, as these will most
likely represent different sectors of the chimera. It is possible,
however, for phenotypically identical mutants to occur in opposite
fruits; these, most likely, originated from independent mutational
events (Figure 4).

Calculation of the Mutation Frequency

Within single fruits, the segregation of mutant and wild-type
embryos is expected to correspond to monogenic Mendelian ratios.
Dominant and recessive mutations may be distinguished if the number
of embryos is sufficiently large. Single fruits may contain up to
30 or 40 embryos, though frequently their number is much reduced
when the chemicals are toxic. Occasionally, recessive mutations
may mimic ratios expected on the basis of dominant inheritance,
because of the small numbers and/or reduced transmission of the
chromosomes (gametes) involved. The simultaneous induction of two
phenotypically similar mutations may result in a theoretical
proportion of 9 wild-type:7 mutant. Poor transmission of the
chromosome carrying the wild-type allele (because of a large

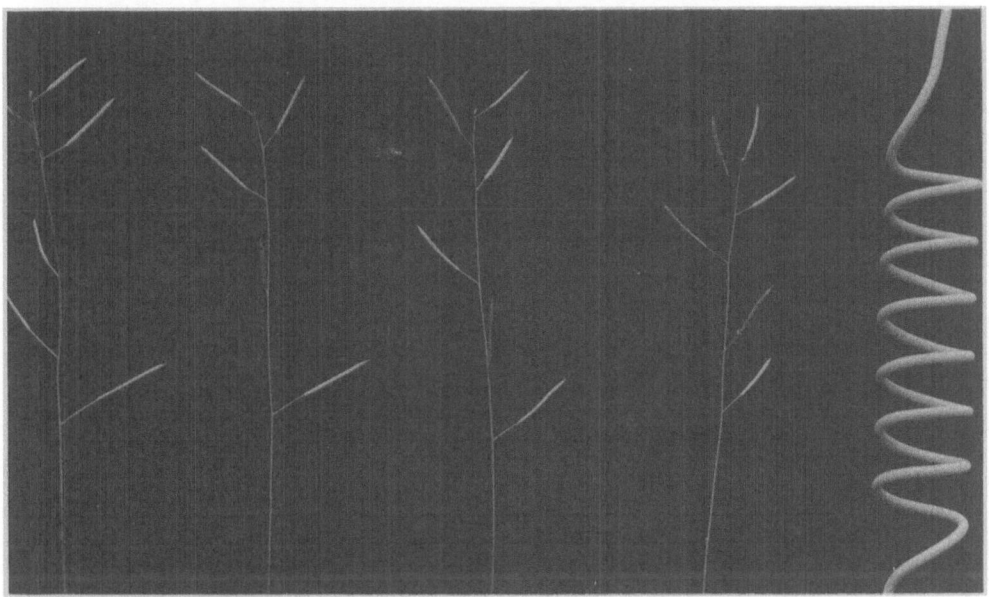

Figure 3. <u>Arabidopsis</u> main stem showing the phyllotaxis of the
fruits (in a right-hand spiral, as illustrated at
right).

deficiency or other gametophyte factors) may give a 2:1 ratio.
It is sometimes not possible to distinguish clearly among these
proportions on the basis of the few embryos in a fruit.

In addition, we may observe a deficiency of the mutant class:
this situation is actually the most common. The reduction of the
mutant class within a fruit can be caused by the poor transmission
of the chromosome carrying the mutant allele or by the early elimination of some of the zygotes homozygous for defective alleles.

When there is normal inheritance of the mutants and not too
high a frequency of mutation (i.e., the majority of plants would
incur only a single mutation), we can determine rather accurately
the number of sectors in the mutant plants. Alternatively (or
additionally), we may harvest all the seeds from individual M_1
plants and determine segregation ratios in the planted M_2
generation. In the case of normal transmission and viability, the
proportion of the wild-type and mutant individuals in the progeny
is expected to reflect the number of cells in the germline that
gave rise to the fruits. (This is what we call the genetically
effective number of cells, or GECN.) Where the germline is

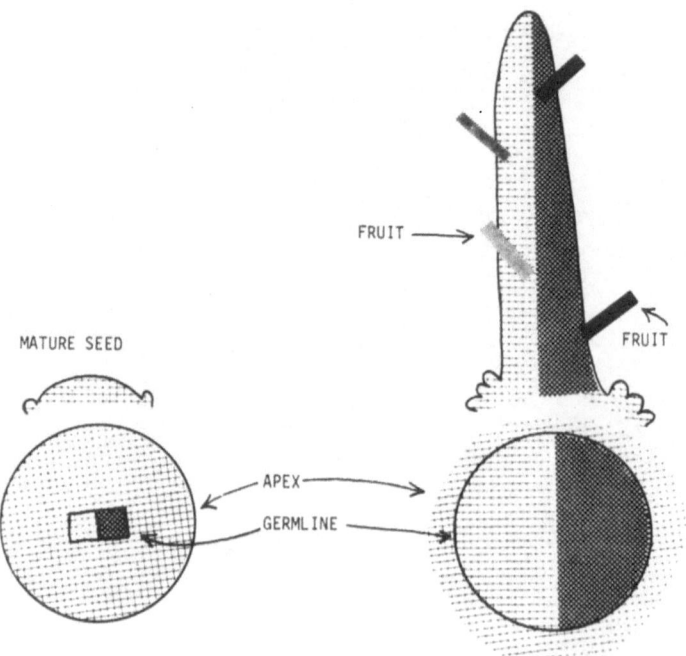

Figure 4.   The time course of the development of the germline.
            In the mature embryo, the germline consists of about
            two cells (left); in the maturing plant, the number
            of cells in the germline increases (right). If one
            of the two cells of the germline in the mature seed
            incurs a mutation, the growing plant becomes a
            chimera containing two identical-size sectors. The
            genetic constitutions of the fruits (embryos) are then
            determined by their locations on the stem.

represented by a single cell at the time of the mutation,
segregation for wild-type and mutant should show a proportion of
3:1 (GECN = 1). Where the germline contains two cells (GECN = 2)
at the time of the mutagenesis and only one of the two cells
suffers a mutation, the expected segregation ratio is 7:1: that is,
one of the two cells segregates 4:0 while the other displays 3:1
proportions, and pooling the data, we obtain the 7:1 ratio. Similarly, if the germline were eight cells, seven of the fruits would
be expected to yield only wild-type embryos, and one would display
3:1 segregation progeny, a proportion of 31:1 ([7 x 4] + 3 : 1).

     Thus, on the basis of genetic data, we can infer the number of
genetically effective cells at the time of the mutagenic exposure
if it is of relatively short duration, that is, lasting for a few

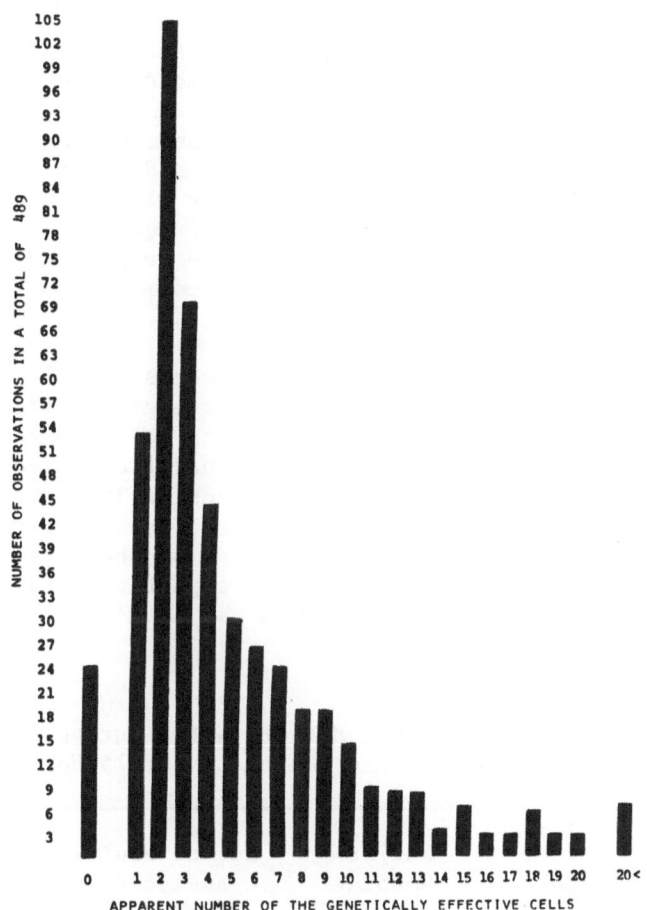

Figure 5. Variations in the apparent number of cells in the germline. The mode (the most frequent class) indicates that the number of genetically effective cells (GECN) is generally two. The classes shown were established by grouping the segregation data of all mutants classified and counted.

hours rather than for many days. The number of genetically effective cells is reasonably consistent at a particular developmental stage. Poor sampling of the seed output of the plants and reduced transmission of certain mutants may, however, cause variation around the most frequent class (Figure 5.)

When the number of genetically effective cells is known, the frequency of mutation can be easily calculated by the following formula:

$$R = \frac{\text{no. of independent mutations observed}}{\text{no. of progenies classified x GECN x ploidy x dose of mutagen.}}$$

This formula expresses mutation frequency on a genome basis, because we take into consideration the diploid nature of all the cells in the germline by substituting 2 for ploidy in the denominator. By this procedure, mutation frequencies can be directly compared with those of other higher organisms or microoganisms. The dose of the mutagen used may be omitted from the formula.

In an experiment, we treated mature seeds with 0.3% ethylmethane sulfonate, and the embryos were classified before maturity as outlined (Table 3). The fruits not used for embryo analysis were harvested, and in the M_2 generation, phenotypically distinguishable mutant classes were counted in 308 families. Some of the families displayed no mutants at all, and others showed one or more types. The frequency distribution of these families is shown in Figure 6, with the theoretical expectation based on the Poission distribution.

Table 3. Frequency of Embryo Mutants after Mutagenic Treatment
with 0.3% Ethylmethane Sulfonate for 15 Hours

Treatment	No. of Plants	No. of Genomes	No. of Fruits Analyzed	Mutation Frequency
Untreated	192	768	599	0.0013
Treated	205	818	485	0.4707

This analysis considered all the mutants that germinated and expressed themselves during early or later stages of development. Many mutations that could be identified in the immature fruits could not be detected after planting the seeds. Apparently, these involved lethality that prevented germination of the individuals affected. A number of additional types of mutants could, however, be identified during the later stages of development.

A closer examination of the classes of families with various or no mutational events revealed some similarities with and some discrepancies from the Poisson distribution. Curiously, the frequency of families with one or more mutations was higher than

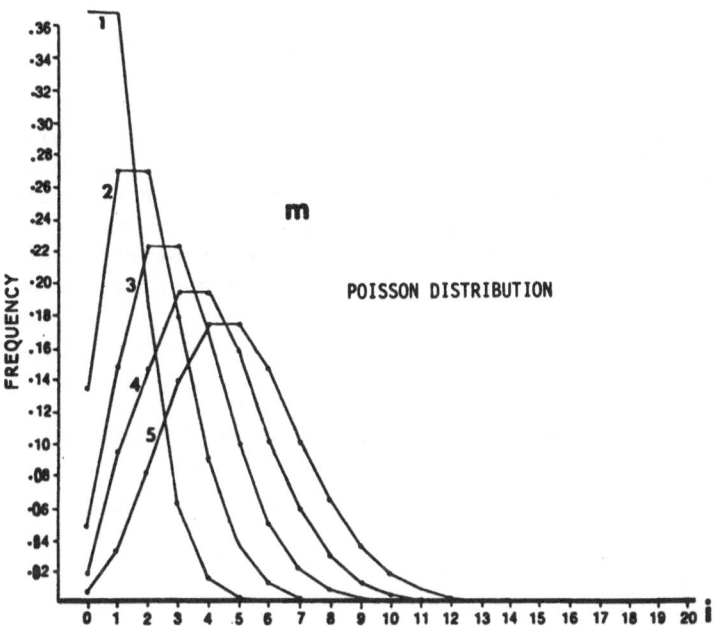

308 Families with Number of Phenotypically Distinguishable Mutations						
	0	1	2	3	4	5	6
Number Observed	22	115	118	45	6	2	0
Frequency	0.071	0.373	0.383	0.146	0.020	0.007	0

Figure 6. Comparison of the distribution pattern of families with multiple mutations with Poisson distributions for averages of one to five independent events (m). The abscissa (i) represents the numbers of events expected to occur at the frequencies given by the ordinate.

expected on the basis of the curves shown. These curves show Poisson distributions with 1, 2, 3, 4, and 5 average mutational events expected per family. On the other hand, the frequencies of classes 1 and 2 were nearly equal, a feature characteristic of the Poisson distributions of integer numbers (Figure 6, top).

In the 308 families, 520 mutations were identified on a phenotypic basis. This classification was obviously loaded with a systematic error, because in the presence of two independent mutations per cell, we would also expect, besides the two single mutants, the double-mutant class. In the case of three mutational events, three single mutants, three different types of double mutants, and one triple-mutant phenotype may be exhibited in a family. Since progeny analysis could not be performed with the lethal or subvital types, the genetic constitution of these phenotypic classes could not be determined.

The average number of mutational events per family can be determined, however, if we consider the zero-mutation class. The frequency of this group is not influenced by the complications just mentioned. The theoretically expected frequency of the zero-mutation class (based on the Poisson distribution) was determined for 2.0 to 3.0 mutational events per progeny, and the frequencies are shown in Table 4.

Table 4. Theoretically Expected Frequency
of the Zero-Mutation Class

Average no. of mutations	2.0	2.5	2.6	2.7	3.0
Expected frequency of zero-mutation class	0.135335	0.082085	0.074274	0.067206	0.049787

Figure 6 (bottom) shows the result of an experiment where the frequency of the zero-mutation class was 0.071, which indicates an average of 2.6 to 2.7 mutations per family (Table 4). Since at the time of the treatment with EMS, the germline of each mature embryo contained four genomes, the average frequency of mutations in this material can be computed as (approximately) 2.6/4 = 0.65. This is considerably higher than the observed 0.47 for mutations expressed at the embryo stage in the immature fruits; it is also higher than the empirically found value of 0.59 for all the mutations observed. We may also conclude that under the conditions of this experiment, approximately 10% of the mutations were missed, either because of misclassification or due to some other accidental causes. The difference between the experimentally observed value of 0.59 and the frequency of 0.65 predicted on the basis of Poisson's exponential binomial limits is not so large as to give basis for serious reservations concerning validity.

Calculation of the Number of Target Loci

From the viewpoint of effectiveness of a mutagen assay, it is important to know how many potential targets can be hit by a mutagenic chemical. It seems that the mutability of the gene loci is not uniform across the entire genome (Table 1). This may be due to differences among the various loci in number of nucleotide pairs (the size of the genes). The genes may be more mutable during periods of replication or transcription, and these differences in state may be reflected in the observed mutabilities. Also, certain sites within a gene may appear as "hot spots" when exposed to one agent but not when another mutagen is applied (Table 2). The nature of particular base pairs and/or the conformation of the DNA may also affect mutability. The metabolic machinery, the genetic repair systems, etc., may also affect the various loci differently.

Assays that are capable of monitoring mutational responses at a large number of loci, presumably representing all of the loci in a fair manner, are particularly attractive. The number of genes cannot be directly counted in higher organisms. Since nucleotide sequencing became practical, the number of genes of a few viruses could be determined (Fiers, 1975; Sanger et al., 1977). Such an approach is still impractical for mammals or higher plants, which may have six to eight orders of magnitude more DNA per cell than the smallest viruses.

The number of genes (cistrons) in Drosophila is believed to be about 5,000 (Garcia-Bellido and Ripoll, 1978), a figure that is close to the number of bands detectable by the most revealing counts on the salivary gland chromosomes. Belling (1928) assumed that the number of chromomeres observed in the lily chromosomes, 2193, indicates the number of genes in this plant. Because of the small size of the chromosomes of Arabidopsis, chromomeric organization cannot be determined (Figure 7), but even if this plant were quite favorable for cytological studies, an estimate of the number of genes on such a basis would not be sufficiently realistic.

The number of gene loci can be calculated more precisely on the basis of overall mutation frequencies if the average rate of mutation per locus is known. In Arabidopsis, estimates of mutation rate are available for specific loci involved in the synthesis of thiamine (vitamin B_1). For the calculations, mutations induced with EMS at the three loci and in the following numbers were used: py 35, tz 5, th 11. The average induced mutation rate at these three loci is 3.2×10^{-5}, which is comparable to the data for mice from the literature (Table 5).

These loci of Arabidopsis were chosen for studies not on the basis of their mutability, but because of our interest in the genetic control of a biosynthetic path. They may represent to a

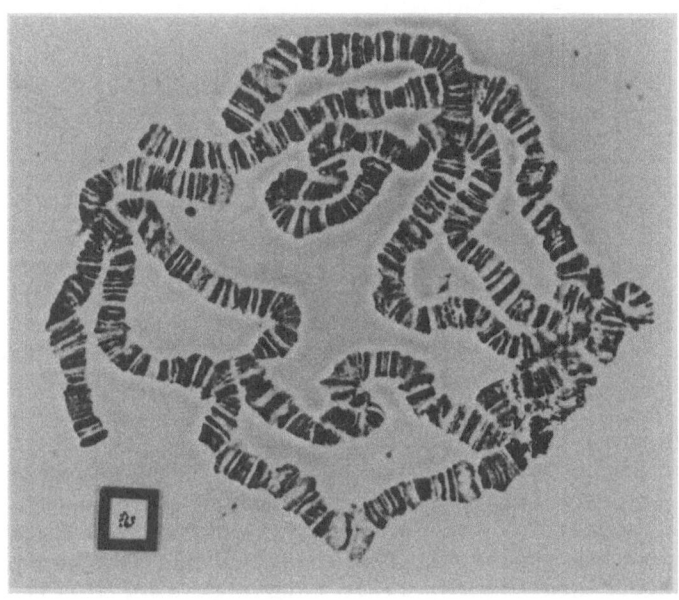

Figure 7. Salivary gland chromosomes of <u>Drosophila</u> compared on
an equal scale with the chromosomes of <u>Arabidopsis</u>,
shown within the box. (Courtesy of Dr. H.K. Mitchell
and Dr. Lotti Sears, respectively.)

Table 5. Induced Mutation Rates in <u>Arabidopsis</u> and Mouse

Organism	Loci	Mutagenic Agent	Rates
<u>Arabidopsis</u>	<u>py</u>	(EMS)	2.0×10^{-4}
	<u>tz</u>	(EMS)	2.5×10^{-5}
	<u>th</u>	(EMS)	7.0×10^{-5}
Mouse	Isozyme loci	X-rays[a]	1.7×10^{-4}
	7 Specific loci	EMS[b]	7.8×10^{-5}

[a]Malling and Valcovic, 1978.
[b]Ehling, 1978.

fair extent the rest of the genes of this plant. This assumption cannot be proven, however. In the absence of better information in this or other systems of higher eukaryotes, we have no other choice.

When the estimated overall frequency of induced mutations is divided by the average frequency of mutations per gene, we obtain an estimate of the number of loci capable of mutation or of the expression of a mutant phenotype at a particular developmental stage. For example, the average frequency of mutations detectable at the embryo stage was found to be approximately 0.47 (Table 3), and the average rate of mutation per locus was estimated to be 3.2 x 10^{-5}. Hence, the minimal number of loci responding with mutations expressed at the stage of immature embryos is 0.47/3.2 x 10^{-5} = 14,687. Similarly, on the basis of mutations expressed at other developmental stages, we can calculate the number of genes expressed in mutant states at early, late, or all developmental stages (Table 6).

Table 6. Calculation of the Number of Loci in <u>Arabidopsis</u>

	Frequencies
Embryo mutations per genome	0.47
Early seedling mutations per genome	0.32
Late seedling mutations per genome	0.12
Frequency of all mutations observed	0.47 + 0.12 = 0.59

Average mutation frequency of 3 loci: 3.2 x 10^{-5}

Estimate of the number of loci with embryo mutations:

$$\frac{0.47}{3.2 \times 10^{-5}} = 14,687$$

Estimate of all the potentially mutable loci (based on Table 4 data):

$$\frac{0.65}{3.2 \times 10^{-5}} \times 20,313$$

Estimate of the number of all loci with mutations observed:

$$\frac{0.59}{3.2 \times 10^{-5}} = 18,438$$

Some objections may be raised to this procedure of calculating gene numbers. For example, these average mutation frequencies may include repeated mutations at some loci and none at others. The repeated occurrences do not seriously bias the data, however, because the probability of including twice a locus with a high, say 10^{-2}, frequency of mutation is only 10^{-4}. Such cases do not much affect the average frequencies, which are in the 10^{-1} range. Thus the expected error in connection with each case is in the 0.0001 range. Similarly, the very stable loci barely bias the figures because the use of approximation shown in Table 5 affords the proper correction.

Since this mutagen test assesses forward mutations, the actual number of sensitive sites substantially exceeds that of the number of genes. If we assume that an average locus consists of one thousand nucleotide pairs, the number of potentially mutable sites per plant may reach several millions.

Critical Population Size

The potencies of various mutagens are very unequal, yet we must evaluate their effectiveness in a reliable manner. Therefore some guidelines are needed as to the size of the populations to be tested in order to find some mutants even if their expected frequency is very low. Also, we must establish some criteria of clearance for an apparently innocuous compound. We need to know how many plants to screen to find at least one mutant or what number of mutations represents a significant increase over the spontaneous rate.

The critical size of a population can be defined as the number of plants to be tested, or better, the number of genomes, which may yield at least one mutation at a chosen level of probability (P). The rationale of the procedure is that we rule out the chance of finding no mutational events at all more frequently than specified by 1-P. Let us use a very simple example: When a plant is heterozygous for a single allelic pair, according to the Mendelian rule, we expect 3/4 of the progeny to have the dominant phenotype and 1/4 of the individuals to display the recessive genotype. How many individuals (n) do we then need in the M_2 generation in order to find at least one recessive plant with a probability (P) of 0.99? Since $P = 1-(3/4)^n$,

$$ n = \frac{\log (1-P)}{\log (3/4)} = \frac{-2}{-0.1249387} = 16.008 \simeq 17. $$

Similarly, if we expect an induced frequency of mutations of 1/200 genomes, the number of genomes to be tested (n) at P = 0.99 should be

$$n = \frac{\log (1-P)}{\log (199/200)} = \frac{-2}{-0.002177} = 918.7$$

to find at least one genome within the germline with a recessive mutation at the 0.99 level of probability. When we treat mature seeds, where the germline is represented by four genomes (as we discussed earlier), we need to test (918.7)/4 = 230 surviving plants.

The question remains whether an "induced" frequency of mutations is higher than the spontaneous rate. Certainly we must use a negative control. The frequencies between the two groups should be compared by determining the appropriate confidence limits. We must remember that even if no mutation is observed in 100 genomes, this may not negate the possibility that we missed up to four according to the 95% confidence limit. Similarly, we may want to know whether the 10% mutations observed among 200 genomes is significantly different from the 15% observed in 100 genomes. Reference to the 0.95 confidence belt convinces us readily that they are not. Were we to have 25% mutations in the latter group, the difference would be significant at the 0.95 level.

CORRELATION BETWEEN MUTAGENIC EFFECTS IN ARABIDOPSIS AND CARCINOGENICITY IN AMIMALS

The embryologist Theodore Boveri attributed cancer to the presence of an abnormal complex of chromatin as early as 1914. Evidence for and against a genetic cause of cancer has been entertained and occasionally negated ever since. In the widely used Ames Salmonella assay system, of more than 200 carcinogens over 90% showed mutagenic effects (Hollstein et al. 1979). A survey (Rinkus and Legator, 1979) of 465 known or suspected carcinogens examined by the Salmonella S-9 method indicated a lower correlation (77%) with mutagenicity. It seems that not all groups of carcinogens are equally efficient mutagens for all organisms.

With Arabidopsis so far, approximately 110 chemical compounds had been tested for mutagenicity, according to a survey of about four dozen publications. Interestingly, this compares favorably with Drosophila, for which about 1000 publications are listed by the Environmental Mutagen Information Center as being concerned with tests of about the same number of compounds (Hollstein et al., 1979).

We have found information concerning carcinogenicity and neoplastic effects for 52 compounds tested for mutagenicity in Arabidopsis (Rédei et al., in press). The correlation between carcinogenicity in animals and mutagenicity in Arabidopsis seems comparable to or better than that for Salmonella (Table 7). It is particularly noteworthy that only one compound, di-2-chloroethylamine phosphamide ester (endoxan, cytoxan), was negative in the mutagenicity assay among those that are listed as category I carcinogens for animals (proven in at least two animal systems) according to the Occupational Safety and Health Administration (OSHA). For this particular carcinogen, only a short negative note is available concerning mutagenicity in Arabidopsis (Müller, 1965a), which may require revision on repetition. This compound (synonym cyclophosphamide), when administered to male mice, showed dominant lethal effects for spermatozoa, but no dominant lethals were induced when spermatocytes or spermatogonia were tested (Röhrborn, 1970). According to Heddle and Bruce (1977), this compound is carcinogenic and shows positive responses in the mouse sperm abnormality, bone marrow micronucleus, and Salmonella tests. Triethylenemelamine is another false negative in Arabidopsis; it was found to be mutagenic in several other assays, including the sex-linked recessive lethal test in Drosophila, in the mouse specific locus test, and in a Salmonella assay (Hollstein et al., 1979). Again, the negative result in Arabidopsis is found in the same undetailed note (Müller, 1965b).

Table 7. Number of Compounds Tested for Mutagenicity in Arabidopsis and for Carcinogenicity in Animal Assays

	Carcinogens	Neoplastic
Mutagenic in Arabidopsis	40 (87%)	4 (66.6%)
Nonmutagenic in Arabidopsis	6	2
Total	46	6

OSHA Categories of Compounds That Are Nonmutagenic in Arabidopsis

Di-2-chloroethylamine phosphamide ester	I
Ethyl alcohol	II
Chloramphenicol	II
Maleic acid hydrazide	III
Sulfathiazole	III
Triethylenemelamine	?

The remaining four nonmutagenic suspected carcinogens are only category II and III compounds, indicating that the evidence for carcinogenicity is fragmentary or insufficient. I was unable to find positive mutagenic information for them in the commonly used mutagenic assay systems.

CONCLUSIONS

The Arabidopsis assay as outlined provides direct evidence concerning the genetic nature of most of the mutagenic alterations. The majority of the mutants detected can be subjected to formal genetic analysis. This system permits the simultaneous study of forward mutation at thousands of loci; therefore, the information obtained must be relevant for practically all the genes of this organism and presumably for other eukaryotes. Because of the very small size of the chromosomes of Arabidopsis, cytological proof for chromosomal aberrations is hard to obtain (except for some very gross ones). The sterility of the fruits can be, however, easily determined by counting the missing embryos in the linear array. A considerable number of these defects, perhaps most, are presumably due to large deficiencies and two- or multiple-hit aberrations of the chromosomes.

The mutation frequencies in Arabidopsis can easily be expressed on the genome basis, and the test results can be compared with those of any prokaryote or eukaryote. Mutagenicity information in Arabidopsis correlates very well with the carcinogenicty data for animal test systems. There is direct evidence (Rédei et al., 1980) that the metabolic system of of Arabidopsis can activate several promutagens into genetically effective compounds.

A mutagen assay with Arabidopsis can be completed within four to five weeks. Because of the small size of the plants, up to 200 or more individuals can be raised per 10-cm-diameter petri plate in inexpensive growth chambers. Therefore, the assays are not only fast and revealing, but they are also very inexpensive. The plants can be grown year round, both in the laboratory and in the environment (this species is winter-hardy even in the North). Though sufficient information is lacking on its utility for in situ testing of environmental pollutants, there is no apparent reason why it could not be employed successfully for this purpose too.

ACKNOWLEDGMENTS

This study was supported by EPA contract D4541 NAET, and it is contribution no. 8497 from the Missouri Agricultural Experiment Station. I appreciate the valuable comments and assistance of

Dr. Shahbeg Sandhu and Dr. William Lower during the course of the investigations. Some of the data were collected with the technical assistance of Patti Reichert and Magdi Redei; their conscientious efforts and enthusiasm are gratefully acknowledged.

REFERENCES

Belling, J. 1928. The ultimate chromomeres of Lilium and Aloe with regard to the number of genes. Univ. California Publ. Bot. 14:307–318.

Ehling, U.H. 1978. Specific-locus mutations in mice. In: Chemical Mutagens: Principles and Methods for Their Detection, Vol. 5. A. Hollaender and F.J. de Serres, eds. Plenum Press: New York. pp. 233–256.

Fiers, W. 1975. Chemical structure and biological activity of bacteriophage MS2 RNA. In: RNA Phages. N.D. Zinder, ed. Cold Spring Harbor Laboratory Press: Cold Spring Harbor, NY. pp. 353–396.

Garcia-Bellido, A., and P. Ripoll. 1978. The number of genes in Drosophila melanogaster. Nature 273:399–400.

Heddle, J.A., and W.R. Bruce. 1977. Comparison of tests for mutagenicity or carcinogenicity using assays for sperm abnormalities, formation of micronuclei, and mutations in Salmonella. In: Origins of Human Cancer, Book C, Human Risk Assessment. H.H. Hiatt, J.D. Watson, and J.A. Winsten, eds. Cold Spring Harbor Laboratory Press: Cold Spring Harbor, NY. pp. 1549–1557.

Hollstein, M., J. McCann, F.A. Angelosanto, and W.W. Nichols. 1979. Short-term tests for carcinogens and mutagens. Mutation Res. 65:133–226.

Malling, H.V., and L.R. Valcovic. 1978. New approaches to detecting gene mutations in mammals. In: Mutagenesis, Advances in Toxicology, Vol. 5. W.G. Flamm and M. Mehlman, eds. Hemisphere: Washington, DC. pp. 149–171.

Müller, A.J. 1963. Embryonentest zum Nachweis rezessiver Letalfaktoren bei Arabidopsis thaliana. Biol. Zbl. 83:133–163.

Müller, A.J. 1965a. A survey on agents tested with regard to their ability to induce recessive lethals in Arabidopsis. Arabidopsis Inf. Serv. 2:22–24.

Muller, A.J. 1965b. In: Induction of Mutations and the Mutation Process. J. Veleminsky and T. Gichner, eds. Publishing House of the Czech. Acad. Sci.: Praha. pp. 46–52.

Prakash, L., and F. Sherman. 1973. Mutagenic specificity: reversion of iso-1-cytochrome c mutants of yeast. J. Molec. Biol. 79:65–82.

Rédei, G.P. 1980. Basic Plant Genetics. University of Missouri: Columbia, MO. p. 567.

Rédei, G.P., M.M. Redei, W.R. Lower, and S.S. Sandhu. 1980. Idendification of carcinogens by mutagenicity for Arabidopsis. Mutation Res. 74:469–475.

Rinkus, S.J., and M.S. Legator. 1979. Chemical characterization of 465 known suspected carcinogens and their correlation with mutagenic activity in the Salmonella typhimurium system. Cancer Res. 39:3289–3318.

Röhrborn, G. 1970. The activity of alkylating agents. I. Sensitive mutable stages in spermatogenesis and cogenesis. In: Chemical Mutagenesis in Mammals and Man. P. Vogel and G. Rohrborn, eds. Springer-Verlag: Berlin. pp. 294–316.

Sanger, F., G.M. Air, B.G. Barrell, N.L. Brown, A.R. Coulson, J.C. Fiddes, C.A. Hutchison III, P.M. Slocombe, and M. Smith. 1977. Nucleotide sequence of bacteriophage φX174 DNA. Nature 265:687–695.

SOYBEAN SYSTEM FOR TESTING THE GENETIC EFFECTS OF INDUSTRIAL EMISSIONS AND LIQUID EFFLUENTS

Baldev K. Vig
Department of Biology
University of Nevada
Reno, Nevada

INTRODUCTION

The testing of complex mixtures found in today's air, water, and soils for harmful effects on man is a gigantic and important task. This paper introduces a eukaryotic test system that yields at least qualitative estimates of genetic damage of the type observable in man. The soybean spot test is based on inducing various types of genetic damage by treating seeds or seedlings with the chemical or mixture of chemicals in question. This review briefly describes the nature of the test system, the kinds of data obtained for various chemicals, their postulated genetic effects, and the potential usefulness of this material for environmental mutagen testing. This summary is only a guideline to understanding the system; the details of previous work can be found in references (for reviews, see Nilan and Vig, 1976; Vig, 1975, 1978).

THE TEST SYSTEM

The Origin of Spots

In 1956, Weber and Weiss described a light-green colored plant of soybean (<u>Glycine</u> <u>max</u> [L.] Merrill) whose progency segregated in a ratio of 1 dark green: 2 light green: 1 golden yellow. The yellow color is due to lack of chlorophyll. The alleles controlling this trait are symbolized by Y_{11} and y_{11}: $Y_{11}Y_{11}$ plants are dark green, $Y_{11}y_{11}$ plants are light green, and $y_{11}y_{11}$ plants are golden yellow. In heterozygotes, the embryonic leaves (which develop into two simple, opposite leaves) and the first compound leaf occasionally show a few dark green spots and about an equal number of yellow spots. Although these spots resemble the leaves of the two homozygotes in color, the intensity of their colors depends on the number of layers of palisade cells involved in the mutational event. An altered embryonic leaf cell usually develops its colony of cells during division and expansion without being outcompeted by the surrounding tissue. The yellow spots survive because the light green background tissue provides the necessary carbohydrates for their growth.

Besides these single spots, the $Y_{11}y_{11}$ leaves also produce twin spots composed of adjacent equal-sized, mirror-image spots, one dark green and one yellow. The cells of such twin spots appear to have complementary genotypes, i.e., $Y_{11}Y_{11}$ and $y_{11}y_{11}$.

The origin of twin spots is best attributed to somatic crossing over in $Y_{11}y_{11}$ cells. The failure of one of these cells to develop into a visible sector results in a single dark green or yellow spot. The frequency of occurrence of these twin spots coincides with the frequency of somatic crossing over in several other organisms (see Vig, 1978) and can be dramatically increased by applying substances known to cause mitotic recombination, such as mitomycin C (German and LaRock, 1969; Holiday, 1954; Vig and Paddock, 1968).

The single yellow spots may also develop by the multiplication of a cell (or cells) that has lost the chromosome segment carrying the Y_{11} allele. Duplication of Y_{11} allele or deletion of the y_{11} allele, followed by cell multiplication, may give single dark green spots (Vig, 1969b). These situations may result from nonhomologous translocations involving the chromosome carrying this gene. Consequently, $Y_{11}y_{11}$ leaves treated with a given mutagen or complex mixtures of mutagens may be analyzed to distinguish between the modes of action of various mutagens. Furthermore, specific locus mutations may be induced (y_{11} to Y_{11}, in $y_{11}y_{11}$ cells) to give light green spots on a yellow background.

The Protocol

A 10- to 15-g sample of seed of variety T219, L65-1237, or L72-1937 is treated for the desired length of time with a solution of the chemical or mixture of chemicals to be tested. The seed is then thoroughly washed in running tap water, sown 6 mm (0.25 in.) deep in washed, coarse sand of no nutritive value in galvanized metal flats, and watered as required, depending on the temperature and humidity of the greenhouse. The protocol can be altered so that the seed is watered with a solution of the suspected mutagen. Agents like mitomycin C, caffeine, or nitrosourea may be used as positive controls.

The plants are ready for analysis in four to five weeks, when the second compound leaf unfolds. For both heterozygous and homogyzous plants, spots are counted on the two simple leaves and the first compound leaf, which is considered equivalent in area to three simple leaves. Usually, the simple leaves have more spots, due to a larger initial number of cells than in the compound leaf. Because the spontaneous background frequency of the control varies with the age of the seed, a statistical procedure such as the t-test is used to confirm the induction of damage by the mutagen. The doses and number of replicates required are determined by the chemical's mutagenic effectiveness.

STUDIES WITH SOME TEST AGENTS

The Induction of Twin Spots

Chemicals whose primary genetic effect is to cause somatic recombination produce a preponderance of twin spots on the heterozygous leaves. Two such chemicals, caffeine and mitomycin C, deserve special mention. Four-hour treatments with caffeine in concentrations ranging from 0.0625% to 0.5% increase the frequencies of all types of spots, especially doubles (Vig, 1973b). In one experiment, the ratio of total spots to twin spots was 11:1 in the control; for seeds treated for 4 h with 0.0625%, 0.125%, 0.25%, and 0.5% caffeine, the ratios were 3.6, 4.0, 3.4, and 2.7, respectively. Several other experiments have confirmed that about one third of the spots induced by caffeine are twin spots.

When seeds are treated with mitomycin C at a wide range of concentrations and at various stages of seed germination, usually one third or more of the spots are twins. The same frequency of twin spots occurs following treatment of seeds with mitomycin C solution and following application of the chemical in lanolin paste to the growing tip of the seedling, including the unexpanded third through fifth compound leaves (Vig and Paddock, 1968). In these studies, concentrations of mitomycin C were as low as 0.00325% (for 24 h) for seed treatments and 0.005% in lanolin paste. Twin spots are also preferentially induced by applying the chemical to the seed at various physiological ages for 4-h periods (i.e., from 0 to 4, 4 to 8, 8 to 12, 12 to 16,..., 32 to 36 hours after germination [Vig, 1973a]).

The alkylating agents diepoxybutane, trenimon, and methylmethane sulfonate affect the frequency of spots similarly. These three agents are potent inducers of spots of all types, about one third of which are twin spots. In the case of diepoxybutane, concentrations as low as 0.5 ppm applied to seed for 24 h are effective, and concentrations of trenimon as low as 0.25 ppm applied under similar conditions increase the total spot frequency to about twice that of the control (Vig and Zimmermann, 1977). Methylmethane sulfonate induces similar types of spots when applied at various phases of germination (Vig et al., 1976) and at concentrations as low as 6 ppm. At high concentrations, however, these alkylating agents affect leaf expansion drastically, resulting in a reduced spot frequency calculated on per leaf basis.

Preponderant Production of Yellow Spots

While ethyl methanesulfonate slightly increases the frequency of twin spots over that of the control, it is far more effective in producing yellow spots than the other two types of spots. In one study (Vig et al., 1976), treatment with a 0.25% solution of this chemical produced about 50% more single yellow spots than single dark green spots and about 2.5 times as many single yellow spots as twin spots. Although ethylmethane sulfonate and methylmethane sulfonate are closely related chemicals, they produce different frequencies of the various spot types at any given concentration (Vig et al., 1976).

Treatment of the seed with cobalt-60-emitted gamma rays gives results similar to those for ethylmethane sulfonate, although the relative frequency of yellow spots is even higher. This has been found over a range of 10 to 750 R for both dry and pre-soaked seed (Vig, 1974); the total frequency of spots is generally higher in leaves from pre-soaked seed. Exposure of seed to beta particles from tritiated water results in equal frequencies of the three types of spots (Vig, 1974; Vig and McFarlane, 1975). Even tritium concentrations as low as 0.01 µC/ml for 96 h (equivalent to 4.5 R) are effective in producing spots. These differences could be due to internal availability of beta particles to the DNA from within the organic fraction of the embryo (Vig and McFarlane, 1975).

A Lack of Production of Twin Spots

Treatment of seeds with sodium azide (NaN_3) greatly increases the frequencies of single dark green and yellow spots (Vig and McFarlane, 1975). However, the frequency of twin spots is barely above that of the control (Vig, 1973c), and the chemical generally does not produce light green spots on yellow leaves. Thus, these spots are probably not due to somatic crossing over, point mutations, or deletions of chromosome segments. We tentatively conclude that NaN_3 causes nondisjunction, giving $Y_{11}Y_{11}y_{11}$ sectors that are dark green and $Y_{11}y_{11}y_{11}$ sectors that are nearly yellow. The monosomic cell lines are presumably inviable or outcompeted by the normal and trisomic lines, and are thus lost. We have not found similar results for any other mutagen that we have tested.

The Induction of Point Mutations from y_{11} to Y_{11}

Light green spots on yellow ($y_{11}y_{11}$) leaves are produced by specific locus mutations, y_{11} to Y_{11}. Not all chemicals that induce spots in the heterozygotes also produce light green spots in yellow homozygotes. Caffeine produces such spots at concentrations of 0.05% or higher (Vig, 1973b). This effect distinguishes caffeine from chemicals like mitomycin C or nitrosoamines, which induce both twin spots and single spots on $Y_{11}y_{11}$ leaves, but have no effect on $y_{11}y_{11}$ leaves.

The alkylating agents methylmethane sulfonate, ethylmethane sulfonate, methylethane sulfonate, and methylbutane sulfonate cause mutation of y_{11} to Y_{11} at concentrations as low as 0.02% when applied to seed for 20 h (Vig et al., 1976). Diepoxybutane and trenimon also induce light green spots on yellow leaves. For these chemicals, a concentration of 0.25 ppm applied to seed for 24 h is mutagenic (Vig and Zimmermann, 1977); thus, these chemicals induce not only somatic recombination and chromosome deletions (as indicated by the induction of spots on $Y_{11}y_{11}$ leaves), but also point mutations of the allele y_{11} to Y_{11}.

No other chemical has been found to induce the mutation of y_{11} to Y_{11}. However, as in other such test systems, beta particles and gamma rays induce this point mutation at about the same frequency as they do spots on $Y_{11}y_{11}$ leaves (Vig, 1974, 1978; Vig and McFarlane, 1975).

MUTAGENS REQUIRING METABOLIC ACTIVATION

In recent years, increasing attention has been focussed on the mutagenic action of chemicals requiring metabolic activation. The S-9 fraction of rat liver homogenate is commonly used to activate promutagens. Recent studies indicate that liver is not the only system with the enzymatic machinery needed for such activation. In 1968, Veleminsky and Gichner demonstrated the mutagenic activity of some promutagens in plant systems without mammalian metabolic activation (see Arenaz and Vig, 1978; Klekowski and Levin, 1979).

We have treated soybean seeds with aqueous solutions of dimethylnitrosoamine at concentrations as low as 1.25 ppm for 24 h (Arenaz and Vig, 1978). At this dose, we found a 2.8-fold increase in the frequency of twin spots, a 2.6-fold increase in dark green spots, and 1.7-fold increase in yellow spots, demonstrating that this plant system can activate this chemical. Treatments with concentrations from 60 to 500 ppm appeared to cause maximal conversion of the chemical into true mutagen. Methylnitrosourea, a much more toxic, in that interference with leaf expansion (yielding an artificially low spot frequency) occurs at 125 ppm. In comparison, dimethylnitrosoamine is tolerated by the plant at doses as high as 500 ppm (Arenaz and Vig, 1974).

Despite data demonstrating metabolic activation in the plant, it remains to be shown that metabolites produced through the mediation of plant extracts (e.g., from nitrosoamines, atrazine, and other such chemicals) can cause mutations or chromosome changes in any mammalian system.

CONCLUSIONS

The soybean spot test is well suited for assaying both pure chemicals and complex mixtures. The results for several agents tested in this system are summarized in Tables 1 and 2. The system allows not only quick and inexpensive assessment of the genetic damage in a eukaryotic system, but also discrimination among different genetic end points and thus among modes of action of the agents. Thus, a chemical like methylmethane sulfonate, which can cause DNA crosslinks, induces twin spots and single spots on $Y_{11}y_{11}$ leaves and light green spots on $y_{11}y_{11}$ leaves, unlike NaN_3, which produces only single spots on $Y_{11}y_{11}$ leaves and apparently has no other effect. This discriminatory ability of the system is advantageous in preliminary screening.

The soybean test system is suitable for use with complex mixtures. Klekowsky and Levin (1979) recently showed that effluent from paper mills induces spots on $Y_{11}y_{11}$ leaves. The system should be adaptable to testing of liquid, solid, or gaseous effluents, as long as the seed or the seedlings (as in Vig and Paddock, 1968) can be treated at appropriate stages of development.

REFERENCES

Arenaz, P. 1977. Ineffectiveness of hycanthone methanesulfonate in inducing somatic crossing over and mutations in Glycine max. Mutation Res. 48:187-190.

Arenaz, P. 1978. Inability of aminoazotoluene to induce somatic crossing over and mutation in Glycine max (L.) Merrill. Mutation Res. 50:295-598.

Arenaz, P., and B.K. Vig. 1976. Induction of somatic mosaicism in the soybean by some carcinogens. Genetics 83:93.

Table 1. Summary of Qualitative Genetic Effects of Agents Tested for Induction of Spots on the Leaves of the Soybean (Glycine max)

Agent	Y11Y11 Plants		Y11y11 Plants	Postulated Primary Mechanism	Reference
	Twin Spots	Single Spots	Single Spots		
Mitomycin C	+	+	−	a,b	Vig and Paddock, 1968
Caffeine	+	+	+	a,b,c	Vig, 1973b
Actinomycin	+	+	+	a,b	Vig, 1973b
Daunomycin	−	±	−	b	Vig and Paddock, 1968
5-Fluorodeoxy uridine	−	+	−	b	Vig, 1973b
N-methyl, N-nitro, N-nitrosoguanidine	+	+	−	b	Arenaz and Vig, 1976
Ethylmethane sulfonate	±	+	+	b,c	Vig et al., 1976
Methylethane sulfonate	(not tested)	+	c		Vig et al., 1976
Methylmethane sulfonate	(not tested)	+	a,b,c		Vig et al., 1976
Methylbutane sulfonate	+	+	−	c	Vig et al., 1976
Colchicine	+	+	−	a	Ashley, 1978; Vig, 1969b
Puromycin	−	±	−	b	Vig, 1973b
Sodium azide	±	+	−	d	Vig, 1973c
N-methyl-N-nitrosurea	+	++	−	b	Arenaz and Vig, 1978
Dimethyl nitrosamine	+	++	−	b	Arenaz and Vig, 1978
Trenimone	+	+	+	a,c	Vig and Zimmermann, 1977
Diepoxybutane	+	+	+	a,c	Vig and Zimmermann, 1977
Carofur (1-[5-nitro-2-furyl]-2-[6-amino-3-pyridazyl]-ethanehydrochloride)	+	+	−	a	Vig and Zimmermann, 1977

[a] Somatic crossing over usually with accompanying chromosome breakage.
[b] Chromosome breakage as primary cause.
[c] Point mutation.
[d] Nondisjunction (?).

Table 2.   Agents That Did Not Induce Spots on the Leaves of the
Soybean (Glycine max)

Agent	Reference
Aminoazo toluene	Arenaz, 1978
Nucleosides, d-A	Vig, 1972
d-C	
d-G	
d-T	
Aluminum potassium nitrate	Vig and Mandeville, 1972
Copper sulfate	Vig and Mandeville, 1972
Ferrous sulfate	Vig and Mandeville, 1972
Hycanthone	Arenaz, 1977
Ammonium thiocyanide	Vig, 1975
1-nitroso-2-naphthyl-2, 6-disulfonic acid	Vig, 1975
Cytosine arabinoside	Vig, 1975
2-Chloroethanol	Vig, 1975
Hydroxylamine hydrochloride	Vig, 1975
Urea	Vig, 1975
N, N-dinitrosopiperazine	Vig, 1975
N-nitroso-methyl urea	Vig, 1975

Arenaz, P., and B.K. Vig. 1978. Somatic crossing over in Glycine
    max (L.) Merrill: activation of dimethyl nitrosoamine by
    plant seed and comparison with methyl nitrosourea in inducing
    somatic mosaicism. Mutation Res. 52:367-380.

Ashley, T. 1978. Effect of colchicine on somatic crossing over
    induced by mitomycin C in soybean (Glycine max). Genetica
    49:87-96.

German, J., and J. LaRock. 1969. Chromosomal effects of
    mitomycin, a potential recombinogen in mammalian cell
    genetics. Texas Rep. Biol. Med. 27:409-418.

Holiday, R. 1954. The induction of mitotic recombination by
    mitomycin C in Ustilago and Saccharomyces. Genetics
    50:323-335.

Klekowski, E., and D.E. Levin. 1979. Mutagens in a river heavily polluted with paper recycling waste: results of field and lab mutagen assays. Environ. Mutagen. 1:209-219.

Nilan, R.A., and B.K. Vig. 1976. Plant test systems for detection of chemical mutagens. In: Chemical Mutagens, Their Principles and Methods of Detection, Vol. 4. A. Hollaender, ed. Plenum Press: NY. pp. 143-170.

Vig, B.K. 1969a. Increase induced by colchicine in the incidence of somatic crossing over in Glycine max. Theoret. Appl. Genet. 41:145-149.

Vig, B.K. 1969b. Relationship between mitotic events and leaf spotting in Glycine max. Canadian J. Cytol. 11:147-152.

Vig, B.K. 1972. Suppression of somatic crossing over in Glycine max (L.) Merrill by deoxyribose cytidine. Molec. Gen. Genet. 116:158-165.

Vig, B.K. 1973a. Mitomycin C induced mosaicism in Glycine max (L.) Merrill in relation to postgermination age of the seed. Theoret. Appl. Genet. 43:27-30.

Vig, B.K. 1973b. Somatic crossing over in Glycine max (L.) Merrill: effect of some inhibitors of DNA synthesis on the induction of somatic crossing over and point mutations. Genetics 73:583-596.

Vig, B.K. 1973c. Somatic crossing over in Glycine max (L.) Merrill: mutagenicity of sodium azide and lack of synergistic effect with caffeine and mitomycin C. Genetics 75:265-277.

Vig, B.K. 1974. Somatic crossing over in Glycine max (L.) Merrill: differential response to ^{3}H-emitted β-particles and ^{60}Co-emitted γ-rays. Radiat. Bot. 14:127-137.

Vig, B.K. 1975. Soybean (Glycine max): a new test system for study of genetic parameters as affected by environmental mutagens. Mutation Res. 31:49-56.

Vig, B.K. 1978. Somatic mosaicism in plants with special reference to somatic crossing over. Environ. Hlth. Perspect. 27:27-36.

Vig, B.K., and W.F. Mandeville. 1972. Ineffectivity of metallic salts in induction of somatic crossing over and mutation in Glycine max (L.) Merrill. Mutation Res. 16:151-155.

Vig, B.K., and J.C. McFarlane. 1975. Somatic crossing over in
Glycine max (L.) Merrill: Sensitivity to and saturation of
the system at low levels of tritium emitted β-radiation.
Theoret. Appl. Genetics 46:331-337.

Vig, B.K., R.A. Nilan, and P. Arenaz. 1976. Somatic crossing
over in Glycine max (L.) Merrill: induction of somatic
crossing over and specific locus mutations by methyl
methanesulfonate. Environ. Exp. Bot. 16:223-234.

Vig, B.K., and E.F. Paddock. 1968. Alteration by mitomycin C of
spot frequencies in soybean leaves. J. Hered. 59:225-229.

Vig, B.K., and E.F. Paddock. 1970. Studies on the expression of
somatic crossing over in Glycine max L. Theoret. Appl.
Genetics 40:316-321.

Vig, B.K., and F.K. Zimmermann. 1977. Somatic crossing over in
Glycine max. An induction of the phenomena by carofur,
diepoxyoutane and trenimon. Environ. Exp. Bot. 17:113-120.

MUTAGENICITY OF NITROGEN COMPOUNDS FROM SYNTHETIC CRUDE OILS: COLLECTION, SEPARATION, AND BIOLOGICAL TESTING

T.K. Rao, J.L. Epler, M.R. Guerin, B.R. Clark, and
C.-h. Ho
Biology and Analytical Chemistry Divisions
Oak Ridge National Laboratory
Oak Ridge, Tennessee

INTRODUCTION

Short-term mutagenesis assays have been used to test complex
environmental mixtures, in order to 1) serve as predictors of
long-range health effects, 2) guide chemical separation procedures
for the isolation and concentration of biologically active
materials, 3) identify chemical agents responsible for biological
activity, and 4) determine priorities for further, extensive
testing. Organic extraction coupled with chemical-class
fractionation is a prerequisite for most of these assays.

Our emphasis has been on evaluating various test materials
from the newly emerging synfuel technologies. We have previously
used the class chemical separation procedure developed by Swain et
al. (1969) to fractionate certain coal-derived (Epler et al., 1978)
and shale-derived (Epler et al., 1979b) oils for mutagenicity
testing. To minimize chemical reactivity during the fractionation
procedure, several synfuels were separated with Sephadex LH-20 gel
chromatography (Rao et al., in press). The bacterial mutagenicity
assay developed by Ames (Ames et al., 1975) employs certain
well-characterized histidine-auxotrophic mutants of Salmonella
typhimurium. Mutagenicity results indicate that the alkaline
fractions containing azaarenes, primary amines, and nitro
polyaromatics are mutagenic, along with the neutral-fraction-
containing polycyclic aromatic hydrocarbons (PAH) and nitrogen-
containing PAH (Ho et al., in press). The biological activity of
these organic compounds is characterized by the ability to revert
the frameshift strains TA1538 and TA98 (Rao et al., 1978) and
dependence on specific activation systems.

Source of Samples

Samples used in this study were supplied by U.S. Environmental Protection Agency/U.S. Department of Energy Synfuel Research Materials Facility (Coffin et al., 1979). Process operating conditions at the time of sampling, sampling conditions, and sample histories are not sufficiently defined to allow process-specific conclusions. Samples (with repository numbers) and their sources are listed in Table 1. A detailed process description has been presented elsewhere (Guerin et al., in press).

Table 1.   Synfuel Samples and Their Sources

Sample[a]	Source
Petroleum	
Wilmington crude oil (5301)	Bartlesville Energy Technology Center
Recluse crude oil (5305)	Bartlesville Energy Technology Center
Shale-derived oils	
Shale oil (in situ) (4101)	Laramie Energy Technology Center
Paraho shale oil (4601)	US Navy/Standard Oil Co. of Ohio
HDT-Paraho shale oil (4602)	US Navy/Standard Oil Co. of Ohio
Coal-derived oils	
SRC II fuel oil (1701)	Pittsburgh and Midway Mining Co.
H-coal dist.--raw (1601)	Mobil Research/EPRI
H-coal dist.--HDT low severity (1602)	Mobil Research/EPRI
H-coal dist.--HDT medium severity (1603)	Mobil Research/EPRI
H-coal dist.--HDT high severity (1604)	Mobil Research/EPRI
$ZnCl_2$ dist. (1801)	Conoco Coal Development Co.

[a]Respository numbers in parentheses.

Purpose

The objective of this study was to identify mutagenic activity in fractionated synfuel samples and to isolate and identify the mutagenic chemical agents.

MATERIALS AND METHODS

Chemical Fractionation

The chemical fractionation procedure (Figure 1) has been
described by Guerin et al. (in press). Acidic and basic fractions
were separated by liquid/liquid partitioning into ether-soluble
and -insoluble acids and bases (water-soluble fractions were
generally inactive in the mutagenicity assays). The neutral
fraction was separated using Sephadex LH-20 into aliphatic,
aromatic, polyaromatic, and polar fractions by isopropanol elution.
Solvents were removed by rotary evaporation; the residue was
dissolved in dimethylsulfoxide for mutagenicity assays. The basic
fraction was subfractionated by use of basic alumina and Sephadex
LH-20. The ether-soluble base fraction was loaded onto the column
and eluted with 500 ml benzene (benzene subfraction) followed by
700 ml ethanol. Ethanol was removed by rotary evaporation, and the
residue was separated further on a Sephadex LH-20 gel column. The
column was eluted sequentially with 250 ml of isopropanol
(isopropanol subfraction) and 600 ml of acetone (acetone
subfraction), and the eluting solvent was removed by rotary
evaporation.

Mutagenicity Assay

Histidine-auxotrophic strain TA98 of S. typhimurium, obtained
through the courtesy of Dr. B.N. Ames (University of California at
Berkeley), was used for these studies. The procedure (based on the
work of Ames et al., 1975) was to overlay minimal-medium agar
plates with soft agar containing the fraction being tested,
bacterial cells (2×10^8), and liver homogenate (S-9 mix) from
Aroclor-1254-induced rats (for metabolic activation). Activity
in revertants per milligram of test substance was derived from the
slope of the induction curve. Total mutagenic activity of a
starting material was computed from the activities of its
subfractions corrected for the percentages by weight contributed
by the subfractions to the starting material.

RESULTS AND DISCUSSION

The crude oils could not be tested for mutagenicity because of
their toxicity; when tested, they yielded questionable results. To
overcome this problem and also to obtain a more homogeneous
distribution of sample in the test medium, the samples were
fractionated by the general procedure described above (see Figure
1). Recovery and reproducibility have been tested by using
triplicate samples of shale oil and crude oil. Chemical recovery

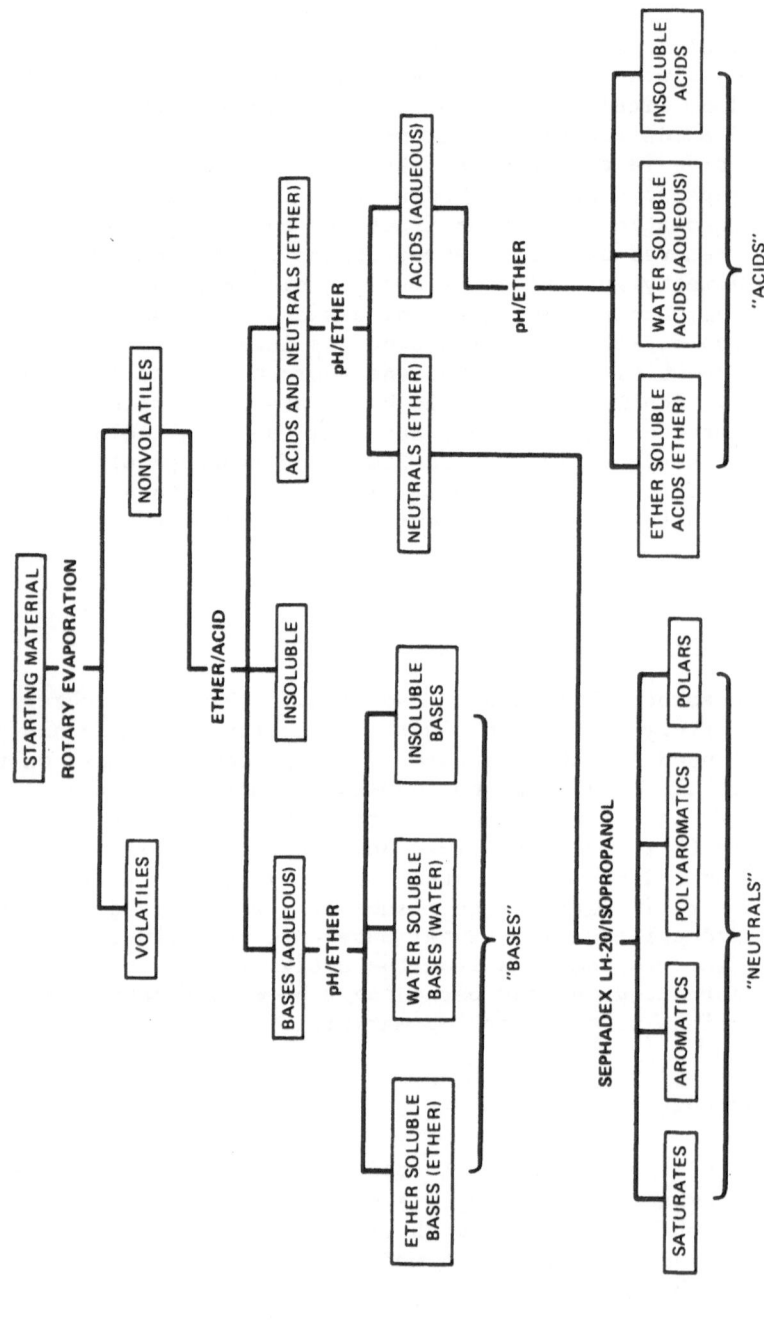

Figure 1. Chemical fractionation procedure (from Guerin et al., in press). Materials are shown in boxes, their phases are indicated in parentheses, and the steps of the procedure are given in boldface type.

and reproducibility are generally adequate for biological testing
purposes, even though recoveries are unacceptably low with certain
samples. The losses are caused by volatile matter, which is
difficult to use in bioassays.

Table 2 gives the distribution of mutagenic activity in
petroleum oils, shale oils, and coal-derived oils. The acidic
fractions were inactive in the mutagenicity assays. All of the
activity was found in the basic and neutral fractions; the activity
levels varied from sample to sample. The petroleum samples showed
activity in the neutral fraction, which was weakly mutagenic. In
the shale oil samples, high in nitrogen, both the neutral and basic
fractions were mutagenic. Significant mutagenic activity was also
observed in both the neutral and basic fractions of the coal-
derived oils. The shale oils and coal-derived oils were more
mutagenic than were the petroleum samples.

Hydrotreatment (HDT) seemed to reduce mutagenic activity, as
seen in the results obtained with HDT-hydrogenated coal (H-Coal)
distillates. The high severity hydrotreatment completely
eliminated mutagenic activity from the H-Coal sample, while the
medium and low severity treatments were less effective. Zinc
chloride- ($ZnCl_2$) catalyzed distillation apparently eliminated
alkaline mutagens from the coal-derived oil.

Mutagenic activity in the basic and neutral fractions led us
to examine these fractions to isolate and identify chemical agents
responsible for the biological activity. When the basic fraction
was subfractionated, biological activity was found in the ultimate
acetone fraction that comprised approximately 10% of the starting
ether-soluble basic (ESB) fraction. Results for subfractionated
SRC-II ESB and shale oil ESB are given in Table 3. The acetone
fractions with high specific activities can be used for biological
assays in other genetic systems as well as for chemical analysis
(Epler et al., 1979a). Azaarenes and primary aromatic amines are
the organic constituents of this fraction suspected of causing the
mutagenic activity.

Chemical separation and analysis of the neutral fraction
suggested PAH, alkylated PAH (Griest et al., 1979), and nitrogen-
PAH (Ho et al., in press) as the possible mutagenic agents.
Mutagenic activity of PAH and alkylated PAH appeared to increase
with the ring size (see Figure 2). The aliphatic constituents of
coal-derived materials were not mutagenic, while the nitrogen-PAH
fraction obtained from coal oil (Ho et al., in press) had a
specific activity more than twice that of the PAH fraction (Table
4). The nitrogen-PAH fraction from a shale-derived oil was not
mutagenic.

Table 2. Distribution of Mutagenic Activity in Synfuels[a]

Sample	Total Mutagenic Activity (revertants/mg)[b]	Distribution of Activity (%)			
		Neutral	Acids	Bases	Other
Petroleum					
Wilmington crude oil	5	100	0	0	0
Recluse crude oil	6	100	0	0	0
Shale-derived oils					
Shale oil (in situ)	178	54	2	42	2
Paraho shale oil	390	31	0	69	0
HDT-Paraho shale oil	0	0	0	0	0
Coal-derived oils					
SRC-II fuel oil	1000	65	0	35	0
H-coal dist.--raw	350	63	0	37	0
H-coal dist.--HDT low severity	540	100	0	0	0
H-coal dist.--HDT medium severity	210	-	-	-	-
H-coal dist.--HDT high severity	0	0	0	0	0
ZnCl2 dist.	530	100	0	0	0

[a]From Guerin et al. (1980, in press).
[b]Determined from the linear portion of a dose-response curve with strain TA98.

Table 3.   Distribution of Mutagenic Activity in the
Ether-soluble Basic Fraction (ESB)

Test Substance	Relative Weight (%)	Specific Activity (rev/mg)	Weighted Activity (rev/mg)
SRC-II ESB	–	0	14,000
Benzene	74	0	0
Isopropanol	6	400	24
Acetone	15	68,000	10,200
Total	95		10,224
Shale oil ESB	–	–	2,500
Benzene	77	0	0
Isopropanol	12	0	0
Acetone	9	20,000	1,800
Total	98		1,800

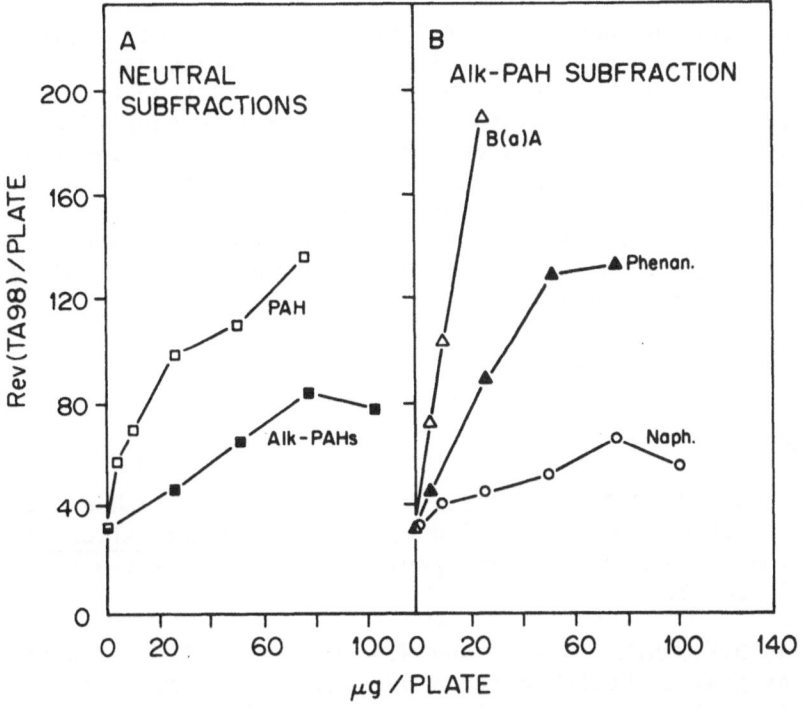

Figure 2.   Mutagenicity of the polycyclic aromatic hydrocarbon
(PAH) and alkylated (PAH) subfractions, including
benz(a)anthracene, phenanthrene, and naphthalene.

Table 4.   Mutagenic Activities of Neutral Subfractions of a
Coal-derived Oil and a Shale Oil[a]

	Specific Activity (rev/mg)	
Subfraction	Coal-derived Oil	Shale Oil
Aliphatic (AL)	0	0
PAH (I)	1390	120
Neutral N-PAH (II)	3250	0
Polar (III)	3380	1100

[a]From Ho et al. (in press).

CONCLUSIONS

Our conclusions are as follows: 1) Short-term bioassays such
as the Ames Salmonella histidine reversion assay can be effectively
applied to complex environmental samples. 2) Proper chemical
extraction and fractionation methods should be coupled to the
assay. 3) The shale- and coal-derived oils were relatively more
mutagenic than was petroleum crude oil. 4) Mutagenic activity was
mainly associated with the basic and neutral fractions.
5) Azaarenes, aromatic amines, PAH, alkylated PAH, and nitrogen-PAH
were the organic constituents of these fractions suspected of
causing the mutagenic activity.

REFERENCES

Ames, B.N., J. McCann, and E. Yamasaki. 1975. Methods for
detecting carcinogens and mutagens with the Salmonella/
mammalian-microsome mutagenicity test. Mutation Res.
31:347-364.

Coffin, D.L., M.R. Guerin, and W.H. Griest. 1979. The interagency
program in health effects of synthetic fossil fuels
technologies: operation of a materials repository. In:
Proceedings of the Symposium on Potential Health and
Environmental Effects of Synthetic Fossil Fuel Technologies,
CONF-780903. U.S. Department of Energy: Oak Ridge, TN.

Epler, J.L., B.R. Clark, C.-h Ho, M.R. Guerin, and T.K. Rao.
    1979a. Short-term bioassay of complex organic mixtures:
    Part II. Mutagenicity testing. Environ. Sci. Res.
    15:269-290.

Epler, J.L., T.K. Rao, and M.R. Guerin. 1979b. Evaluation of
    feasibility of mutagenic testing of shale oil products and
    effluents. Environ. Hlth. Perspect. 30:179-184.

Epler, J.L., J.A. Young, A.A. Hardigree, T.K. Rao, M.R. Guerin,
    I.B. Rubin, C.-h. Ho, and B.R. Clark. 1978. Analytical and
    biological analyses of test materials from synthetic fuel
    technologies. I. Mutagenicity of crude oils determined
    by the Salmonella typhimurium/microsomal activation system.
    Mutation Res. 57:265-276.

Griest, W.H., B.A. Tomkins, J.L. Epler, and T.K. Rao. 1979.
    Characterization of multialkylated polycyclic aromatic
    hydrocarbons in energy-related materials. In: Polynuclear
    Aromatic Hydrocarbons. P.W. Jones and P. Leber, eds. Ann
    Arbor Science Publishers, Inc: Ann Arbor, MI. pp. 395-409.

Guerin, M.R., C.-h. Ho, T.K. Rao, B.R. Clark, and J.L. Epler.
    1980. Polycyclic aromatic primary amines as determinant
    chemical mutagens in petroleum substitutes. Environ. Res.
    23:42-53.

Guerin, M.R., I.B. Rubin, T.K. Rao, B.R. Clark, and J.L. Epler.
    (in press). Distribution of mutagenic activity in petroleum
    and petroleum substitutes. Fuel.

Ho, C.-h, M.R. Guerin, B.R. Clark, T.K. Rao, and J.L. Epler. (in
    press). Preparative-scale isolation of alkaline mutagens from
    complex mixtures. Environ. Sci. Technol.

Rao, T.K., B.E. Allen, D.W. Ramey, J.L. Epler, I.B. Rubin, M.R.
    Guerin, and B.R. Clark. (in press). Analytical and
    biological analyses of test materials from the synthetic
    fuel technologies. III. Fractionation with LH-20 gel for
    the bioassay of crude synthetic fuels. Mutation Res.

Rao, T.K., J.A. Young, A.A. Hardigree, W. Winton, and J.L. Epler.
    1978. Analytical and biological analyses of test materials
    from the synthetic fuel technologies. II. Mutagenicity of
    organic constituents from the fractionated synthetic fuels.
    Mutation Res. 54:185-191.

Swain, A.P., J.E. Cooper, and R.L. Stedman. 1969. Large scale
    fractionation of cigarette smoke condensate for chemical and
    biologic investigations. Cancer Res. 29:579-583.

THE DETECTION OF POTENTIAL GENETIC HAZARDS IN COMPLEX ENVIRONMENTAL MIXTURES USING PLANT CYTOGENETICS AND MICROBIAL MUTAGENESIS ASSAYS

Milton J. Constantin and Karen Lowe
Comparative Animal Research Laboratory
University of Tennessee
Oak Ridge, Tennessee

T.K. Rao, Frank W. Larimer, and James L. Epler
Biology Division
Oak Ridge National Laboratory
Oak Ridge, Tennessee

INTRODUCTION

The recent realization that industrial effluents and wastes could pose health hazards to future generations has led to considerable research on the health effects of these substances and to efforts to regulate their release. The Research Conservation and Recovery Act (RCRA) specifically addresses the potential health and environmental hazards of solid wastes. Evaluation of long-term consequences of exposure to the huge number of potentially hazardous compounds and complex mixtures would tax the scientific community's whole-animal testing capabilities. Therefore, many short-term bioassays have been developed to rapidly screen compounds for toxicity, mutagenicity, teratogenicity, and carcinogenicity. It is hoped that these short-term tests will predict health hazards early in the development of new technologies, allowing technological changes to be made at early stages, and resulting in timely whole-animal testing where screening indicates potential hazards. Time and money should thus be saved in both technological development and biological testing.

In the approach that we have taken with an array of complex mixtures from the technological world, chemical characterization and preparation are coupled with short-term bioassays. Other such investigations have usually involved chemical analyses and preparation for biological testing (e.g., cytotoxicity, mutagenicity, and carcinogenicity assays). The genetic test for Salmonella histidine-mutant reversion (Ames et al., 1975) has been widely used to screen potential mutagens or carcinogens. Subsequent fractionation procedures are carried out to isolate and identify the mutagens in the material; the bioassay is used to

trace the biological activity and guide the separations. The
biological tests function as 1) predictors of long-range health
effects such as mutagenesis, teratogenesis, or carcinogenesis; 2)
predictors of toxicity to man and his environment; 3) a mechanism
to rapidly isolate and identify hazardous agents in complex
material; and 4) indicators of relative biological activity,
through the correlation of control data with changes in
environmental or process conditions.

Plant systems may offer an additional means of evaluating
complex mixtures. According to Kihlman (1971), root tips are the
ideal plant tissue in which to study the effects of chemicals on
chromosomes because 1) they are readily available throughout the
year; 2) they are inexpensive; 3) they are easy to handle; 4)
they provide data within a few days; 5) they are directly exposed
to the chemical in aqueous solutions of known concentrations; 6)
they provide numerous dividing cells for analysis; and 7) they
have few, relatively large chromosomes.

Barley (Hordeum vulgare [L] emend Lam.) has 2n = 14
chromosomes, ranging from 6 to 8 μm in length. The barley embryo
has from five to seven seminal roots that yield numerous dividing
cells for analysis within 24 to 48 h after germination starts.
Following seed treatments, the effects of chemical and physical
agents can be assessed in terms of cytogenetic and genetic end
points in the same population of plants. This unique advantage has
made barley a useful mutagenesis test organism.

The most frequently used method for studying the effects of
chemicals on the chromosomes of barley root tip cells is the
analysis of anaphases for detectable aberrations (Nicoloff and
Gecheff, 1976). The assay has been used to test chemicals either
suspected or known to be mutagens; generally, the results have
agreed with those from other organisms tested with the same
compounds. The assay has not been used widely to test complex
mixtures.

In this report, we present data from a study comparing
microbial mutagenicity assays (Salmonella and yeast) with a plant
cytogenetic assay. The materials used were aqueous extracts from a
fly ash sample and an arsenic- (As) contaminated groundwater sample
(provided by the U.S. Environmental Protection Agency). The
purpose of this research was to assess the potential of complex
environmental mixtures (extracts of solid wastes) to produce
cytogenetic effects in a higher plant and gene mutations in
microbes. A complex environmental mixture capable of inducing both
end points warrants close scrutiny as a potential health hazard to
humans.

MATERIALS AND METHODS

Preparation of Liver Homogenate (S-9)

The microsomal preparation was made according to Ames et al. (1975). The livers of Sprague-Dawley rats (induced with either Aroclor 1254 [Ar S-9] or phenobarbital [ψB S-9]) were washed in an equal volume of 0.15 $\underline{M}$ NaCl, minced with sterile scissors, and homogenized with a Potter-Elvehjem apparatus in 3 vol of 0.15 $\underline{M}$ KCl. The homogenate was centrifuged at 4°C for 10 min at 9000 x g. The supernatant was collected and stored at -80°C. The activation system (S-9) mix contained, per milliliter, 0.3 ml of liver homogenate, 8 μmol magnesium chloride, 33 μmol potassium chloride, 5 μmol glucose-6-phosphate, and 4 μmol nicotinamide adenine dinucleotide phosphate in 100 μmol of sodium phosphate buffer (pH 7.4).

Bacterial Mutagenicity Assay

Among bacterial mutagenicity test systems, the assay developed by Ames using the histidine auxotrophic strains of Salmonella typhimurium is widely used as a screening test to detect potential genetic and carcinogenic hazards. Of the four standard tester strains generally used in the assays, the missense strains TA100 and TA1535 detect base-pair substitutions, while strains TA1537 and TA98 detect frameshift mutations. The general procedure was described by Ames et al. (1975). A bacterial suspension (2 x 10^8/ml) was added to 2 ml of molten top agar (45°C) containing test substance. S-9 mix (0.5 ml/plate) was added, when required, to provide metabolic activation. The top agar was overlayed on Vogel-Bonner (1956) minimal medium, and revertant colonies were counted after two days of incubation at 37°C. Known chemical mutagens as positive controls and solvent (negative) controls were routinely run.

Yeast Assays

The Saccharomyces assay measures both forward and reverse mutation. Forward mutation is detected by the inactivation of the arginine permease gene (CAN1), leading to resistance (can^r) to the toxic antimetabolite canavanine. Reverse mutation is monitored by use of a histidine auxotroph (his1-7) that reverts by base-pair substitution.

The test strain used in these experiments was constructed from stocks of Saccharomyces cerevisiae maintained at Oak Ridge National Laboratory (ORNL) and from stocks obtained from the Berkeley

collection.   Strain XL7-10B has the genotype α p[+] CAN1 his1-7
lys1-1 ural.

    Details of the preparation of media, rat-liver homogenates,
and the mutagenicity assay have been described by Larimer et al.
(1978, 1980).   The test material was mixed with yeast cells in
buffer or rat-liver homogenate (S-9) in buffer and incubated for 3
or 24 h at 30°C with shaking before it was plated on selective
media.   Mutant clones were scored on selective plates after
incubation for five days at 30°C.   To determine survival, dilutions
were plated on yeast-extract-peptone-dextrose plates and scored
after two days at 30°C.

Plant Cytogenetic Assay

    Seeds of Himalaya barley from R.A. Nilan (Washington State
University, Pullman, WA) were stored under refrigeration and
hand-picked for quality just prior to each experiment.
Approximately 25 seeds were sown embryo-side-up on Whatman #1
filter paper in glass petri plates.   The filter paper was
thoroughly saturated by adding 7.5 ml of either double-distilled
water or a test solution at pH 7.0.   Each treatment was done in
triplicate; plates were placed in sealed polyethylene bags and
cultured for 42 to 46 h at 25°C under 10 to 15 $\mu E\ m^{-2} sec^{-1}$
fluorescent light.

    Germinating seeds were killed and fixed in 3:1 ethanol-
glacial acetic acid and stored in the refrigerator until the roots
were processed.   Excised roots were hydrolyzed for 9 min in 1 $\underline{N}$ HCl
at 60°C, reacted with Schiff's Reagent for at least 15 min, treated
with pectinase (0.4% in water, pH 4.0) for at least 30 min, and
squashed in aceto-carmine; Deckglaskitt was used to seal around the
coverslip.   Cells were observed at 1000X magnification for the
presence of bridges and fragments at anaphase.

    Data were expressed as aberrations per hundred anaphases and
as percentage of aberrant anaphases.   Statistical inferences were
drawn on the bases of analysis of variance and chi-square analysis
of 2 x 2 contingency tables for aberrant vs. normal anaphases in
control vs. chemical treatments.

Preparation of Samples

    Aqueous extracts of the fly ash samples were prepared
according to the extraction procedure (EP) given in the Federal
Register (1978).   The As-contaminated sample was used as provided
by the U.S. Environmental Protection Agency.   Resin concentration
(XAD-2; Isolab, Inc.) procedures were described by Epler et al.

(1980). A 500-ml aqueous extract was passed through 4 g XAD-2 at a
flow rate of 1.2 ml/min. The column was rinsed with deionized
water and then eluted with acetone. The acetone eluate was
evaporated to dryness, and the sample was dissolved in 2 ml
dimethylsulfoxide.

RESULTS AND DISCUSSION

Salmonella Assay

    Aqueous samples from the As-contaminated groundwater and its
XAD-2 concentrate (12.5-fold) were tested for mutagenicity over a
nontoxic dose range. The results are given in Figure 1. The
groundwater sample was not mutagenic (Figure 1A), even with
metabolic activation, for strains TA98 and TA100. However, XAD-2
concentrate of groundwater (Figure 1B) exhibited a clear dose
response with the frameshift strains TA1537 and TA98 and the
highly sensitive TA100 strain. The missense strain TA1535 was not
reverted. These results suggest the presence of a mutagenic
constituent(s) in the aqueous sample whose activity was evident
only when the sample was concentrated by an appropriate method.

    The EP extract and its XAD-2 concentrate from fly ash sample
were not mutagenic when tested with the basic set of tester strains
(see Table 1). Addition of the metabolic activation system did not
influence the result. Lack of mutagenic activity of fly ash from
power plants conflicts with earlier reports of Chrisp et al. (1978)
and Kubitschek and Venta (1979), who used serum extraction, and
Hobbs et al. (1979), who used dimethylsulfoxide for extraction.
The EP extraction procedure is probably not adequate to extract
mutagenic agents from fly ash. When benzene extraction was used
with a similar fly ash sample from a different power plant, a
dose-dependent increase in the induction of histidine revertants
was observed (unpublished data). Results obtained with known
chemical mutagens (positive controls) are given in Table 1B. The
alkylating agents ethylmethane sulfonate (EMS) and methylmethane
sulfonate (MMS) were very specific in reverting TA1535 and TA100,
respectively. The frameshift strains were reverted by 8-amino-
quinoline (8-AmQ), specific to TA1537, and benzo(a)pyrene (B[a]P),
which is mutagenic only when activated with Aroclor-1254-induced
rat-liver homogenate (Ar S-9 mix).

Saccharomyces Mutation Assay

    The As-contaminated groundwater sample was not mutagenic (see
Table 2). The XAD-2 concentrate of this sample was mutagenic
without metabolic activation for a 24-h exposure, giving a dose-
dependent response. Metabolic activation appeared to reduce the

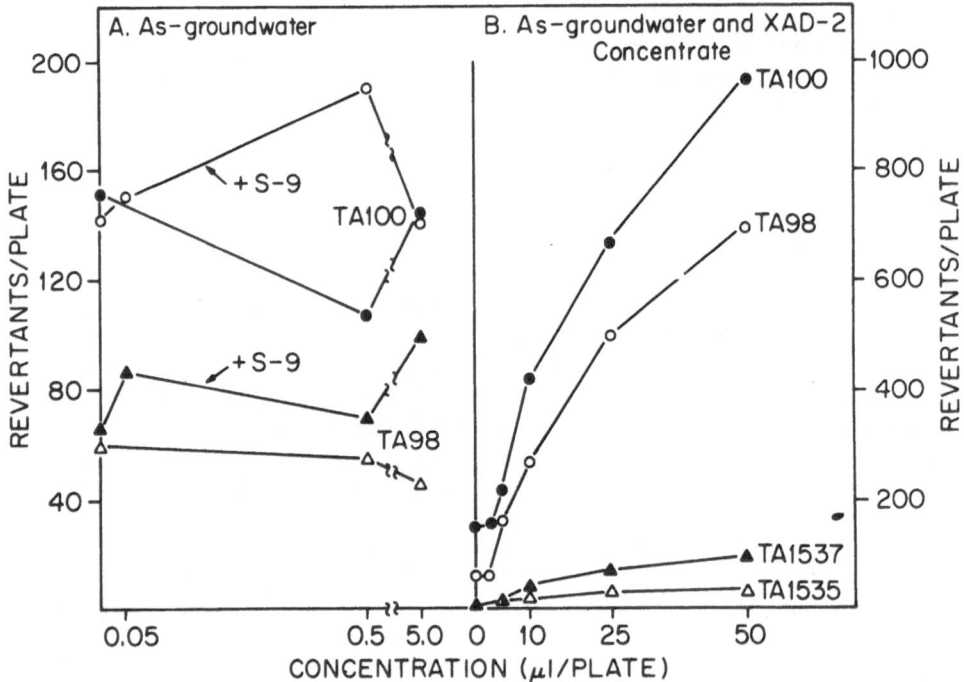

Figure 1. Mutagenicity of As-contaminated groundwater and its
           XAD-2 concentrate in the <u>Salmonella</u> assay. Each point
           represents an average from at least three independent
           experiments.

mutagenic potential of the XAD-2 concentrate. Neither of these
test materials was toxic. More induced forward mutations to can[r]
were seen than his1-7 base-pair substitutions. This is the typical
response of this system to a frameshifting agent. The fly ash EP
extract and its XAD-2 concentrate were not toxic or mutagenic to
yeast (Table 3).

Plant Cytogenetic Assay

     The cytogenetic effects observed in barley root tip cells are
presented as percentage of aberrant anaphases in Table 4, which
shows the distribution of anaphases into normal and aberrant
classes for negative, positive, and solvent controls and complex
environmental mixtures. The probability values were determined
through chi-square tests for independence of the treatments and the
proportions of aberrant anaphases, in 2 x 2 contingency tables.

Table 1. Salmonella Mutation Assay: Fly Ash—Aqueous
Extract (EP) and its Concentrate (XAD-2)

Treatment and Concentration (μm/plate)	his^+ Revertants/Plate					
	TA1535[a]	TA1537[a]	TA98		TA100	
	EP	EP	EP	XAD-2	EP	XAD-2
No activation						
Solvent control	21	14	32	57	136	231
50	18	15	27	30	141	180
Ar S-9 activation						
Solvent control	15	16	46	33	192	209
10	15	14	35	42	108	203
25	8	15	34	28	141	206
50	16	15	53	26	123	206
75	16	18	42	38	172	208
Positive controls[b]						
50 μl (5%) EMS	704	11	48	–	386	–
50 μl (5%) MMS	42	13	48	–	1495	–
20 μg 8–AmQ	6	102	56	–	270	–
20 μg B(a)P + S-9	7	122	625	–	1700	–

[a]XAD-2 not tested with these strains.
[b]Underscoring represents positive response to mutagen.

The solvent controls (phosphate buffer and acetic acid extraction
solution) yielded aberrant anaphases at frequencies not
significantly different from those of the negative control,
distilled water. In contrast, the frequency of aberrant anaphases
was increased in the seeds treated with EMS, a positive control,
and in those treated with the fly ash extract and the As-
contaminated groundwater. The fly ash extract was used as we
received it, except for the adjustment to pH 7.0, whereas the As-
contaminated groundwater was diluted 1:8 and 1:16 with distilled
water (1.25 and 0.625 ml extract diluted to 10 ml of solution),
and these results were pooled. The more concentrated solutions
were toxic to the germinating seed; i.e., germination was delayed
and root growth was inhibited.

Table 2. Yeast Mutation Assay: Arsenic-Contaminated Groundwater and its XAD-2 Concentrate

| Concentration (µl) | % Survival | | | | $can^r/10^7$ Survivors | | | | $his^+/10^7$ Survivors | | | |
| | 3 h | | 24 h | | 3 h | | 24 h | | 3 h | | 24 h | |
	Sample	XAD-2	Sample	XAD-2	Sample	XAD-2	Sample	XAD-2	Sample	XAD-2	Sample	XAD-2
No activation												
Control	100	100	100	100	27	24	19	15	6	13	11	6
0.1	71	–	94	–	23	–	16	–	16	–	8	–
1.0	84	–	81	–	20	–	19	–	8	–	8	–
10	75	116	85	95	13	19	19	17	13	8	9	9
20	–	127	–	98	–	15	–	34	–	8	–	24
50	–	138	–	93	–	17	–	69	–	6	–	37
100	89	156	99	86	16	17	13	176	9	9	11	55
YB S-9 activation												
Control	100	100	100	100	14	17	17	15	12	10	12	8
0.1	107	–	103	–	21	–	14	–	10	–	9	–
1.0	59	–	87	–	27	–	21	–	10	–	10	–
10	82	105	103	94	21	18	14	19	14	11	9	12
20	–	98	–	86	–	20	–	31	–	13	–	6
50	–	103	–	79	–	16	–	89	–	8	–	29
100	91	112	119	71	18	21	18	107	12	7	7	24
Ar S-9 activation												
Control	100	100	100	100	17	15	21	14	9	9	11	6
0.1	97	–	94	–	18	–	20	–	11	–	9	–
1.0	92	–	96	–	13	–	27	–	8	–	7	–
10	87	107	91	108	21	12	22	13	12	10	10	8
20	–	111	–	122	–	25	–	41	–	11	–	9
50	–	122	–	91	–	13	–	74	–	5	–	17
100	88	90	103	82	24	13	22	118	11	8	13	38
EMS, 1% v/v	22				448				1270			

Table 3. Yeast Mutation Assay: Fly Ash EP Extract and its XAD-2 Concentrate

Concentration (µl)	% Survival				$can^r/10^7$ Survivors				$his+/10^7$ Survivors			
	3 h		24 h		3 h		24 h		3 h		24 h	
	EP	XAD-2	EP	XAD-2	EP	XAD-2	EP	XAD-2	EP	XAD-2	EP	XAD-2
No activation												
Control	100	100	100	100	12	26	19	24	13	10	13	6
0.1	104	–	101	–	18	–	18	–	5	–	10	–
1.0	109	–	96	–	14	25	13	27	6	6	15	8
10	94	103	97	96	14	27	12	25	9	6	16	7
20	–	105	–	98	–	23	–	22	–	8	–	8
50	–	102	–	86	–		–		–		–	
100	97	96	99	92	11	23	16	23	9	.9	10	10
YB S-9 activation												
Control	100	–	100	100	17	24	16	23	11	9	9	8
0.1	106	–	107	–	20	–	11	–	9	–	8	–
1.0	100	–	102	–	12	25	13	19	3	6	14	9
10	101	–	98	102	17	23	17	19	9	8	6	11
20	–	–	–	107	–	18	–	22	–	11	–	7
50	–	–	–	96	–		–		–		–	
100	94	–	109	89	10	29	13	27	6	9	8	11
Ar S-9 activation												
Control	100	100	100	100	21	22	14	25	13	6	10	11
0.1	95	–	93	–	17	–	11	–	16	–	8	–
1.0	93	–	103	–	13	20	10	19	8	9	14	9
10	103	90	92	102	13	21	19	20	12	8	5	12
20	–	91	–	94	–	28	–	25	–	8	–	7
50	–	86	–	92	–		–		–		–	
100	109	80	97	84	14	23	13	18	11	6	7	10
B(a)P, 100 µg Ar-activation			89				181				52	

Table 4.   Chromosome Aberrations in Barley Embryos
Treated with Extracts of Complex Environmental Mixtures:
Numbers of Aberrant Anaphases

Treatments	Normal Anaphases N	Aberrant Anaphases		
		N	%	p[a]
Distilled water	2182	48	2.15	
Phosphate buffer[b]	2197	75	3.30	0.05 to 0.10
EMS[c]	2241	184	7.59	< 0.001
Acetic acid extraction solution[d]	3784	88	2.27	0.30 to 0.50
Fly ash extract	2874	437	13.20	< 0.001
As-contaminated groundwater	1221	190	13.47	< 0.001

[a]P = probability that difference was due to chance, according to chi-square test for independence.
[b]Monobasic and dibasic phosphate buffer at pH 7.0.
[c]EMS at 0.025 $\underline{M}$ in phosphate buffer at pH 7.0; seeds soaked aerobically in water at ~ 1°C for 16 h and in EMS for 2 h at ~ 1°C plus 6 h at 24°C, rinsed, and cultured on a water-saturated Whatman #1 filter.
[d]Weak acetic acid extraction solution as described in Federal Register (1978).

Table 5 shows the same experimental data as in Table 4, but expressed as aberrations per hundred cells (the mean for each treatment).   An arcsine transformation of the data (arcsin p, where p is a proportion) was done prior to the analysis of variance, to reduce the heterogeneity of the treatment variances. A Student-Newman-Keuls comparison of treatment means (Steel and Torrie, 1960) showed no difference between the negative control (distilled water) and either of the solvent controls (phosphate buffer and acetic acid extraction solution), whereas the positive control (EMS) and the two complex environmental mixtures (fly ash extract and the As-contaminated groundwater) induced a significantly greater number of aberrations per hundred anaphases.

Although the two methods of data analysis address different aspects of the cell population's response to seed treatment, the conclusion is the same: the barley root tip system responded to an unknown mutagenic substance(s) in the two complex environmental

Table 5.   Chromosome Aberrations in Barley Embryos
Treated with Extracts of Complex Environmental Mixtures:
Aberrations per Hundred Cells

Treatments[a]	Embryos	Mean Number of Aberrations/ 100 Cells[b]	Standard Deviation
Distilled water	60	2.75	3.04
Acetic acid extraction solution	60	2.57	2.85
Fly ash extract	60	16.30	9.10
Phosphate buffer	59	4.37	4.23
EMS (0.025 M)	63	11.01	9.73
As-contaminated groundwater	63	16.19	9.10

[a]See footnotes to Table 4 for details.
[b]Not arcsine-transformed.

mixtures much as it did to EMS, a known mutagen. This response was observed as increases in the percentage of aberrant anaphases and in the number of aberrations per hundred anaphases.

Our results with the barley root tip cytogenetic aberration assay agree with expectations. According to Brewen and Preston (1978), structural changes in chromosomes constitute a significant proportion of mutagenic events. Chemicals that induce mutations in eukaryotes invariably also induce chromosomal structural changes (Evans, 1976; Kunzel, 1971). Fly ash from coal combustion is mutagenic in bacteria; known mutagens have been isolated and identified. Arsenic and heavy metals are known to induce both mutations and cytogenetic effects.

Chrisp et al. (1978) reported that horse serum, phosphate-buffered saline, and cyclohexane filtrates of fly ash of 2.2-μm mass median diameter (the finest particle size fraction tested) induced histidine revertants in S. typhimurium strains TA98 and TA1538. The order of activity was horse serum >> cyclohexane > saline. More recently, Fisher et al. (1979) reported that serum filtrates of the most respirable stack-collected fly ash are mutagenic in S. typhimurium strains TA98, TA100, and TA1538. However, after heating to 350°, these serum filtrates are not mutagenic in the Salmonella assays. The authors hypothesized that the mutagenic activity of fly ash is associated with organic compounds. Lee et al. (1980) have found dimethylsulfate and its

hydrolysis product monoethylsulfate at concentrations as high as
830 ppm in fly ash and airborne particulate matter from coal
combustion. These compounds are known mutagens.

In the case of the As-contaminated groundwater, numerous
metals were present (e.g., in µg/l: cadmium, 485; nickel, 935,
lead, 117; antimony, 297; thallium, 7720, and zinc, 251; Epler et
al., 1980). Arsenic, especially in the arsenite state, is known to
induce mutations and chromosome aberrations (Rossner, 1977). Some
of the heavy metals are known to be mutagenic and/or carcinogenic
(Freese, 1971; Miller and Miller, 1971).

CONCLUSIONS

The following conclusions were reached: 1) The Salmonella and
Saccharomyces assays indicated the presence of mutagenic activity
in the XAD-2 concentrate of the As-contaminated groundwater but not
in the aqueous extract of the fly ash sample. 2) Both assays
implicated frameshift mutagenesis as the mechanism involved. 3)
The Hordeum root tip assay indicated mutagenic activity in both
complex mixtures tested. 4) Chemical analyses of both complex
mixtures showed the presence of heavy metals, implicating them as
the possible cause of chromosomal aberrations.

ACKNOWLEDGMENTS

The authors gratefully acknowledge the staff of the Analytical
Chemistry and the Environmental Sciences Divisions of ORNL who
contributed to the research that led to this comparative study.

REFERENCES

Ames, B.N., J. McCann, and E. Yamasaki. 1975. Methods for
      detecting carcinogens and mutagens with the Salmonella/
      mammalian-microsome mutagenicity test. Mutation Res.
      31:347-364.

Brewen, J.G., and R.J. Preston. 1978. Analysis of chromosome
      aberrations in mammalian germ cells. In: Chemical Mutagens:
      Principles and Methods For Their Detection. Vol. 5. A.
      Hollaender, ed. Plenum Press: New York. pp. 127-150.

Chrisp, C.E., G.L. Fisher, and J.E. Lammert. 1978. Mutagenicity
      of filtrates from respirable coal fly ash. Science 199:73-75.

Epler, J.L., F.W. Larimer, T.K. Rao, E.M. Burnett, W.H. Griest, M.R. Guerin, M.P. Maskarinec, D.A. Brown, N.T. Edwards, C.W. Gehrs, R.E. Milleman, B.R. Parkhurst, B.M. Ross-Todd, D.S. Shriner, and H.W. Wilson, Jr. 1980. Toxicity of Leachates. Final Report for Office of Research and Development, U.S. Environmental Protection Agency, Cincinnati, OH. Oak Ridge National Laboratory: Oak Ridge, TN.

Evans, H.J. 1976. Cytological methods for detecting chemical mutagens. In: Chemical Mutagens: Principles and Methods For Their Detection, Vol. 4. A. Hollaender, ed. Plenum Press: New York. pp. 1-29.

Federal Register. 1978. 43 (Dec.): FR58946.

Fisher, G.L., C.E. Chrisp, and O.G. Raabe. 1979. Physical factors affecting the mutagenicity of fly ash from a coal-fired power plant. Science 204:879-881.

Freese, E. 1971. Molecular mechanisms of mutations. In: Chemical Mutagens: Principles and Methods For Their Detection, Vol. 1. A. Hollaender, ed. Plenum Press: New York. pp. 1-56.

Hobbs, C.H., C.R. Clark, L.C. Griffis, R.O. McClellan, R.F. Henderson, J.O. Hill, and R.E. Royer. 1979. Inhalation toxicology of primary effluents from fossil fuel conversion and use. ORNL Publication Conf-780903. Oak Ridge National Laboratory: Oak Ridge, TN.

Kihlman, B.A. 1971. Root tips for studying the effects of chemicals on chromosomes. In: Chemical Mutagens: Principles and Methods For Their Detection, Vol. 2. A. Hollaender, ed. Plenum Press: New York. pp. 489-514.

Kubitschek, H.E., and L. Venta. 1979. Mutagenicity of coal fly ash from electric power plant precipitators. Environ. Mutagen. 1:79-83.

Kunzel, G. 1971. The ratio of chemically induced chromosome aberrations to gene mutations in barley. Mutation Res. 12:397-409.

Larimer, F.W., A.A. Hardigree, W. Lijinsky, and J.L. Epler. 1980. Mutagenicity of N-nitrosopiperazine derivatives in Saccharomyces cerevisiae. Mutation Res. 77:143-148.

Larimer, F.W., D.W. Ramey, W. Lijinsky, and J.L. Epler. 1978. Mutagenicity of methylated N-nitrosopiperidines in Saccharomyces cerevisiae. Mutation Res. 57:155-161.

Lee, M.L., D.W. Later, D.K. Rollins, D.J. Eatough, and L.D. Hansen. 1980. Dimethyl and monomethyl sulfate: presence in coal fly ash and airborne particulate matter. Science 207:186–188.

Miller, E.C., and J.A. Miller. 1971. The mutagenicity of chemical carcinogens: Correlations, problems, and interpretations. In: Chemical Mutagens: Principles and Methods For Their Detection, Vol. 1. A. Hollaender, ed. Plenum Press: New York. pp. 83–119.

Nicoloff, H., and K. Gecheff. 1976. Methods of scoring induced chromosome structural changes in barley. Mutation Res. 34:233–244.

Rossner, P. 1977. Mutagenic effect of sodium arsenite in Chinese hamster cell line Dede. Mutation Res. 46 (3):234–235.

Steel, R.G.D., and J.H. Torrie. 1960. Principles and Procedures of Statistics. McGraw–Hill: New York. p. 481.

Vogel, H.J., and D.M. Bonner. 1956. Acetylornithase of E. coli: partial purification and some properties. J. Biol. Chem. 218:97–106.

SESSION 4

MOBILE SOURCES

SHORT-TERM CARCINOGENESIS AND MUTAGENESIS BIOASSAYS OF MOBILE-SOURCE EMISSIONS

Joellen Lewtas Huisingh
Health Effects Research Laboratory
U.S. Environmental Protection Agency
Research Triangle Park, North Carolina

INTRODUCTION

The combustion emissions from mobile sources, including both gases and particles, are very complex and may have thousands of separate components. Qualitative and quantitative identification of all of these individual components is a tremendous task. The analytical challenge is facilitated if the number of compounds requiring identification can be reduced.

Short-term bioassays can be used to narrow the compounds requiring identification to those potentially responsible for adverse health effects. Initial screening of complex mixtures is useful to

1) indicate particular emissions or portions of an emission that are potentially toxic, mutagenic, or carcinogenic and that should be evaluated in confirmatory and, possibly, long-term bioassays;

2) biologically direct the fractionation and identification of hazardous components and specific chemicals in complex mixtures; and

3) compare the relative biological activity of similar emissions that result from different sources, fuels, control technologies, or operating conditions.

The introduction of increasing numbers of light-duty diesel automobiles has stimulated environmental concern over the health effects of diesel particulate emissions. Currently, diesel

269

automobiles emit over one hundred times the particles (grams per mile) emitted by gasoline-powered, catalyst-equipped (gasoline-catalyst) automobiles. Diesel particles, emitted as carbonaceous soot, serve as condensation nuclei for higher-molecular-weight organic combustion vapors, which condense onto the soot particles as the exhaust is diluted and cooled to ambient temperature. The diesel particles emitted into the ambient air contain 10 to 50% extractable organic constituents.

The gaseous organics that do not adsorb onto particles are currently regulated only as total hydrocarbon emissions. Components of this general class of emissions that are of potential concern, such as aldehydes, nitrosamines, phenols, and cyanides, are not specifically regulated. Research to apply short-term bioassays to these gaseous emissions is being initiated; much of the research completed to date, however, has focussed on the organic compounds extracted from diluted particulate emissions.

APPLICATION OF MICROBIAL ASSAYS TO MOBILE SOURCE EMISSIONS

Mutagenic activity resulting from organics extracted from diesel particulate emissions was first detected using microbial mutagenesis assays. Particles collected from two heavy-duty diesel engines were subjected to extraction and fractionation techniques (Huisingh et al., 1979). The resulting organic fractions (acidic, basic, and neutral) were then screened using bioassays that employed bacteria (Salmonella typhimurium) to detect gene mutations and mammalian cells to detect cellular toxicity. None of the diesel organic fractions was found to be highly cytotoxic in the mammalian cell assays. All but one of the fractions showed some mutagenicity in the S. typhimurium plate-incorporation assay for gene mutations.

The neutral components of the diesel extract accounted for 84% of the mass and were fractionated into four subfractions (paraffins, aromatics, and transitional and oxygenated polar neutrals). The paraffinic fraction (39% by weight) was not mutagenic, and the aromatic fraction (13% by weight) accounted for only 1.5% of the mutagenic activity in the TA98 strain of S. typhimurium. The two polar neutral fractions, transitional and oxygenated, were the most mutagenic. These two fractions accounted for one third of the mass of the extractable organics and over 90% of the mutagenic activity in both TA98 and TA1538 strains of S. typhimurium.

These results suggest that there is more than one mutagen present in the polar neutral fractions of organics bound to diesel particles. These mutagens are not artifacts of the extraction or fractionation processes (Huisingh et al., 1979), but appear to be products of the combustion process, since fractions of uncombusted fuel were not mutagenic.

Various fuels appear to differ in the mutagenicity of their particle-bound combustion organics. Studies comparing the mutagenic activity of combustion emission organics from two passenger cars operated with five different fuels show that the poorest quality fuel (No. 2 diesel fuel) generated the largest quantity of mutagenic particle-bound organics (Huisingh et al., 1979). This minimum-quality fuel had the lowest Cetane index (41.8), highest aromatic content, and highest nitrogen and sulfur contents.

The effects of engine, fuel, and operating conditions on the mutagenicity of automotive emissions were studied using short-term bioassays. These conditions are variable and may affect not only the mutagenic activity of the organic fractions but also the amount of extractable organics present on the diesel particles and the particulate emission rate. These factors can be accommodated by calculations that determine the mutagenic activity on a per-mile- or per-kilogram-fuel-consumed basis.

In comparing different diesel automobiles, Claxton and Kohan (1980) found as much as a three-fold difference in their mutagenic emission rate. Although the extractable organics from the gasoline-catalyst automobile emissions were more mutagenic than many of the diesel organics, the amount of extractable organics and the particle emission rate were so low for the gasoline-catalyst automobile that the net mutagenic activity per mile was two orders of magnitude less than that from a comparable diesel automobile.

The total mutagenic activity resulting from automotive emissions depends on the release of the mutagenic organics from the particles. The ability of physiological fluids (serum, lung-cell cytosol, and lung-lavage fluid) to release mutagens from diesel particles has been compared with the extraction capability of solvents. Serum and lung cytosol were found to remove 80 to 85% of the solvent-extractable mutagenic activity from the diesel particles (King et al., in press). The serum- and cytosol-associated mutagens were essentially undetectable when the serum itself was tested in the S. typhimurium mutagenesis bioassay. This effect is possibly due to binding of the mutagens by the serum. Other studies have shown that whole diesel particles are engulfed by mammalian cells in vitro and are capable of causing gene mutations (Chescheir et al., 1980).

APPLICATION OF MAMMALIAN CELL BIOASSAYS TO MOBILE-SOURCE EMISSIONS

The extractable organics from diesel particles, although showing a low cellular toxicity in the microbial bioassays, were mutagenic in a microbial (S. typhimurium) assay and positive in a yeast (Saccharomyces cerevisciae) assay for DNA damage (mitotic recombination). These results indicated the presence of potentially mutagenic or carcinogenic chemicals in diesel emission organics (Huisingh et al., in press b).

Mammalian cell bioassays were initiated to verify the microbial screening results; mammalian cells are much more similar to human cells in cellular and chromosomal organization than are microbes. Diesel organics gave positive results in two forward mutational assays using mammalian cells. Two assays for DNA damage--unscheduled DNA synthesis (UDS) and sister chromatid exchange (SCE) assays--were also used. The UDS assay was negative and the SCE assay positive with the diesel organics tested. The carcinogenesis assay for morphological oncogenic transformation in the mammalian (BALB/c 3T3) cells was positive.

Additional research is needed to determine which bioassays are most useful in evaluating automotive emissions and to develop new methods to expose these test systems to "difficult" samples, such as gases and insoluble organics.

COMPARATIVE BIOASSAYS OF MOBILE SOURCE EMISSIONS

A matrix of in vitro and in vivo bioassays is currently being used to quantitatively compare the effects of a series of mobile-source emissions (extractable organics from particulate emissions: Huisingh et al., in press a). The normalized rankings for four bioassays are compared in Table 1. The quantitative results from the mobile-source samples show a general overall consistency (Nesnow and Huisingh, in press). The Cat sample was very weak in all of the assays. The Nissan sample showed the highest activity in these assays, while the other three mobile sources showed intermediate activity.

In theory, gene mutation and skin tumor initiation arise from similar mechanisms and, thus, should give similar results (assuming equal toxicity and mutagen or carcinogen transport and activation by the various cell types). A comparison of the results of the microbial and mammalian cell mutation assays with the results of the rodent skin tumor initiation assay seems to support this hypothesis.

Table 1. Activity Rankings for Mobile-source Emissions[a],[b]

Activity	Heavy-duty Diesel Cat	Light-duty Diesel			Gasoline-catalyst Mustang
		Nissan	Olds	VW Rab	
Microbial mutation[c]	4.3	100	23	22	25
Sister chromatid exchange[d]	0	100	0	50	1
Mammalian cell mutation[e]	1	100	64	50	36
Rodent skin tumor initiation[f]	0	100	45	1	35

[a]All data are expressed as a percentage of the Nissan diesel activity, which was assigned a value of 100.
[b]Cat is the Caterpillar 3208, 4-stroke cycle engine; Olds is Oldsmobile; VW Rab is Volkswagen Rabbit.
[c]S. typhimurium histidine reversion assay; TA98 with S-9 activation (Aroclor-induced).
[d]Chinese hamster ovary cell assay with Aroclor-induced S-9 activation.
[e]L5178 mouse lymphoma forward mutation asssay at the thymidine kinase locus with Aroclor-induced S-9 activation.
[f]SENCAR mouse assay using TPA (12-0-tetradecanoylphorbol-13-acetate as the tumor promoter.

Other comparative source samples (roofing tar, coke oven emissions, and cigarette smoke condensate) were also evaluated in this study (Huisingh et al., in press a; Nesnow and Huisingh, in press). The quantitative results for these samples, which required metabolic activation, showed less agreement between these bioassays. Thus, it may not be possible to quantitatively extrapolate from in vitro to in vivo results for all types of complex mixtures.

CONCLUSIONS

Short-term carcinogenesis and mutagenesis bioassays, now being widely applied to the evaluation and characterization of mobile source emissions, show that the organics associated with both diesel and gasoline-catalyst particulate emissions exhibit mutagenic and carcinogenic activity. The relative potency of different mobile sources varies significantly.

Current research is focussing on the following areas: 1) comparative potency of the emissions from a variety of mobile sources, 2) comparative evaluation of a battery of bioassays for mobile-source applications, 3) identification of the hazardous components in diesel emissions, and 4) determination of the effective dose and target for those hazardous components.

REFERENCES

Chescheir, G.M., N.E. Garrett, J. Lewtas Huisingh, M.D. Waters, and J.D. Shelburne. 1980. Mutagenic effects of environmental particulates in the CHO/HGPRT system. Presented at the U.S. Environmental Protection Agency Second Symposium on the Application of Short-term Bioassays in the Fractionation and Analysis of Complex Environmental Mixtures, Williamsburg, VA.

Claxton, L., and M. Kohan. 1980. Bacterial mutagenesis and the evaluation of mobile source emissions. Presented at the U.S. Environmental Protection Agency Second Symposium on the Application of Short-term Bioassays in the Fractionation and Analysis of Complex Environmental Mixtures, Williamsburg, VA.

Huisingh, J.L., R. Bradow, R. Jungers, L. Claxton, R. Zweidinger, S. Tejada, J. Bumgarner, F. Duffield, V.F. Simmon, C. Hare, C. Rodriguez, L. Snow, and M. Waters. 1979. Application of bioassay to the characterization of diesel particle emissions. Part I. Characterization of heavy duty diesel particle emissions. Part II. Application of a mutagenicity bioassay to monitoring light duty diesel particle emissions. In: Application of Short-term Bioassays in the Fractionation and Analysis of Complex Environmental Mixtures. Plenum Press: New York. pp. 382-418.

Huisingh, J. Lewtas, R. Bradow, R. Jungers, B. Harris, R. Zweidinger, K. Cushing, B. Gill, and R. Albert. (in press a). Mutagenic and carcinogenic potency of extracts of diesel and related environmental emissions: study design, sample generation, collection, and preparation. In: Proceedings of the International Symposium on Health Effects of Diesel Engine Emissions, December, 1979. U.S. Environmental Protection Agency: Cincinnati, OH.

Huisingh, J. Lewtas, S. Nesnow, R. Bradow, and M. Waters. (in
    press b). Application of a battery of short-term mutagenesis
    and carcinogenesis bioassays to the evaluation of soluble
    organics from diesel particles. In: Proceedings of the
    International Symposium on Health Effects of Diesel Engine
    Emissions, December, 1979. U.S. Environmental Protection
    Agency: Cincinnati, OH.

King, L., M. Kohan, A. Austin, L. Claxton, and J. Huisingh. (in
    press). Evaluation of the release of mutagens from diesel
    particles in the presence of physiological fluids. Environ.
    Mutagen.

Nesnow, S., and J. Lewtas Huisingh. (in press). Mutagenic and
    carcinogenic potency of extracts of diesel and related
    environmental emissions: summary and analysis of the results.
    In: Proceedings of the International Symposium on Health
    Effects of Diesel Engine Emissions, December, 1979. U.S.
    Environmental Protection Agency: Cincinnati, OH.

TUMORIGENESIS OF DIESEL EXHAUST, GASOLINE EXHAUST, AND RELATED EMISSION EXTRACTS ON SENCAR MOUSE SKIN

Stephen Nesnow
Health Effects Research Laboratory
U.S. Environmental Protection Agency
Research Triangle Park, North Carolina

Larry L. Triplett and Thomas J. Slaga
Biology Division
Oak Ridge National Laboratory
Oak Ridge, Tennessee

INTRODUCTION

Recent advances in the study of particulate emissions have brought to light several facts concerning their health effects. Many emission sources produce respirable particles with associated organic substances (Waters et al., 1979). These organic substances may be unburned fuel or they may result from pyrosynthetic reactions at or near the combustion source and photosynthetic and oxidative processes that occur after their initial formation (Crittenden and Long, 1976). Some of these organic materials contain known carcinogens and are mutagenic in short-term bioassays (Huisingh et al., 1979). Previous work by Kotin et al. (1966) and by Mittler and Nicholson (1957) gave conflicting results on the mouse skin tumorigenicity of diesel exhaust components. Similar studies with gasoline exhaust revealed a positive tumorigenic response from multiple application of condensates and extracts to mouse skin (Kotin et al., 1964; Mittler and Nicholson, 1957; Hoffmann and Wynder, 1963; Hoffmann et al., 1965). The present study was performed to examine the tumorigenicity of the organics associated with diesel exhaust particulate emissions using a sensitive mouse skin tumorigenesis model (SENCAR) and to compare the tumorigenic potency of the organics from particulate emissions of diesel, gasoline, and related emission sources.

The SENCAR mouse is a relatively new stock of carcinogen-sensitive animals, which up to this time has not been used extensively in bioassay programs. A description of the SENCAR system and of mouse skin tumorigenesis in general follows, to explain the strengths and weaknesses of this short-term in vivo carcinogenesis bioassay.

The SENCAR mouse stock has been selected for its increased
sensitivity to two-stage carcinogenesis using 7,12-dimethylbenz(a)-
anthracene (DMBA) as the initiator and 12-0-tetradecanoylphorbol-
13-acetate (TPA) as the promotor. This system is also more
sensitive to other polycyclic aromatic hydrocarbons (PAH) such as
benzo(a)pyrene (B[a]P) (Slaga et al., in press a). In addition to
its well-documented response to PAH (Slaga et al., 1978b), the
mouse skin tumorigenesis bioassay system has identified many
chemicals other than PAH as potential carcinogens (Table 1). These
chemicals represent a wide variety of structural classes, including
aldehyde, carbamate, epoxide, haloalkyether, haloaromatic,
haloalkylcarbonyl, hydroxylamine, lactone, nitrosamide, sulfonate,
sultone, and urea. This list of 32 chemicals includes such well-
known chemical carcinogens as aflatoxin B1, bis(chloromethyl)ether,
chloromethyl methyl ether, urethane, N-acetoxy-2-acetamidofluorene,
β-propiolactone, N-methyl-N'-nitro-N-nitrosoguanidine,
1,3-propanesultone, N-nitrosomethyl urea, triethylenemelamine, and
4-nitroquinoline-N-oxide. The mouse skin tumorigenesis bioassay
can also detect chemicals that cause tumors in the respiratory
tract of animals (Table 2). Of 11 known animal respiratory
carcinogens, the mouse skin tumorigenesis system has to date
detected PAH, quinolines, and carbamates. Of 11 highly suspect
occupational respiratory carcinogens, the mouse skin tumorigenesis
system has to date detected chloromethyl ethers and coke oven
emissions. These results indicate that the mouse skin
tumorigenesis bioassay can detect both dermal and nondermal
carcinogens.

The two basic protocols that can be employed to detect
chemical carcinogens in the mouse skin tumorigenesis assay are
illustrated in Figure 1. Multiple application of the test agent
for up to 60 weeks will give rise primarily to malignant carcinomas
of the skin. This protocol for complete carcinogens is a test for
agents exhibiting both tumor-initiating and tumor-promoting
activities. The bioassay protocol for tumor initiators is a single
application of test agent followed one week later by multiple
applications of a potent tumor promoter. Tumor initiation is one
step in the multistep carcinogenic process and involves the
conversion of a normal cell into a preneoplastic one. In the case
of chemical carcinogens, it involves the interaction of chemicals
or their activated forms with cellular DNA. These initiated cells
remain dormant for periods of up to one year, or until they are
stimulated to progress into hyperplastic or neoplastic lesions.
This stimulation is called tumor promotion and is accomplished by
applying croton oil or its most active component, TPA. An
initiated cell is, therefore, an irreversibly formed preneoplastic
lesion that can be stimulated to express the transformed phenotype.

Table 1.   Chemicals Other Than PAH Detected by Mouse Skin Bioassay

Class	Chemical	Reference
Aldehyde	Malonaldehyde	Shamberger et al., 1974
Carbamate	Urethane	Salaman and Roe, 1953 Slaga et al., 1973
	Vinyl carbamate	Dahl et al., 1978
	Ethyl N-phenylcarbamate	Roe and Salaman, 1955
Epoxide, diepoxide	Glycidaldehyde	Shamberger et al., 1974 Van Duuren et al., 1965
	1,2,3,4-Diepoxybutane	Van Duuren et al., 1965
	1,2,4,5-Diepoxypentane	Van Duuren et al., 1965
	1,2,6,7-Diepoxyheptane	Van Duuren et al., 1965
	Chloroethylene oxide	Zajdela et al., 1980
Haloalkylether	Bis(chloromethyl)ether	Van Duuren et al., 1969 Zajdela et al., 1980 Slaga et al., 1973
	Chloromethyl methyl ether	Slaga et al., 1973 Van Duuren et al., 1969
Haloaromatic	2,3,4,5-Tetrachloronitrobenzene	Searle, 1966
	2,3,4,6-Tetrachloronitrobenzene	Searle, 1966
	2,3,5,6-Tetrachloronitrobenzene	Searle, 1966
	Pentachloronitrobenzene	Searle, 1966
Haloalkylcarbonyl	Chloroacetone	Searle, 1966
	3-Bromopropionic acid	Searle, 1966
Hydroxylamine	N-Acetoxy-4-acetamidobiphenyl	Scribner and Slaga, 1975
	N-Acetoxy-2-acetamidofluorene	Scribner and Slaga, 1975 Slaga et al., 1978b
	N-Hydroxy-2-aminonaphthalene	Clayson and Garner, 1976
	N-Acetoxy-2-acetoamidophenanthrene	Scribner and Slaga, 1975
	N-(4-Methoxy)benzoyloxypiperidine	Scribner and Slaga, 1975
	N-(4-Nitro)benzoyloxypiperidine	Scribner and Slaga, 1975
	N-Acetoxy-4-acetamidostilbene	Scribner and Slaga, 1975
Lactone	β-Propiolactone	Roe and Salaman, 1955 Slaga et al., 1973 Hennings and Boutwell, 1969
Multifunctional	Triethylenemelamine	Roe and Salaman, 1955
	4-Nitroquinoline-N-oxide	Hennings and Boutwell, 1969
Natural products	Aflatoxin B1	Lindenfelser et al., 1974
Nitrosamide	N-Methyl-N'-nitro-N-nitrosoguanidine	Hennings et al., 1978 Fujii, 1976
Sulfonate	Allyl methylsulfonate	Roe, 1957
Sultone	1,3-Propanesultone	Slaga et al., 1973
Urea	N-Nitrosomethylurea	Graffi and Hoffman, 1966

Table 2. Response of Carcinogens in Humans, Animals, and Mouse Skin

Sample	Occupational Respiratory Carcinogen[a]	Animal Respiratory Carcinogen[a]	Mouse Skin Tumorigen[b]
Arsenic	+		
Asbestos	+	+	
Beryllium	+	+	
Carbamates		+	+
Chloromethyl ethers	+	+	+
Chromium	+		
Coke oven	+		+
Isopropyl oil	+		
MOCA[c]	+	+	
Mustard gas	+	+	
Nickel	+	+	
Nitrosamines		+	
PAH		+	+
Quinolines		+	+
Vinyl chloride	+	+	

[a]Frank, 1978.
[b]Slaga et al., 1978b, in press; Van Duuren, 1976.
[c]Methylene bis(ortho-chloroaniline).

    Relationships between tumor initiators and complete
carcinogens have been previously described. Various structurally
diverse chemicals (Table 3) are both complete carcinogens and tumor
initiators in mouse skin from CD-1 and the genetically related
SENCAR mouse. Some agents, however, appear to have only tumor-
initiating activities in mouse skin (Table 4). The correlation
between potencies as complete carcinogens and as tumor initiators
is excellent for the 12 chemicals that show both kinds of activity
(Table 5). The relationship between the production of papillomas
and the production of carcinomas in the same animals treated with
the skin tumor initiator DMBA or B(a)P is shown in Table 6. These
results indicate that the number of papillomas per mouse at 15 to
20 weeks correlates well with the number of malignant carcinomas
formed at 50 weeks, for animals treated with these two strong skin
tumor initiators.

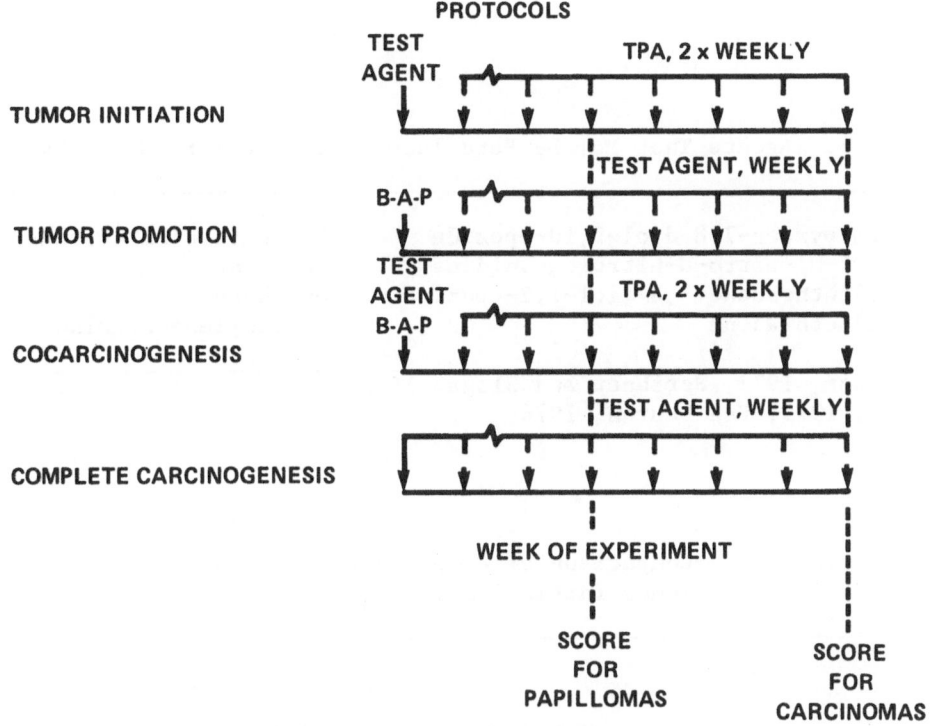

Figure 1. Protocols for bioassays of test agents as tumor
initiators, tumor promoters, cocarcinogens, and complete
carcinogens.

Table 3. Compounds That Are Both Complete Carcinogens and
Tumor Initiators in CD-1 and SENCAR Mouse Skin[a]

7,12-Dimethylbenz(a)anthracene	β-Propiolactone
3-Methylcholanthrene	Bis(chloromethyl)ether
Benzo(a)pyrene	2-Hydroxybenzo(a)pyrene
7-Methylbenz(a)anthracene	Benzo(a)pyrene-7,8-oxide
Dibenz(a,h)anthracene	Benzo(a)pyrene-7,8-diol
5-Methylchrysene	7,12-Dimethylbenz(a)anthracene-3,4-diol

[a]Hecht et al., 1979; Slaga et al., 1978b, in press b.

Table 4.   Agents That May Be Pure Tumor Initiators in Mouse Skin[a]

Benzo(a)pyrene-7,8-diol-9,10-epoxide	Dibenz(a,c)anthracene
N-Methyl-N'-nitro-N-nitrosoguanidine	Chrysene
Benz(a)anthracene-3,4-diol-1,2-epoxide	Urethane
Benz(a)anthracene	Triethylenemelamine

---

[a]Scribner, 1973; Scribner and Slaga, 1975; Slaga et al., 1973, 1978a, 1979; Van Duuren, 1976.

Table 5.   Comparison of Complete Carcinogenesis and
           Tumor Initiation in Mouse Skin

Compound	Relative Potency[a]	
	Complete Carcinogenesis (carcinomas)	Tumor Initiation (papillomas)
7,12-Dimethylbenz(a)anthracene	100	100
3-Methylcholanthrene	50	50
Benzo(a)pyrene	30	30
2-Hydroxybenzo(a)pyrene	30	30
7-Bromomethyl-12-methylbenz(a)anthracene	20	20
Benzo(a)pyrene-7,8-oxide	20	20
Dibenz(a,h)anthracene	20	20
Benz(a)anthracene	5 ± 5	5
Dibenz(a,c)anthracene	0	3
Pyrene	0	0
Benzo(a)pyrene-4,5-oxide	0	0
Anthracene	0	0

---

[a]Relative potency was determined from dose-response data.   DMBA was given a maximum value of 100 (Slaga et al., in press b).

Table 6.   Dose-response Studies on the Ability of DMBA and B(a)P
to Initiate Skin Tumors in SENCAR Mice[a]

Initiator	Dose (nmol)	No. of Papillomas per Mouse at 15 Weeks[b]	% of Mice With Papillomas at 15 Weeks	% of Mice With Carcinomas at 50 Weeks[b]
DMBA	100.0	22.0 (100)	100	100
DMBA	10.0	6.8 (32)	100	40
DMBA	1.0	3.2 (15)	93	22
DMBA	0.1	0.5 (2)	20	5
B(a)P	200.0	7.5 (100)	100	55 (100)
B(a)P	100.0	3.2 (43)	78	30 (55)
B(a)P	50.0	1.4 (19)	60	18 (33)

[a]Mice were treated one week after initiation with twice weekly
applications of 5 μg TPA.
[b]Values in parentheses represent percent normalized to the highest
dose tested of each agent (Slaga et al., in press b).

MATERIALS AND METHODS

Sample Generation and Isolation

The details of sample generation and isolation have been
reported elsewhere (Huisingh et al., in press). Briefly, the
mobile-source samples consisted of particulate emissions from two
diesel-fueled vehicles, one gasoline-fueled vehicle, and one diesel
engine (Table 7):   a heavy-duty Caterpillar 3304 engine mounted on
an engine dynamometer at 2200 rpm steady state with an 85-1b load;
a Datsun-Nissan 220-C; an Oldsmobile 350; and a 1978 Mustang II-302
V-8 catalyst engine (with emission controls and using unleaded
gasoline) mounted on a chassis dynamometer with a repeated highway
fuel economy cycle of 10.24 mi, an average speed of 48 mph, and a
running time of 12.75 min. The Caterpillar, Datsun-Nissan, and
Oldsmobile engines were fueled with the same batch of No. 2 diesel
fuel. Particulate samples were collected using a dilution tunnel
in which the hot exhaust was diluted, cooled, and filtered through
Pallflex Teflon-coated fiberglas filters.

The comparative sources employed were cigarette smoke
condensate, coke oven samples, and roofing tar emissions.
Cigarette smoke condensate was obtained by condensing smoke from an
85-mm nonfilter Kentucky reference cigarette 2R1. Condensate was
collected in acetone and refrigerated in Dry-Ice-isopropanol bath.
Cigarette smoke condensate acetone suspension was adjusted with

Table 7. Mobile Source Sample Generation

Sample	Description	Fuel	Driving Cycle
Diesel			
Cat	Caterpillar 3304	Diesel No. 2	Mode II[a]
Nissan	Nissan Datsun 220C	Diesel No. 2	HWFET[b]
Olds	Oldsmobile 350	Diesel No. 2	HWFET
Gasoline			
Mustang	1978 Mustang, II-302, V-8 catalyst and EGR	Unleaded gasoline	HWFET

[a]Mode II cycle was conducted at 2200 rpm steady state with an 85-lb load.
[b]Highway fuel economy cycle (HWFET) was a 10.24-mi cycle averaging 48 mph and taking 12.75 min.

appropriate amounts of acetone and water. Coke oven samples were collected from the top of a coke oven battery at Republic Steel, Gadston, AL, using the Massive Air Volume Sampler. Due to local wind conditions, various types of aerosols were sampled; thus, an unknown but significant portion of the emission sample may have been from the urban environment. The roofing tar emission sample was collected using a conventional tar pot with external propane burner. Pitch-based tar was heated to 360 to 380°F (182 to 193°C), and emissions were collected using a 6-ft (1.8-m) stack extension and Teflon socks in a baghouse.

The mobile source, coke oven, and roofing tar emission samples were Soxhlet-extracted with dichloromethane. The dichloromethane was removed by evaporation under dry nitrogen, and the samples were shipped in coded form in dry ice to Oak Ridge National Laboratories where the animal experiments were conducted. Table 8 shows the amount of organic material extracted from the particles with dichloromethane and the amount of B(a)P per milligram extract or per milligram particle in each sample. The B(a)P analysis was performed according to the method of Snook et al. (1976) or Swanson et al. (1978). Percent extractable of organic material from the particles varied from 8% of the Nissan sample to a maximum of 99% for the roofing tar sample. Since cigarette smoke condensate was not a particulate sample per se, the complete sample was used in

Table 8. Benzo(a)pyrene Analysis[a]

Sample	Extractable (%)	B(a)P (ng/mg extract)	B(a)P (ng/mg particle)
Diesel:			
Cat	27	2	0.5
Nissan	8	1173	96.2
Olds	17	2	0.4
Gasoline:			
Mustang	43	103	44.1
Comparative Sources:			
Cigarette	--	<1	--
Coke	7	478	31.5
Roofing tar	>99	889	889

[a]B(a)P analysis was performed according to Swanson et al. (1978), except for analysis of cigarette smoke condensate, which was performed according to Snook et al. (1976).

the biological analysis. B(a)P in the extracts varied from less than 1 ng/mg extract for the cigarette sample to a high of 1173 ng/mg extract for the Nissan sample.

Animals

SENCAR mouse stock, selected for its increased sensitivity to carcinogenesis (Boutwell, 1964) was used in this study. These mice were derived by breeding Charles River CD-1 mice with male STS (skin-tumor-sensitive) mice that were originally derived from Rockland mice. Mice were selected for sensitivity to the DMBA-TPA two-stage system of tumorigenesis for eight generations. These mice were initially obtained from Dr. R. Boutwell (McArdle Laboratory for Cancer Research, University of Wisconsin, Madison, WI) and now are being raised at the Oak Ridge National Laboratory, Oak Ridge, TN.

Chemicals

TPA was obtained from Dr. P. Borchert (University of Minnesota Minneapolis, MN) and B(a)P from Aldrich Chemical Co. All the agents were prepared under yellow light immediately before use and applied topically in 0.2 ml of spectral-quality acetone.

Tumor Experiments

These studies employed 80 mice per treatment group (40 of each sex). All the mice were shaved with surgical clippers two days before the initial treatment, and only those mice in the resting phase of the hair cycle were used. Five dose levels were used for the tumor-initiating activities of the various samples, except for the Mustang sample, which was tested at four dose levels. B(a)P was used as the standard for the tumor-initiation studies, using four dose levels. One week after application, the tumor promoter TPA was administered twice weekly. All samples at all doses were applied as a single treatment, except for the 10-mg dose, which was administered in five daily doses of 2 mg. Skin tumor formation was recorded weekly, and papillomas greater than 2 mm in diameter were included in the cumulative total if they persisted for one week or longer. Both the number of mice with tumors and the number of tumors per mouse were determined and recorded weekly. Papillomas and carcinomas were removed randomly for histological verification.

RESULTS AND DISCUSSION

The organic extracts from particulate emissions described previously were applied to the backs of SENCAR mice according to the protocols cited in Materials and Methods. The production of benign papillomas on a weekly basis is depicted in Figure 2 for both the reference standard B(a)P and the Nissan sample. In both cases, after a 7-to 8-week latency period, the percent of animals bearing tumors rose dramatically between weeks 8 to 14, with a 95 to 100% tumor incidence observed in both of these dose groups. Mean number of papillomas per mouse began to rise from control between weeks 6 to 8, increasing much more slowly than did the number of animals with tumors. A plateau was reached during weeks 22 to 25. In both cases, the numbers of papillomas per animal ranged from five to six.

B(a)P exhibited a linear dose response between 2.52 and 100.92 µg (10 to 400 nmol) in both male and female SENCAR mouse skin (Figure 3). The males seemed to be more sensitive than the females to this carcinogen, although this sex difference was not evident for the complex mixture samples evaluated. The most active sample tested in this series was the coke oven extract. The response to

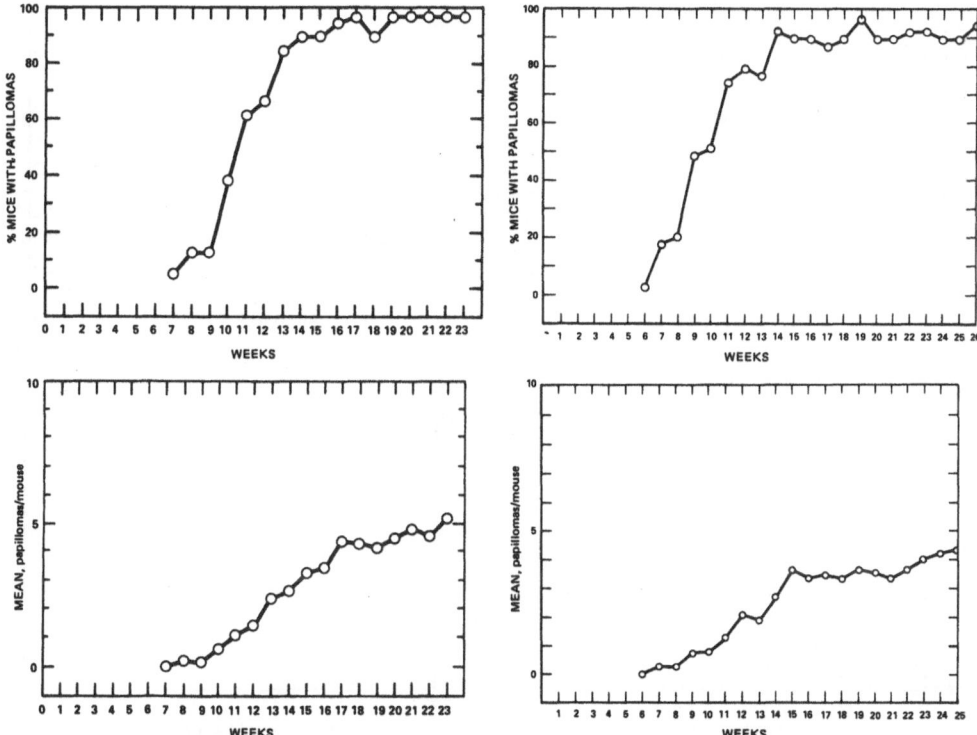

Figure 2. SENCAR mouse skin-tumor initiation. Male SENCAR (40) were initiated with either a single dose of B(a)P (50.4 µg) or five daily treatments with Nissan extract (2 mg). Animals were then treated biweekly with TPA (2 µg). Left: B(a)P. Right: Nissan extract.

this sample in both male and female animals was biphasic. An initial linear dose response was observed between 0.1 and 2 mg extract, with animals carrying an average of five to six papillomas. The roofing tar extract and Nissan extract (Figure 4) also produced a large tumor response in both male and female animals.

The Oldsmobile sample exhibited a linear dose response up to 1 mg and a subsequent loss of activity at 10 mg. The response to the Oldsmobile sample was one tenth that for the Nissan, coke oven, and roofing tar samples. The gasoline-fueled Mustang II sample also produced a weak response in both male and female animals. The "goodness of fit" (R^2) to the linear regression analysis for the female animals was extremely low, 0.686, indicating a lack of

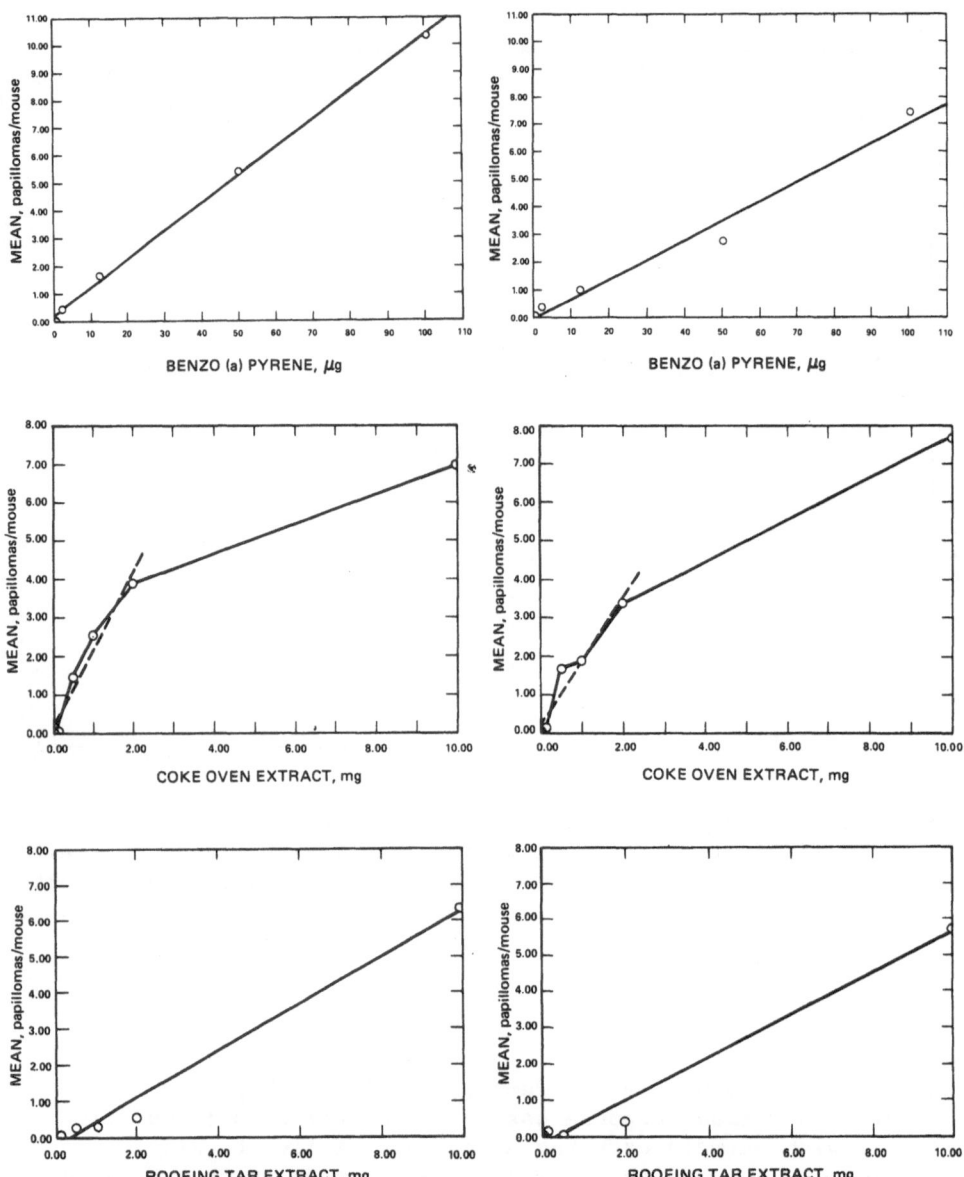

Figure 3.   SENCAR mouse skin tumor initiation dose-response plots
            (mean number of papillomas per mouse, after subtracting
            background level). Graphs on left are for males and
            those on right are for females. There were 40 animals
            per dose group. The numbers of surviving animals at
            scoring were as follows: B(a)P--males, 156; females,
            156; coke oven extract--males, 195; females 197; roofing
            extract--males, 197; females, 196.

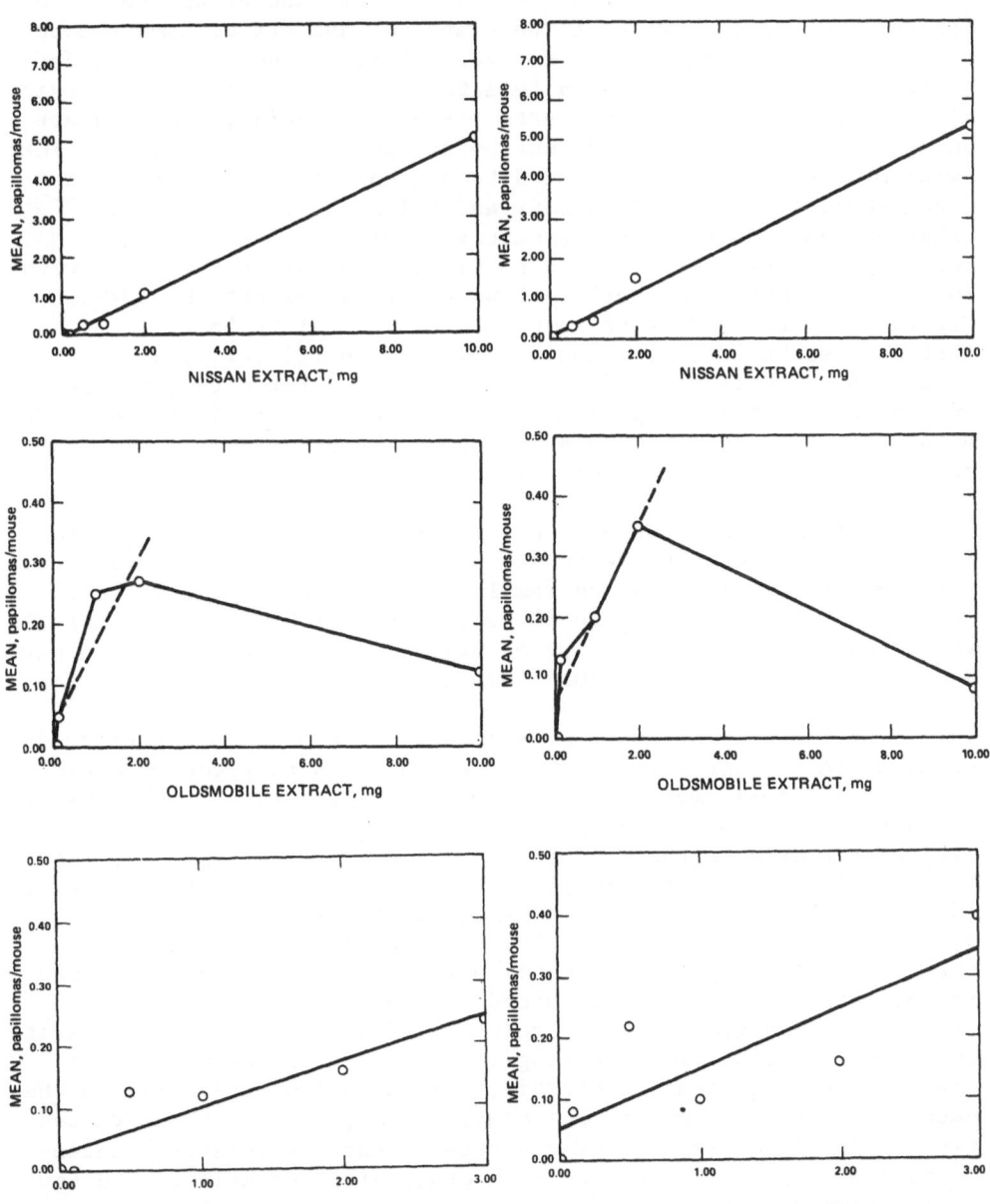

Figure 4.   SENCAR mouse skin tumor initiation dose-response plots
(mean number of papillomas per mouse, after subtracting
background level). See Figure 3 caption for
explanation. Nissan extract--males, 190; females, 198;
Oldsmobile extract--males, 156; females, 157; Mustang
extract--males, 188; females, 195.

linear dose response. The Caterpillar sample and cigarette smoke
condensate produced two to three times the numbers of tumors found
in the controls (Figure 5). However, there was no observable dose
response for the doses tested (0.1 to 10 mg). The lack of activity
of the cigarette smoke condensate was disappointing, although not
unexpected. Cigarette smoke condensate when applied to female ICR
Swiss mice twice weekly produced tumors only at relatively high
doses (Gori et al., 1977; Wynder and Hoffmann, 1967). It was
expected that the increased sensitivity of SENCAR mice to
carcinogens would allow tumors to be observed after treatment with
10 mg whole-smoke condensate. However, this was not the case.
Cigarette smoke condensate is not an extract of isolated
particulates but a suspension of organics, particles, and
volatiles. Therefore, it has not been concentrated to the same
extent as the other samples. The detectability limit of the SENCAR
mouse skin tumorigenesis assay is above the doses and
concentrations tested of the cigarette smoke condensate.

     The formation of spontaneous tumors in animals treated with
acetone and promoted twice weekly with TPA was 0.08 and 0.05
papillomas/mouse in male and female animals, respectively, at 22
weeks after initiation; 7 to 8% of the animals had tumors. Animals
initiated with up to 100.92 µg of B(a)P, followed by promotion with
acetone alone, did not produce tumors.

     A preliminary analysis of the results obtained was performed
using a linear regression statistical analysis to produce potencies
in terms of papillomas per animal per milligram agent. The results
of these calculations are found in Table 9. The R^2 (goodness of
fit) of the data to the linear response was greater than 0.920 for
8 out of 12 of the test groups and greater than 0.84 for 11 out of
12. Potency values ranged from 0 to 101 papillomas/mouse/mg
agent. The higher of these values was obtained from the B(a)P
treatment groups and was an extrapolation from the microgram dose
range, where the data was obtained, to the milligram range.
Obviously, this number is theoretical and based on strict linearity
throughout a 1000-fold dose range, an assumption not yet proven.
Also, it is a physical impossibility to have 100 papillomas on the
back of a mouse. However, for comparative purposes, these values
give a good approximation of the true values. A relative ranking
of each of the test groups to each other after normalizing to the
Nissan sample is also found in Table 9. The ranking indicates
that the potency of B(a)P was greater than that of the coke oven
sample, which was in turn greater than those of the roofing tar and
Nissan samples. The potencies of these samples were greater than
those of the Oldsmobile and Mustang samples. All of these samples
were greater in potency than were cigarette smoke condensate and
the Caterpillar sample, whose potencies were not significantly
different from zero.

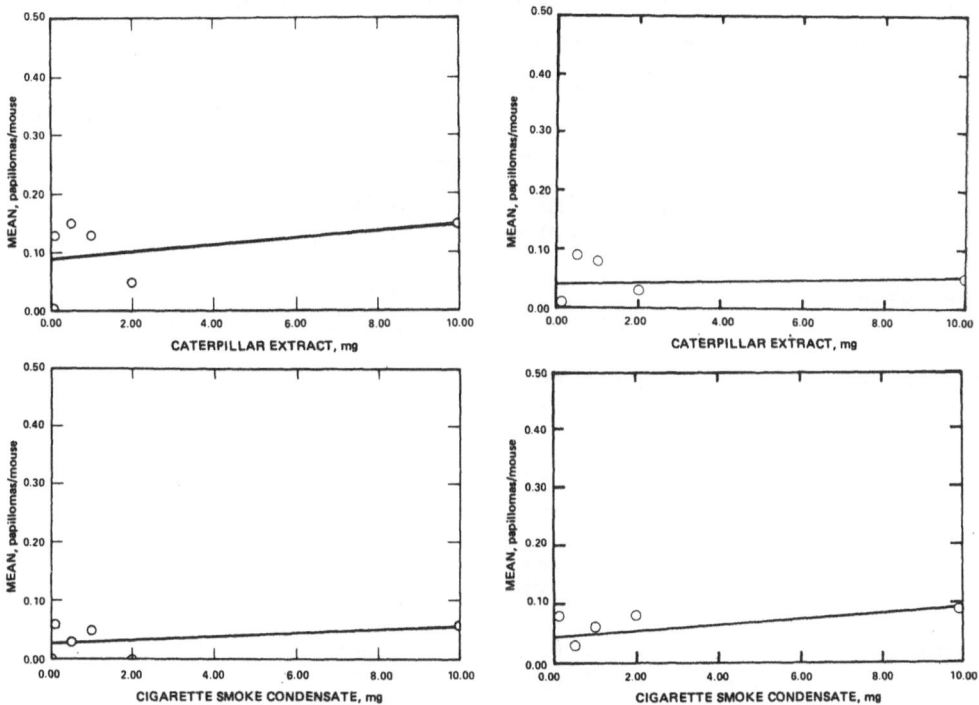

Figure 5.   SENCAR mouse skin tumor initiation dose-response plots
            (mean number of papillomas per mouse, after subtracting
            background level). See Figure 3 caption for
            explanation. Caterpillar extract--males, 196; females
            191; cigarette smoke condensate--males, 187; females,
            194.

        The results presented here confirm and expand the earlier
observations by Kotin et al. (1966) on the tumorigenesis of diesel
exhaust components and clearly indicate the tumorigenic potential
of these materials. The results also indicate a range of response
of diesel engines, presumably due to differences in engine
technology.

        Comparison of the tumor data in Table 9 with the B(a)P
content per milligram extract in Table 8 indicates a lack of
correlation between the two parameters. This result suggests that
B(a)P and possibly other associated PAH are not reliable markers
for tumorigenic activity in these complex mixtures, and that other
non-PAH chemicals in the mixtures make major contributions to their
overall potency.

Table 9. SENCAR Mouse Skin Tumor Initiation: Sample Rankings[a]

Sample	Papillomas/ Mouse/mg		R^2	Relative Ranking
Benzo(a)pyrene	101	(M)	0.999	20000
	71.1	(F)	0.979	13000
Coke oven	2.00	(M)	0.960	400
	1.65	(F)	0.922	310
Roofing tar	0.640	(M)	0.975	130
	0.571	(F)	0.977	110
Nissan	0.532	(F)	0.991	100
	0.507	(M)	0.998	100
Olds	0.148	(F)	0.896	28
	0.135	(M)	0.844	27
Mustang	0.097	(F)	0.686	18
	0.073	(M)	0.842	14
Cigarette	0		--	0
Caterpillar	0		--	0

[a]A linear regression model was applied to the individual data points to obtain both slope potency and R^2. M and F refer to results from male and female animals, respectively.

In conclusion, the SENCAR mouse skin tumorigenesis bioassay for tumor initiation is a quantitative short-term in vivo rodent carcinogenesis system that detects a variety of structurally diverse chemical carcinogens. This bioassay system has also shown its utility in evaluating complex environmental mixtures for tumorigenic potential. It gives excellent dose responses with both pure substances and complex mixtures and has shown utility for comparative potency analysis. Additional statistical models are being evaluated to analyze this data, and the results will be reported elsewhere.

ACKNOWLEDGMENTS

   The authors wish to thank R.L. Bradow, R.H. Jungers, B.D.
Harris, T.O. Vaughan, R.B. Zweidinger, K.M. Cushing, J. Bumgarner,
and B.E. Gill for the sample isolation and characterization, and
Carol Evans for the ADP programming. The research was sponsored by
the U.S. Environmental Protection Agency, contract no. 79D-X0526,
under the Interagency Agreement, U.S. Department of Energy no.
40-728-78, and the Office of Health and Environmental Research,
U.S. Department of Energy, under contract no. 7405 eng-26 with the
Union Carbide Corporation.

REFERENCES

Boutwell, R.K. 1964. Some biological aspects of skin
     carcinogenesis. Progr. Exp. Tumor Res. 4:207-250.

Clayson, D.B., and R.C. Garner. 1976. Carcinogenic aromatic
     amines and related compounds. In: Chemical Carcinogens.
     C.E. Searle, ed. American Chemical Society: Washington, DC.
     pp. 366-461.

Crittenden, B.D., and R. Long. 1976. The mechanisms of formation
     of polynuclear aromatic compounds in combustion systems. In:
     Polynuclear Aromatic Hydrocarbons. R. Freudenthal and P.W.
     Jones, eds. Raven Press: New York. pp. 209-223.

Dahl, G.A., J.A. Miller, and E.C. Miller. 1978. Vinyl carbamate
     as a promutagen and a more carcinogenic analog of ethyl
     carbamate. Cancer Res. 38:3793-3804.

Frank, A.L. 1978. Occupational lung cancer. In: Pathogenesis
     and Therapy of Lung Cancer. C.C. Harris, ed. Marcel Dekker,
     Inc.: New York. pp. 25-51.

Fujii, M. 1976. Carcinogenic effect of N-butyl-N-nitrosourethane
     on CDF-1 mice. Gann 67:231-236.

Gori, G.B., ed. 1977. Toward Less Hazardous Cigarettes: The
     First Set of Experimental Cigarettes. Department of Health,
     Education, and Welfare publication no. 76-905. National
     Institutes of Health: Washington, DC.

Graffi, A., and F. Hoffman. 1966. Strong carcinogenic effect of
     methl nitrosourea on mouse skin in the drop test. Acta Biol.
     Med. Ger. 16:K1-K3.

Hecht, S.S., E. LaVoie, R. Mazzarese, N. Hirota, T. Ohmori, and D. Hoffmann. 1979. Comparative mutagenicity, tumor-initiating activity, carcinogenicity and in vitro metabolism of fluorinated 5-methylchrysenes. J. Natl. Cancer Inst. 63:855-861.

Hennings, H., and R.K. Boutwell. 1969. Inhibition of DNA synthesis by initiators of mouse skin tumorigenesis. Cancer Res. 29:510-514.

Hennings, H., B. Michael, and E. Patterson. 1978. Croton oil enhancement of skin tumor initiation by MNNG. Proc. Soc. Exp. Biol. Med. 158:1-4.

Hoffmann, D., E. Theisz, and E.L. Wynder. 1965. Studies on the carcinogenicity of gasoline exhaust. J. Air Pollut. Contr. Assoc. 15:162-165.

Hoffmann, D., and E.L. Wynder. 1963. Studies on gasoline engine exhaust. J. Air Pollut. Contr. Assoc. 13:322-327.

Huisingh, J. Lewtas. (in press). Mutagenic and carcinogenic potency of extracts of diesel and related environmental emissions: preparation and characterization of the samples. Proceedings of the International Symposium of Health Effects of Diesel Engine Emissions.

Huisingh, J., R. Bradow, R. Jungers, L. Claxton, R. Zweidinger, S. Tejada, J. Bumgarner, F. Duffield, M. Waters, V. Simmon, C. Hare, C. Rodriguez, and L. Snow. 1979. Application of bioassay to the characterization of diesel particle emissions. In: Application of Short-term Bioassays in the Fractionation and Analysis of Complex Environmental Mixtures. M.D. Waters, S. Nesnow, J.L. Huisingh, S.S. Sandhu, and L. Claxton, eds. Plenum Press: New York. pp. 383-418.

Kotin, P., H.L. Falk, and M. Thomas. 1964. Aromatic hydrocarbons, II. Presence in the particulate phase of gasoline-engine exhausts and the carcinogenicity of exhaust extracts. AMA Arch. Ind. Hyg. Occup. Med. 9:164-177.

Kotin, P., H.L. Falk, and M. Thomas. 1966. Aromatic hydrocarbons, III. Presence in the particulate phase of diesel-engine exhausts and the carcinogenicity of exhaust extracts. AMA Arch. Ind. Hyg. Occup. Med. 11:113-120.

Lindenfelser, L.A., E.B. Lillehoj, and H.R. Burmeister. 1974. Aflatoxin and trichothecene toxins: skin tumor induction and synergistic acute toxicity in white mice. J. Natl. Cancer Inst. 52:113-116.

Mittler, S., and S. Nicholson. 1957. Carcinogenicity of
    atmospheric pollutants. Ind. Med. Surg. 26:135-138.

Roe, F.J.C. 1957. Tumor initiation in mouse skin by certain
    esters of methane sulfonic acid. Cancer Res. 17:64-67.

Roe, F.J.C., and M.H. Salaman. 1955. Further studies on
    incomplete carcinogenesis: triethylene melamine (TEM),
    1,2-benzanthracene, and β-propiolactone as initiators of skin
    tumor formation in the mouse. Br. J. Cancer 9:177-203.

Salaman, H.H., and F.J.C. Roe. 1953. Incomplete carcinogens:
    Ethyl carbamate (urethane) as an initiator of skin tumor
    formation in the mouse. Br. J. Cancer 7:472-481.

Scribner, J.D. 1973. Tumor initiation by apparently
    noncarcinogenic polycyclic aromatic hydrocarbons. J. Natl.
    Cancer Inst. 50:1717-1719.

Scribner, J.D., and T.J. Slaga. 1975. Tumor intiation by
    N-acetoxy derivatives of piperidine and N-arylacetamides. J.
    Natl. Cancer Inst. 54:491-493.

Searle, C.E. 1966. Tumor initiatory activity of some
    chloromononitrobenzenes and other compounds. Cancer Res.
    26:12-17.

Shamberger, R.J., T.L. Andreone, and C.E. Willis. 1974.
    Antioxidants and cancer IV. Initiating activity of
    malonaldehyde as a carcinogen. J. Natl. Cancer Inst.
    53:1771-1773.

Slaga, T.J., G.T. Bowden, B.G. Shapas, and R.K. Boutwell. 1973.
    Macromolecular synthesis following a single application of
    alkylating agents used as initiators of mouse skin
    tumorigenesis. Cancer Res. 33:769-776.

Slaga, T.J., W.M. Bracken, A. Viaje, D.L. Berry, S.M. Fischer, D.R.
    Miller, W. Levin, A.H. Conney, H. Yagi, and D.M. Jerina.
    1978a. Tumor initiating and promoting activities of various
    benzo(a)pyrene metabolites in mouse skin. In: Polynuclear
    Aromatic Hydrocarbons. R.I. Freudenthal and W. Jones, eds.
    Raven Press: New York. pp. 371-382.

Slaga, T.J., S.M. Fischer, L. Triplett, and S. Nesnow. (in
    press a). Comparison of complete carcinogenesis and tumor
    initiation in mouse skin: Tumor initiation promotion, a
    reliable short term assay. In: Proceedings of the
    Symposium on Short-term In Vivo Carcinogenesis Bioassays.

Slaga, T.J., G.L. Gleason, J. DiGiovanni, D.L. Berry, M.R. Juchau, and R.G. Harvey. 1979. Tumor initiating activities of various derivatives of benz(a)anthracene and 7,12-dimethylbenz(a)anthracene in mouse skin. In: Proceedings of the Third International Battelle Conference on Polynuclear Aromatic Hydrocarbons. P.W. Jones and P. Leber, eds. Ann Arbor Press: Ann Arbor, MI. pp. 753-764.

Slaga, T.J., A. Sivak, and R.K. Boutwell, eds. 1978b. Carcinogenesis: A Comprehensive Survey, Vol. 2. Mechanisms of Tumor Promotion and Cocarcinogenesis. Raven Press: New York.

Slaga, T.J., L.L. Triplett, and S. Nesnow. (in press b). Mutagenic and carcinogenic potency of extracts of diesel and related environmental emissions: two-stage carcinogenesis in skin tumor sensitive mice (SENCAR). Proceedings of the International Symposium on Health Effects of Diesel Emissions.

Snook, M.E., R.F. Severson, H.C. Higman, R.F. Arrendale, and O.T. Chortyk. 1976. Polynuclear aromatic hydrocarbons of tobacco smoke: Isolation and identification. Zeit. zur Tabak. pp. 250-276.

Swanson, D., C. Morris, P. Hedgecoke, R. Jungers, R. Thompson, and J. Bumgarner. 1978. A rapid analytical procedure for the analysis of benzo(a)pyrene in environmental samples. In: Trends in Fluorescence, Vol. 1, No. 2. Perkin-Elmer Corp.: Norwalk, CT. pp. 22-27.

Van Duuren, B.L. 1976. Tumor-promoting and co-carcinogenic agents in chemical carcinogenesis. In: Chemical Carcinogens. C.E. Searle, ed. American Chemical Society: Washington, DC. pp. 24-51.

Van Duuren, B.L., L. Orvis, and N. Nelson. 1965. Carcinogenicity of epoxides, lactones and peroxy compounds, Part II. J. Natl. Cancer Inst. 35:707-717.

Van Duuren, B.L., A. Sivak, B.M. Goldschmidt, C. Katz, and S. Melchionne. 1969. Carcinogenicity of haloethers. J. Natl. Cancer Inst. 43:481-486.

Waters, M.D., S. Nesnow, J.L. Huisingh, S.S. Sandu, and L. Claxton, eds. 1978. Application of Short-term Bioassays in the Fractionation and Analysis of Complex Environmental Mixtures. Plenum Press: New York.

Wynder, E.L., and D. Hoffmann, eds. 1967. Tobacco and Tobacco Smoke. Academic Press: New York. pp. 133-315.

Zajdela, F., A. Croisy, C. Malaveille, L. Tomatis, and H. Bartsch. 1980. Carcinogenicity of chloroethylene oxide, an ultimate reactive metabolite of vinyl chloride, and bis(chloromethyl)ether after subcutaneous administration and in initiation-promotion experiments in mice. Cancer Res. 40:352-356.

BACTERIAL MUTAGENESIS AND THE EVALUATION OF MOBILE-SOURCE EMISSIONS

Larry Claxton and Mike Kohan
Health Effects Research Laboratory
U.S. Environmental Protection Agency
Research Triangle Park, North Carolina

INTRODUCTION

Interest in developing a rapid, inexpensive means of detecting and evaluating the potential health hazards of mobile-source emissions is increasing. Faced with the staggering numbers of chemicals created through combustion processes that have never been assayed for mutagenicity and/or carcinogenicity, the chemist faces a futile task of identifying and controlling all potential health hazards. This study will demonstrate how bioassay techniques, and particularly the Salmonella assay, can be coupled with the fractionation of chemically complex emissions to identify components requiring more extensive analysis and control.

Emission source organics vary with factors such as time, fuel, and environmental conditions. In this study, several variables influencing mobile source studies are examined: day-to-day variation from a single source; variation among several vehicles of the same make, model, and configuration; and variation between different light-duty mobile sources. Also explored are the effects of storage and the creation of artifacts during the initial collection of sample. By understanding certain characteristics and uses of the available bacterial strains, the investigator can use microbial bioassays to aid in identification of mutagens in complex mixtures, characterize and compare the types of mutagenic components within complex mixtures, and screen various mobile sources for levels of mutagenic compounds.

MATERIALS AND METHODS

Bioassays

The primary mutation assay used was the Salmonella typhimurium plate-incorporation assay as described by Ames et al. (1975). However, the test protocol had the following minor modifications: 1) minimal histidine was added to the base agar in the petri dish rather than to the soft-agar overlay; 2) plates were counted at 48 and 72 h to provide an additional check for toxicity factors; 3) colony counting was performed with an Artek automatic colony counter; and 4) when adequate sample was available, each dose was done in triplicate. Microsomal activation was provided using a 9000 x g supernatant of Aroclor-1254-induced Charles River CD-1 rats, as described by Ames et al. (1975).

Six indicator strains of S. typhimurium were used: TA98, TA100, TA1537, TA1538, TA1535, and TA98-FR1. The TA98-FR1 strain is a nitroreductase-deficient strain (Rosenkranz and Speck, 1975), which was provided by Dr. Herbert Rosenkranz (New York Medical College, Valhalla, NY). TA98-FR1 is deficient in only one of several nitroreductase enzymes (H.S. Rosenkranz, personal communication, 1979). All other strains were provided by Dr. Bruce Ames (University of California at Berkeley). The test results for all six indicator strains are presented in this summary paper; however, the data is given for only one strain, due to the large volume of data collected.

A second bioassay was also conducted: the 8-azaguanine forward mutation assay. Two strains of S. typhimurium, TM677 and TM35, were used, as described by Skopek et al. (1978a, b).

Samples

All samples were organic extracts from automotive exhaust particles. In each case, the vehicle was operated on a chassis dynamometer, the exhaust was diluted and cooled in a stainless steel dilution tunnel, and the particles were collected on Pallflex T60A20 glass fiber filters. The entrapped particles were then extracted with dichloromethane (Huisingh et al., 1978). The remaining organics were solvent exchanged to dimethylsulfoxide (DMSO) to a final concentration of 2 mg exhaust organics/ml DMSO. The DMSO solution was used in the bioassay. Each sample was given a unique identificaiton number to prevent bias in testing and to allow computerization of all data. These identification numbers are used for sample identification in this paper.

Sample CMBX-79-0092 is a Nissan 220C diesel vehicle, and the results for this sample are reported to demonstrate the typical

response in the five major tester strains. Samples CMBX-79-0001 to
CMBX-79-0012 were 12 aliquots of the same sample. Each aliquot was
tested during consecutive months to ascertain any effects of
storage. Multiple samples were collected on three consecutive days
from the same Oldsmobile 350 vehicle and supplied as samples MSER-
78-0122 to MSER-78-0135. A collection of exhaust organics from the
Oldsmobile 350 diesel was chemically fractionated by the Research
Triangle Institute (RTI), Research Triangle Park, NC, and was
assigned numbers MSER-79-0039 to MSER-79-0046. The method of
chemical fractionation is reported elsewhere (Little, 1978; Lee et
al., 1976). MSER-79-0032 to MSER-79-0036 were samples from five
separate gasoline automobiles of the same make, model, and
configuration (Ford LTD, catalyst equipped). The variation among
different makes and models of diesel automobiles was demonstrated
with the TAEB samples. The nitroreductase-deficient strains were
used with the following samples: an Oldsmobile 350 diesel (MSER-
79-0074), a 1978 Datsun 810 gasoline (MSER-79-0064), an Oldsmobile
260 diesel (TAEB-79-0029), and a VW Dasher diesel (TAEB-79-0034)
vehicle. The comparison of forward and reverse mutation systems
using a preincubation protocol used an Oldsmobile 260 sample (TAEB-
79-0030) and a VW Dasher sample (TAEB-79-0036). Table 1 summarizes
the samples.

RESULTS

The qualitative response of the five Ames tester strains to
exhaust extracts from various internal combustion engines was very
consistent. TA1535 generally gave either a very low or a negative
response. Each of the frameshift tester strains has given positive
results. Since TA100 responds to frameshift mutagens as well as to
other mutagens, it also gave a positive response. Figure 1 shows
this typical response of the five tester strains to the organics
from the exhaust of a Nissan 220C automobile.

Tables 2 and 3 summarize the results of bioassay data from
multiple Highway Fuel Economy Test (HWFET) cycles run on three
consecutive days with the same Oldsmobile 350 diesel vehicle. The
bioassay data is expressed for the organics as revertants per
microgram organic (slope of the linear regression line). By using
the percent extractable mass and the particulate emission rate
(PER) for each automobile, the revertants per gram particulate and
the revertants per mile were calculated from the given slope. The
revertants per microgram ranged from 2.86 to 4.33, and the
revertants per mile ranged from 187,000 to 284,000. Even with the
variation of test runs and bioassays, there is less than a twofold
difference in the bioassay values for different filter extracts
from the same vehicle operated on the HWFET cycle. The overall
coefficient of variation is approximately 11%.

Table 1. Sample Information

Sample	Sample Number
Oldsmobile 350 diesel[a]	
Total particle extract	CMBX-79-0001 to 0012
	MSER-78-0122 to 0135
Acid I fraction	MSER-79-0039
Acid II fraction	MSER-79-0040
Base I fraction	MSER-79-0041
Base II fraction	MSER-79-0042
Insoluble tars	MSER-79-0043
Polar neutral fraction	MSER-79-0044
Polynuclear aromatics	MSER-79-0045
Nonpolar neutral fraction	MSER-79-0046
Ford LTD gasoline automobiles[a]	
Total particle extract	MSER-79-0032 to 0036
Nissan 220C diesel[a]	
Baseline study for strains	CMBX-79-0092
Comparative vehicles[a,b]	
Oldsmobile 350 diesel[a,b]	TAEB-78-0501-0507
	MSER-79-0074
Oldsmobile 260 diesel[b]	TAEB-78-0502, 0503
	TAEB-79-0029, 0030
Mercedes Benz 300D diesel[b]	TAEB-78-0504, 0505
Open Record E diesel[b]	TAEB-78-0506, 0510
Chevrolet truck 350 diesel[b]	TAEB-78-0511, 0512
VW Dasher wagon diesel[b]	TAEB-78-0513, 0514
1978 Datsun 810 gasoline[a]	MSER-79-0064
VW Dasher diesel[b]	TAEB-79-0034, 0036

[a]Supplied by Ron Bradow and Roy Zweidinger, U.S. Environmental
Protection Agency (EPA), Research Triangle Park (RTP), NC.
Fractionation by Edo Pellizzari, RTI, RTP, NC.
[b]Supplied by Tom Baines, Emission Control Technology Division,
EPA, Ann Arbor, MI.

    The same diesel automobile (Oldsmobile 350) was used to
generate multiple samples that were pooled subsequently for a large
single sample. Twelve of the aliquots (CMBX-79-0001 to CMBX-79
-0012) from this sample were stored in glass vials at -80°C in the
dark. Each month for 12 months, one aliquot was tested in the
microbial assay, and the results are recorded in Table 4. For

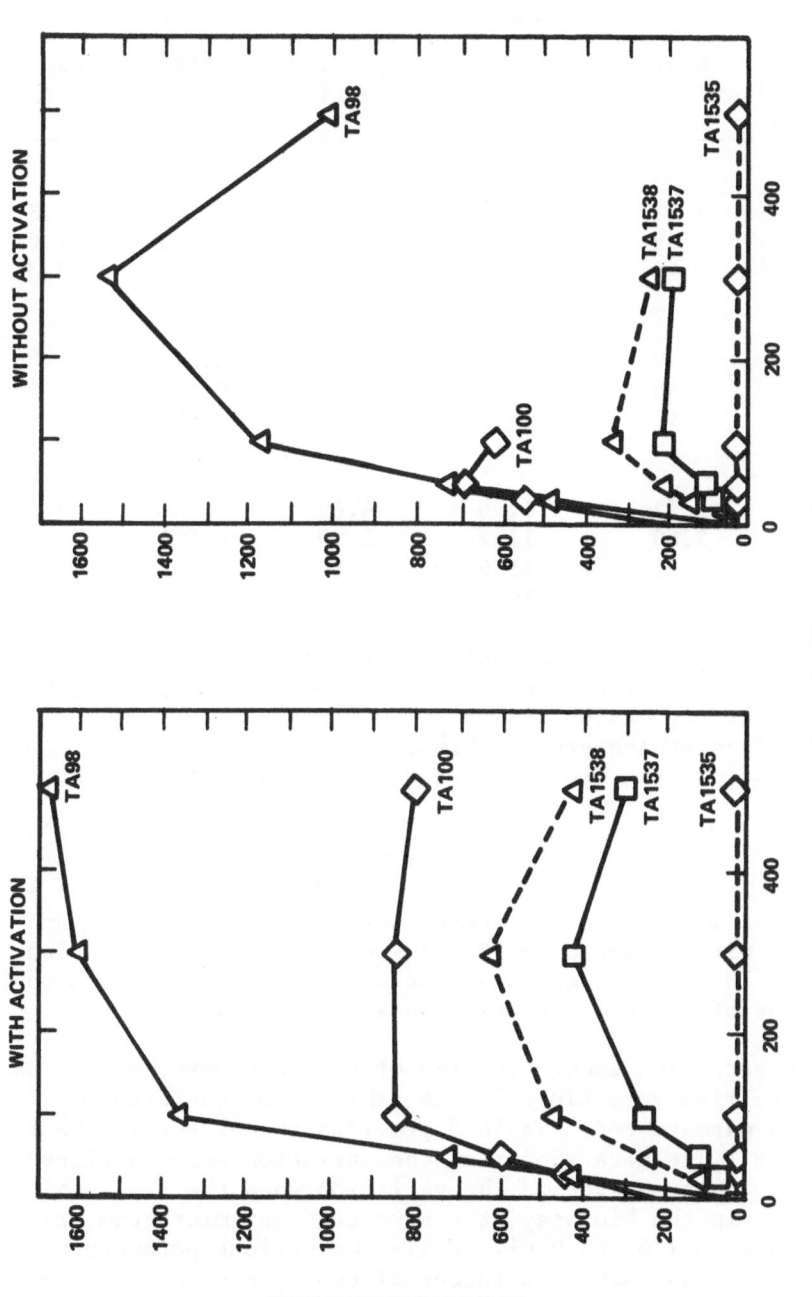

Figure 1. Exhaust organics from a Nissan 220C vehicle tested in the five tester strains of S. typhimurium.

Table 2. Comparative Results of Exhaust Organics from an Oldsmobile
350 Diesel Tested with S. typhimurium TA98 Without Activation

Sample No. (MSER-78-)[a]	Slope[b] (Rev/plate/µg)	% Ext.[c]	Rev x 10[5] /g Partic.	PER[d] (g/mile)	Rev x 10[5]/mile
Day 1					
0122	4.33	10.8	4.68	0.502	2.35
0123	3.99	10.8	4.31	0.484	2.09
0124	3.82	12.4	4.74	0.495	2.34
Day 2					
0126	3.99	13.8	5.51	0.516	2.84
0129	3.86	11.4	4.40	0.517	2.28
0130	3.41	12.8	4.36	0.487	2.13
Day 3					
0132	3.72	12.6	4.69	0.518	2.43
0133	2.86	11.2	3.20	0.584	1.87
0134	3.35	11.6	3.89	0.569	2.21
0135	3.46	10.8	3.74	0.568	2.12

[a]Numbers assigned to each automobile exhaust sample collected by
 EPA, Environmental Sciences Research Laboratory (ESRL), RTP, NC.
[b]Slope of linear regression line.
[c]Percent dichloromethane-extractable mass.
[d]Particle emission rate (courtesy of Roy Zweidinger, EPA, ESRL, RTP,
 NC).

these storage samples, the revertants per microgram ranged from
4.05 to 7.23, with a mean of 5.52. The coefficient of variation
for slope was 18.1%. Although obvious variance was seen from month
to month, none of the differences were significant.

    To test variation among vehicles of the same make, model, and
configuration, five Ford LTD gasoline vehicles of the same
configuration were tested. Table 5 provides the bioassay data for
these vehicles. The data show that the variation among different
vehicles was much greater than the variation when the same vehicle
was retested. In the bioassay, the revertant per microgram organic
levels varied from 1.02 to 9.61. While the various parameters of a
single vehicle could vary by a factor of two, the same parameters
for different individual vehicles could show a 5- to 10-fold
difference. It can be seen from Table 5 that a single sample
(i.e., sample MSER-79-0036) could be responsible for most of the
variation, when small numbers of samples were used. When the

Table 3. Daily and Total Statistics of Results for Exhaust Organics
from an Oldsmobile 350 Diesel Tested with S. typhimurium TA98
Without Activation

Sample No. (MSER-78-)[a]	Slope[b] (Rev/plate/μg)	% Ext.[c]	Rev x 10^5 /g Partic.	PER[d] (g/mile)	Rev x 10^5/mile
Day 1					
Mean	4.05	11.3	4.58	0.494	2.26
Standard dev.	0.26	0.9	0.23	0.009	0.15
Coeff. of var.	0.06	0.08	0.05	0.02	0.07
Day 2					
Mean	3.75	12.7	4.76	0.507	2.42
Standard dev.	0.30	1.2	0.65	0.017	0.37
Coeff. of var.	0.08	0.08	0.14	0.03	0.15
Day 3					
Mean	3.35	11.6	3.88	0.560	2.16
Standard dev.	0.36	0.8	0.62	0.029	0.23
Coeff. of var.	0.11	0.07	0.16	0.05	0.11
All days combined					
Mean	3.68	11.8	4.35	0.524	2.27
Standard dev.	0.42	1.0	0.64	0.037	0.26
Coeff. of var.	0.11	0.09	0.15	0.07	0.11

[a]Numbers assigned to each automobile exhaust sample collected by
EPA, Environmental Sciences Research Laboratory (ESRL), RTP, NC.
[b]Slope of linear regression line.
[c]Percent dichloromethane-extractable mass.
[d]Particle emission rate (courtesy of Roy Zweidinger, EPA, ESRL, RTP,
NC).

revertants per microgram organic were normalized to revertants per
mile, the variation was reduced from a 10-fold to a 5-fold
difference.

The TAEB samples (Table 6) demonstrated the degree of
variation in mutagenic response found between diesel vehicles of
different makes and models. Among the diesel vehicles tested, the
revertants per microgram organic ranged from 1.15 to 3.23, while
the revertants per mile ranged from 250,000 to 878,000.

Results from bioassay techniques can be used to guide the
chemical fractionation of complex exhaust organics. This

Table 4. Mutagenicity of Diesel Exhaust Organic Samples Stored Over Varying Periods of Time Tested with S. typhimurium TA100

Mean Revertants per Plate for Monthly Samples

Compound	Act[a]	Dose (µg)	JAN Mean	JAN SD[b]	FEB Mean	FEB SD	MAR Mean	MAR SD	MAY Mean	MAY SD	JUN Mean	JUN SD
Positive control	+		734.50	62.93	741.00	43.02	778.67	36.02	669.33	10.12	993.67	43.13
	−		291.00	2.83	305.00	2.83	84.67	7.51	466.66	43.13	510.67	25.70
Negative control												
Dimethylsulfoxide	+	100.00 µl	95.00	7.07	90.00	7.07	81.67	3.21	94.00	18.08	85.67	4.73
Dimethylsulfoxide	−	100.00 µl	89.00	8.49	101.50	6.36	88.00	17.35	90.67	4.51	85.00	9.54
Oldsmobile 350 diesel	+	60.00	227.67	11.37	250.67	8.50	242.33	4.51	248.00	9.64	310.00	19.47
	+	100.00	318.33	40.50	393.33	28.02	362.50	12.02	305.00	34.39	421.33	19.22
	+	200.00	504.67	87.05	—	—	564.33	46.61	483.67	4.73	660.33	6.51
	+	600.00	—	—	—	—	1137.00	14.42	917.00	12.12	1185.00	21.07
	−	60.00	473.00	30.05	453.67	47.88	471.33	4.93	494.67	18.61	532.00	56.15
	−	100.00	633.33	29.14	591.00	78.48	628.33	27.39	605.33	12.22	739.33	20.03
	−	200.00	—	—	—	—	1077.67	20.03	934.00	8.72	1117.33	16.92
	−	600.00	—	—	—	—	1176.33	44.00	1125.00	48.12	1298.33	82.50

Compound	Act[a]	Dose (µg)	JUL Mean	JUL SD[b]	AUG Mean	AUG SD	SEP Mean	SEP SD	OCT Mean	OCT SD	NOV Mean	NOV SD
Positive control	+		674.00	49.43	118.67	21.83	338.33	20.50	1065.67	74.10	1368.00	37.72
	−		477.67	22.50	365.67	18.50	662.67	21.03	591.67	3.79	675.67	17.21
Negative control												
Dimethylsulfoxide	+	100.00	92.67	5.51	72.33	8.08	101.00	8.72	48.00	5.57	111.67	19.63
Dimethylsulfoxide	−	100.00	103.67	13.61	70.67	16.01	87.33	10.41	49.33	2.89	98.00	1.00
Oldsmobile 350 diesel	+	60.00	587.33	10.07	286.00	16.64	167.00	18.25	188.67	13.05	229.00	43.84
	+	100.00	746.67	14.29	422.33	31.56	210.33	6.03	288.33	15.95	334.33	13.65
	+	200.00	1074.67	13.58	649.33	19.40	360.33	29.40	503.67	11.02	564.00	13.23
	+	600.00	1186.33	79.05	909.00	121.62	730.00	40.34	1034.67	59.65	1186.67	70.69
	−	60.00	262.67	12.42	528.67	26.54	589.67	3.06	494.33	20.40	685.00	44.84
	−	100.00	366.00	11.14	701.00	33.96	735.33	59.80	767.33	28.57	1041.00	20.66
	−	200.00	685.67	44.38	926.33	44.74	1182.33	75.39	1230.33	16.17	1565.67	25.74
	−	600.00	1119.00	25.36	961.33	16.62	1227.33	146.17	1436.67	19.30	1920.00	59.23

[a] Metabolic activation.
[b] Standard deviation.

Table 5. Comparison of Exhaust Organics from Gasoline Vehicles of the Same Make, Model, and Configuration (Ford LTD Automobiles) in S. typhimurium TA98 Without Activation[a]

Sample No. (MSER-79-)		Slope[b] (Rev/plate/μg)	% Ext.[c]	Rev x 10^5 /g Partic.	PER[d] (g/mile)	Rev/ Mile
0032	Lt. blue	7.54	5.8	4.37	0.0079	3455
0033	Mid. blue	7.51	3.4	2.55	0.0100	2553
0034	Gray	9.49	3.2	3.04	0.0180	5466
0035	Silver	9.61	3.8	3.65	0.0100	3652
0036	Brown	1.02	21.4	2.18	0.0050	1091
Mean		7.03	7.52	3.16	0.0102	3243
Standard dev.		3.51	7.83	0.87	0.0048	1602
Coeff. of var.		0.50	1.04	0.28	0.47	0.49

[a]Numbers assigned to each automobile exhaust sample collected by EPA, ESRL, RTP, NC.
[b]Slope of linear regression line.
[c]Percent dichloromethane-extractable mass.
[d]Particle emission rate.

possibility was demonstrated with exhaust organic samples from an Oldsmobile 350 diesel vehicle chemically fractionated at RTI under the direction of Edo Pellizzari. When adequate sample was available, each fraction was bioassayed with all five tester strains. Results (Table 7) were similar to previous results with exhaust organics from heavy-duty diesel engines (Little, 1978). (A more complete summary is in preparation.) TA1535 gave negative results with all fractions except for the polynuclear aromatic (PNA) fraction when exogenous activation was used. The basic and nonpolar neutral fractions were negative; however, the basic fraction could not be adequately tested due to a lack of sample. The acid I, acid II, polar neutral, and PNA fractions gave positive results with each strain that responds to frameshift mutagens. This activity was demonstrated both with and without activation.

When sample CMBX-79-0047 was tested with TA98 and TA98-FR1 (Table 8), the nitroreductase-deficient strain demonstrated approximately one half the activity of TA98. With cigarette smoke condensate, both strains provided equal responses. When the results of other mobile-source organics simultaneously tested with these two strains were compared (Table 9), this relationship was not maintained.

Table 6. Comparison of Organic Exhausts from Diesel Vehicles Using S. typhimurium TA98 Without Activation

Sample No. (TAEB-78-)	Vehicle	HC (ppm)	NO_x (ppm)	CO (ppm)	Slope[b] (Rev/plate /μg)	Mean % Ext.[c]	Rev x 10^5 /g Partic.	PER[d] (g/mile)	Rev x 10^5 /mile
501 + 0507	Olds 350	0.593	1.49	1.508	1.85	21.1	3.90	0.846	3.30
502 + 503	Olds 260	0.584	1.59	1.504	1.51	16.5	2.49	1.005	2.50
511 + 512	Chev. truck 350 diesel	0.784	1.53	1.590	2.64	53.4	14.10	0.623	8.78
504 + 505	M.B. 300D	0.251	1.38	1.377	3.23	23.3	7.53	0.840	6.32
506 + 510	Opel Record E	0.454	2.08	1.540	1.15	51.6	5.93	0.483	2.87
513 + 514	VW Dasher	0.503	0.97	1.170	1.45	53.7	7.79	0.322	2.51
Mean		0.528	1.51	1.448	1.97	36.6	6.96	0.687	4.38
Standard deviation		0.176	0.36	0.153	0.80	18.0	4.06	0.256	2.59
Coefficient of variation		0.33	0.24	0.11	0.41	0.49	0.58	0.37	0.59

[a] Provided by Tom Baines, Emission Control Technology Division, EPA, Ann Arbor, MI.
[b] Slope of linear regression line.
[c] Percent dichloromethane-extractable mass.
[d] Particle emission rate.

Table 7. Activity of Organic Fractions from an Oldsmobile 350 Diesel Exhaust Sample Tested in S. typhimurium[a]

| | Specific Activity for Strain[b] | | | | | | | | | |
| Sample | TA100 | | TA98 | | TA1537 | | TA1538 | | TA1535 | |
	+S-9	-S-9	+S-9	-S-9	+S-9	-S-9	+S-9	-S-9	+S-9	-S-9
Acid I fraction	+	+	+	+	+	+	+	+	-	-
Acid II fraction	+	+	+	+	+	+	+	+	-	-
Base I fraction[c]	-	-	-	-	NT	NT	-	-	NT	NT
Base II fraction[c]	-	-	-	-	NT	NT	-	-	NT	NT
Insoluble tars	+	+	+	-	-	-	-	-	-	-
Polar neutrals	+	+	+	+	+	+	+	+	-	-
PNA	+	+	+	+	+	+	+	+	+	-
Non-polar neutrals	-	-	-	-	-	-	-	-	-	-

[a] positive activity indicates a dose-dependent response with at least one dose giving a two-fold increase over spontaneous activity.
[b] NT = not tested; + = positive; - = negative.
[c] Incomplete testing.

Table 8. Mutagenicity of a Diesel Exhaust Organic and a Cigarette Smoke Condensate Tested in S. typhimurium TA98-FR1 and TA98

Compound	Metabolic Activation	Sample (µg/Plate)	TA98-FR1 (rev/plate)		Sample (µg/plate)	TA98 (rev/plate)	
			Mean	St. Dev.		Mean	St. Dev.
Positive control							
2-Nitrofluorine	−	3.0	64.33	11.72	3.0	308.33	11.68
2-Aminoanthracene	+	0.5	1062.33	57.07	0.5	1609.00	73.26
Negative control							
Dimethylsulfoxide	−	100.0 µl	22.00	7.81	100.0	23.67	5.51
Dimethylsulfoxide	+	100.0 µl	46.00	6.24	100.0	44.00	1.73
Cigarette smoke condensate	−	60.0	27.33	2.08	30.0	25.00	9.64
	−	100.0	20.33	9.07	50.0	29.67	6.43
	−	200.0	27.00	1.73	100.0	28.33	1.15
	−	600.0	21.00	6.24	300.0	24.33	4.04
	−	1000.0	8.00[a]	−	500.0	26.00	6.56
	+	60.0	58.67	7.64	30.0	55.00	7.21
	+	100.0	67.33	12.01	50.0	68.67	6.66
	+	200.0	93.33	9.81	100.0	99.67	2.89
	+	600.0	96.33	16.17	300.0	102.67	18.18
	+	1000.0	16.00[a]	−	500.0	85.00	4.58
Nissan diesel	−	60.0	263.0	24.24	30.0	479.67	37.74
	−	100.0	421.0	25.50	50.0	729.33	31.50
	−	200.0	672.67	13.20	100.0	1175.67	10.02
	−	600.0	902.33	27.00	300.0	1543.33	73.79
	−	1000.0	575.67	76.79	500.0	1011.33	101.75
	+	60.0	231.67	27.57	30.0	421.00	55.43
	+	100.0	426.67	29.37	50.0	723.00	50.39
	+	200.0	740.00	12.17	100.0	1363.00	70.19
	+	600.0	901.00	75.36	300.0	1607.33	35.91
	+	1000.0	797.67	24.68	500.0	1680.33	38.18

Table 9. Comparison of Exhaust Organics from Different
Light-Duty Automobiles with Nitroreductase-deficient (TA98-FR1)
and -competent (TA98) Strains of S. typhimurium

Sample	Metabolic Activation	Specific Activity (rev/100 μg)	
		TA98	TA98FR1
MSER-79-0064	−	154	210
(Datsun 810 gasoline auto.)	+	341	490
MSER-79-0074	−	79	221
(Olds 350 diesel auto.)	+	81	207
TAEB-79-0029	−	46	104
(VW Dasher diesel wagon)	+	65	106
TAEB-79-0034	−	108	330
(VW Dasher diesel auto.)	+	156	257

The preliminary study comparing the response of complex
mixtures in a forward and reverse mutation system was completed
using four strains of S.typhimurium: TM677, TM35, TA100, and
TA1535. Each of the four strains were used for forward mutation to
8-azaguanine resistance. Strains TA100 and TA1535 were used for
reverse mutation to histidine prototrophy. Throughout this study,
the preincubation assay as described by Skopek et al. (1978a, b)
was used. Two samples, from an Oldsmobile 260 and a VW Dasher,
were used for comparison. Table 10 gives a summary of the results.
The mutagenic activity of the two samples was detected by the
forward mutation system using strains TM677 and TA100; however,
TA100 was not as sensitive as strain TM677. The preincubation
protocol also was used for the reverse mutation assay in which
strain TA100 provided a positive response. TA1535 was negative for
the reversion assay.

DISCUSSION

Microbial mutagenesis assays can be used for rapid initial
evaluation of combustion organics and emissions. Previously
published reports (Huisingh et al., 1978) stated that most of the
mutagenic activity associated with diesel samples from heavy-duty

Table 10. Comparison of Forward and Reverse Mutation Systems With the Preincubation Protocol Using Two Diesel Exhaust Samples

Sample	Dose	8-Azaguanine Resistance (mutants/10^5 survivors)				Histidine Reversion (mutants/10^5 survivors)	
		TM677	TA100	TM35	TA1535	TA100	TA1535
Spontaneous	-	4.4	1.2	1.2	0.6	4.1	0.2
Control: MNNG	2.04 µM	37.9	95.0	42.0	27.0	7.9	1.9
Olds 260	12.5 µg/ml	5.4	2.7	1.1	0.9	4.6	0.3
	25.0 µg/ml	6.1	4.2	1.5	1.3	6.1	0.2
	50.0 µg/ml	7.9	4.6	1.3	1.7	5.4	0.4
	100.0 µg/ml	14.6	9.3	1.2	2.0	10.8	0.3
VW Dasher	12.5 µg/ml	7.0	9.6	1.6	1.2	4.4	0.1
	25.0 µg/ml	10.7	19.1	1.6	1.8	5.9	0.1
	50.0 µg/ml	18.4	34.3	1.4	1.8	5.8	0.3
	100.0 µg/ml	34.2	43.4	1.6	1.5	9.5	0.3

engines was detected by the indicator strains that respond to frameshift mutagens. This paper provides data on light-duty diesel passenger automobiles. Each diesel and gasoline organic sample for which testing has been completed in all five strains to date produced a negligible response in TA1535, but positive responses in TA98, TA100, TA1537, and TA1538. Figure 1 shows the response of the five tester strains to Nissan 220C diesel automotive exhaust organics. Although the magnitude of response varied, the overall qualitative response was similar for all automobiles tested to date.

A great concern in testing complex environmental samples is the variability expected from multiple mixtures of chemicals. Therefore, the difficulty in sorting out the factors that cause this variability was explored. Tables 2, 3, 4, and 6 show the variability of bioassay and chemical data with different types of comparisons. Tables 2 and 3 compare the results for samples collected from one automobile at various times. Both the bioassay data and chemical data showed less than a twofold difference in all comparisons and a coefficient of variation less than 20%. A comparison of samples from vehicles of the same make, model, and configuration (Table 5) demonstrates an increased variability over multiple samples taken from one vehicle. There was nearly a tenfold difference in revertants per microgram of organic material between two automobiles. In this comparison, one automobile (the brown Ford LTD, sample MSER-79-0036) introduced significantly different values for most of the parameters measured; however, the variability for the bioassay was normalized to a large extent with the calculation of revertants per mile. Clearly, differences in emission characteristics can result in widely different mutagenic activities in emission organics. The brown Ford LTD, for example, may have emitted an unusual amount of unburned fuel or oil that would have diluted the concentration of mutagens and increased the percent extractable. However, since the amount of fuel burned for a predetermined distance would not be altered, the calculation of revertants per mile would normalize the data. Table 6 is the averaged data from replicate experiments for six different light-duty diesel vehicles. The variation among different diesel vehicles is similar to the variation between the different Ford LTD gasoline automobiles. There are, however, two dramatic differences First, the organics of the gasoline vehicles demonstrate more mutagenicity on a per weight of particulate basis. Second, when normalized to revertants per mile, there is an approximately 50- to 100-fold difference between the diesel vehicles and the Ford LTDs. The mean for the different diesel vehicles is 438,000 rev/mile, and the mean for the Ford LTDs is 3,200 rev/mile. This difference is attributable mainly to the difference in PER and demonstrates the need for adjusting the data for specific needs and comparisons. Table 11 summarizes the data from different comparisons.

Table 11.   Comparison of Summary Data Demonstrating the Effect of Differing Sampling
Parameters (Derived from Tables 4, 7, and 9)

	HC (ppm)	NO_x (ppm)	CO (ppm)	Slope (rev/plate/µg)	% Ext.	Rev x 10^5 /g Partic.	PER[b] (g/mile)	Rev x 10^5/mile
Different runs within same automobile (diesel)								
Mean	0.249	1.550	0.899	3.68	11.8	4.35	0.524	2.27
Standard deviation	0.032	0.10	0.070	0.42	1.0	0.64	0.037	0.26
Coeff. of var.	0.13	0.07	0.08	0.11	0.09	0.15	0.07	0.11
Vehicles of same make, model, and configuration (gasoline)								
Mean	0.23	1.74	0.37	7.03	7.52	3.16	0.0102	0.032
Standard deviation	0.20	0.66	0.10	3.51	7.83	0.87	0.0048	0.016
Coeff. of var.	0.89	0.38	0.27	0.50	1.04	0.28	0.47	0.49
Different diesel vehicles								
Mean	0.528	1.51	1.448	1.98	36.6	6.96	0.687	4.38
Standard deviation	0.176	0.36	0.153	0.80	18.0	4.06	0.256	2.59
Coeff. of var.	0.33	0.24	0.11	0.41	0.49	0.58	0.37	0.59

[a]Slope of linear regression line.
[b]Particle emission rate.

Decisions on pollution control devices, alterations in engine
design, criteria for fuel characteristics, and some environmental
regulations rely upon an understanding of which specific organics
are likely to have a detrimental health effect. Bioassay-guided
chemical fractionation has the potential to speed the identification
of biologically active compounds in this category. Table 7
indicates that the chemist needs to place a higher priority on
chemical identification of the polar neutral fraction than the
nonpolar neutral fraction. Also, by examining the specific
activity of the different fractions, one notices that the PNA
fraction gives a positive response without activation. This
response may be due to spillover of polar neutral chemicals into
the PNA fraction with this particular fractionation scheme.

Nitroreductase-deficient strains of bacteria cannot metabolize
the nitrogen components of a chemical; therefore, a decreased
response in the nitroreductase-deficient strain supports the
conclusion that active nitro compounds are present in the organic
mixture. If activity of the compound does not depend on the
reduction of the nitrogen group, activity in a nitroreductase-
deficient strain should not differ from that of its parental
strain. This decreased response occurred with some samples, though
not all (Tables 8 and 9). This demonstrates that nitroaromatics
may be one class of mutagens prominent in mobile source emissions.
Whether these nitro compounds are true artifacts or are also
created under normal environmental conditions is yet to be
demonstrated.

The results of the 8-azaguanine forward mutation bioassay are
consistent with earlier results in the Ames reversion assay.
Although the Ames tester strains TA100 and TA1535 can be used for
this forward mutation assay, the strains developed by Skopek et al.
(1978a, b) were more sensitive and gave fewer technical problems in
the performance of the assay. Although the assay detects a range
of mutagens equivalent to the range detected by all five of the
routinely used Salmonella strains, these forward mutation strains
cannot be used in the same diagnostic manner. In other words, the
forward mutation system detects a variety of mutagenic insults,
whereas a specific type of DNA damage must occur to be detected by
reverse mutation systems. As the forward mutation assay is more
thoroughly validated, it becomes increasingly feasible to use the
8-azaguanine system for general screening and to resort to Ames
tester strains for better characterization of the substance(s)
being tested.

These samples demonstrate the uses of microbial mutagenesis
assays for both qualitative and semi-qualitative assessment of
mobile source emissions. This paper has demonstrated a means of
comparing and evaluating the potential health hazard of polluting
soot material from various mobile sources and from new technological

developments for each of these sources. The classical chemical examination of exhaust components cannot be used effectively to quantitate and evaluate the thousands of organic chemicals derived from each combustion source. Whole-animal evaluation of multiple mobile sources and technological alterations would involve extremely slow evaluation and high costs. Although still controversial in many aspects, the Salmonella assay for mutagenicity provides a rapid, inexpensive means to assess genetically toxic effects. Recent results (Nesnow and Huisingh, in press) indicate that the plate-incorporation test correlates well with other tests for genotoxicity of combustion organics. By assuming that increased response in the plate-incorporation test represents greater potential health hazard, the Salmonella assay can be used to compare various sources, evaluate technological developments, and guide chemical characterization. Since mutation assays have some power to predict heritable effects, carcinogenesis, and teratogenesis (Hollaender and de Serres, eds., 1978), this assumption does not rely on direct correlation of Salmonella assay results with any one end point.

REFERENCES

Ames, B.N., J. McCann, and E. Yamasaki. 1975. Methods for detecting carcinogens and mutagens with the Salmonella/ mammalian microsome mutagenicity test. Mutation Res. 31:347-364.

Claxton, L.D., and C.C. Evans. (MS). The microbial mutagenicity of various environmental substances. U.S. Environmental Protection Agency: Research Triangle Park, NC.

Hollaender, A., and F.J. de Serres, eds. 1978. Chemical Mutagens: Principles and Methods for Their Detection. Vol. 5. Plenum Press: New York.

Huisingh, J., R. Bradow, R. Jungers, L. Claxton, R. Zweidinger, S. Tejada, J. Bumgarner, F. Duffield, M. Waters, V.F. Simmon, C. Hare, C. Rodriguez, and L. Snow. 1978. Application of bioassay to the characterization of diesel particle emissions. M. Waters, S. Nesnow, J. Huisingh, S. Sandhu, and L. Claxton, eds. In: Application of Short-term Bioassays in the Fractionation and Analysis of Complex Environmental Mixtures. Plenum Press: New York. pp. 381-418.

Lee, M.L., M. Novotny, and K.D. Bartle. 1976. Gas chromatography/ mass spectrometric and nuclear magnetic resonance determination of polynuclear aromatic hydrocarbons in airborne particulates. Anal. Chem. 48:1566.

Little, L. 1978. Microbiological and chemical testing of air samples for potential mutagenicity: First annual report. EPA/68-02-2724. U.S. Environmental Protection Agency: Research Triangle Park, NC.

Nesnow, S., and J. Lewtas Huisingh. (in press). Mutagenic and carcinogenic potency of extracts of diesel and related environmental emissions: Summary and discussion of results. In: International Symposium on the Health Effects of Diesel Engine Emissions. U.S. Environmental Protection Agency: Cincinnati, OH.

Rosenkranz, H.S., and W.T. Speck. 1975. Mutagenicity of metranidazole: Activation by mammalian liver microsomes. Biochem. Biophys. Res. Commun. 66:520-525.

Skopek, T.R., H.L. Liber, D.A. Kaden, and W.G. Thilly. 1978a. Proc. Natl. Acad. Scie. USA 75:4465-4469.

Skopek, T.R., H.L. Liber, J.J. Krolewski, and W.G. Thilly. 1978b. Proc. Natl. Acad. Sci. USA 75:410-414.

COMPARISON OF THE MUTAGENIC ACTIVITY IN CARBON PARTICULATE MATTER AND IN DIESEL AND GASOLINE ENGINE EXHAUST

Göran Löfroth
Radiobiology Department
University of Stockholm
Stockholm, Sweden

INTRODUCTION

Airborne carbon particulate matter is a variable and complex mixture of components, including a variety of organic compounds. Its origin in urbanized and industrialized areas is primarily through various combustion processes. Motor vehicles are often viewed as a major source.

Talcott and Wei (1977) and Pitts et al. (1977) first showed that the Salmonella/microsome mutagenicity test, as developed by Ames et al. (1975), can be used to detect mutagenic--and thus potentially carcinogenic--activity in airborne particulate matter. Exploratory studies of motor vehicle exhaust (Tokiwa et al., 1978) and of emissions from a stationary combustion plant (Löfroth, 1978) also demonstrated the feasibility of the Salmonella assay for this purpose.

The present report summarizes some of the results obtained in an ongoing study of the mutagenicity, as detected by the Salmonella assay, of airborne particulate matter collected in the Stockholm area. The results are compared with mutagenicity data for motor vehicle exhaust samples obtained in a previous study (Löfroth, 1979) and by additional analyses of these samples.

MATERIALS AND METHODS

Motor Vehicle Exhaust Samples

Exhaust samples from gasoline and indirect injection (IDI) diesel passenger cars were the same as those described by Löfroth (1979). They were obtained during U.S. Federal Test Procedure 1973 cycles by sampling the undiluted exhaust and by collecting the particulate matter on the filter at 25 to 50°C and also the formed aqueous condensate containing compounds that had not adsorbed onto particulate matter at the point of sampling. Particulate matter was Soxhlet-extracted overnight with acetone, and after removal of the major part of the acetone, the residue was dissolved in dimethylsulfoxide (DMSO). The aqueous condensates were treated in various ways, including extraction with n-pentane followed by removal of the pentane and dissolution of the residue in DMSO. These pentane-extracted samples, shown to contain the major part of the mutagenicity of the condensates, are the condensate samples that have been further analyzed in the present study, together with the samples of particulate matter. All samples dissolved in DMSO were stored frozen at -20°C. No detectable change in their mutagenic activity was noted over a period of about one year.

Airborne Particulate Matter

Airborne particulate matter was collected with Sierra 305 High Volume Samplers or similar equipment on 20.3- x 25.4-cm (8- x 10-in) glass fiber filters (Stora Kopparberg Special-produkter, Sweden) with a flow rate of 68 m^3/h (40 f^3/min). Sampling sites were located on the roof of a ten-story building in the northern part of the inner city of Stockholm and on the roof of a two-story house in a suburban community 22 km NNW from the center of Stockholm. Sampling was usually performed for 24 h, starting and ending at 6 to 7 AM. Night samples were collected between 10 PM and 6 AM.

To study the mutagenic activity of size-fractionated particles, samples were taken in the inner city of Stockholm. Particulate matter was collected with a Sierra 305 High Volume Sampler equipped with five stages of cascade impactors with slotted glass fiber collection substrates and the regular glass fiber filter as back-up filter. Simultaneous sampling at the same flow rate (68 m^3/h) was done using a second 305 Sampler without impactors. Sampling was done over four days; substrates and filters were changed every 24 h. Filters were extracted individually; the samples were first assayed separately and then combined and assayed as one sample. The four substrates from each impactor stage were extracted and assayed as one sample.

Collected filters were wrapped in aluminum foil and stored at
-20°C until they were extracted (between one and eight days). The
filters were Soxhlet-extracted for 16 h with 250 ml acetone. The
acetone was first evaporated under vacuum to about 10 ml and then
under a stream of nitrogen on a heating block at < 40°C to 0.3 to
0.5 ml. This residue was diluted with DMSO to a known volume,
usually about 4 ml/filter. All samples were stored frozen at -20°C
prior to and between mutagenicity assays.

Meteorological data were acquired from three official
stations. One is located in the inner city of Stockholm, and
records are made from three daily observations. Two are located at
airports, in the western part of the city and north of Stockholm,
and observations are made every 30 min.

Mutagenicity Assay

Mutagenicity was determined by the <u>Salmonella</u> plate-
incorporation method with bacterial cultures fully grown overnight,
as described by Ames et al. (1975). Assays generally were
performed with strains TA98 and TA100, obtained from Dr. B.N. Ames
(University of California, Berkeley, CA), and with the
nitroreductase-deficient strains TA98 NR and TA100 NR, obtained
from Dr. H.S. Rosenkranz (New York Medical College, Valhalla, NY).
The microsome-containing rat-liver supernatant (S-9) was prepared
from Aroclor-1254-induced male Sprague-Dawley rats and was used
with the necessary cofactors.

All assays included tests with positive control compounds.
The bacterial strains were routinely checked for the presence of
known characteristics, including spontaneous reversion frequency,
sensitivity to ultraviolet light and crystal violet, and
sensitivity or resistance to ampicillin. Benzo(a)pyrene was used
as a positive control for the S-9; in TA98, a 5-μg dose of this
compound yielded between 300 and 600 revertants/plate with S-9 at
20 and 50 μl/plate.

The sources of the chemicals used were as follows:
2-nitropropane was purchased from Merck-Schuchardt; FRG and
2-nitronaphthalene from EGA-Chemie; and FRG and 1-nitropyrene
(labeled 3-nitropyrene) from Koch-Light, England. Di- and
tetranitropyrenes were obtained from Dr. R. Mermelstein, Xerox
Corporation, USA.

During the course of this investigation, it was found that the
strain TA100 could vary in its response to large molecules and that
the change in characteristics was not readily detected with crystal
violet. The quantitative monitor presently used is 1-nitropyrene
(Koch-Light); 1 μg of the commercial product gives between 1000 and

1500 revertants/plate in TA100. Other samples of 1-nitropyrene may give different responses, due to the presence of variable amounts of more mutagenic impurities.

Each sample was assayed with one plate per dose level at two or more dose levels in at least three independent tests. The mutagenic response, expressed as revertants per cubic meter of air for airborne particulate matter and as revertants per gram of consumed fuel for motor vehicle exhaust, was calculated from the linear part of the dose-response curve.

RESULTS AND DISCUSSION

Mutagenicity Pattern of Motor Vehicle Exhaust

Results presented in a previous report (Löfroth, 1979) showed both similarities and differences in the mutagenic responses of Salmonella to gasoline exhaust and diesel exhaust from normal passenger cars. All samples were mutagenic in the absence of mammalian metabolic activation, showing that the mutagenic compounds present were either directly acting mutagens or were converted to ultimate mutagens by bacterial metabolism. Both gasoline and diesel exhausts were more mutagenic in TA100 than in TA98.

Enhancement of the mutagenic activity by mammalian metabolic activation was observed for only one type of sample: particulate matter from gasoline exhaust. All other samples were less mutagenic in the presence of S-9 than in its absence. The decrease in mutagenic activity depended on the amount of S-9; it is thus not appropriate to report the mutagenic activity in the presence of S-9 for such samples. For this reason, these results cannot be compared with those of Ohnishi et al. (1980), who reported Salmonella mutagenicity data for several gasoline and diesel exhaust samples assayed in the presence of mammalian metabolic activation.

Diesel exhaust was far more mutagenic than was gasoline exhaust. Irrespective of the manner of comparison--revertants per volume of exhaust, revertants per test cycle, or revertants per amount of fuel consumed--diesel exhaust was more than ten times as mutagenic as gasoline exhaust.

Assays involving anaerobic incubation during the first 16 h of the 48-h incubation indicated that the exhaust samples contained no or undetectable amounts of certain nitro compounds that become more mutagenic when assayed under anaerobic conditions.

Mutagenicity Pattern of Urban Particulate Matter

Sampling of urban particulate matter for mutagenicity studies was attempted in the early part of 1976, but the low volumes sampled did not yield conclusive results. High-volume sampling was started at the end of 1977 to test the feasibility of the assay, and regular sampling was begun during 1978.

Several different Soxhlet-extraction solvents--acetone, benzene, cyclohexane, dichloromethane, and methanol--were tested in the early phase of the study. Acetone extraction gave the highest yield of mutagenic activity. Extraction with benzene, cyclohexane, dichloromethane, and methanol followed by further extraction with acetone resulted in samples containing additional mutagenic activity; however, extraction with acetone followed by further extraction with any of the solvents resulted in samples with no additional detectable mutagenicity.

The samples collected above the rooftops were generally tested in the strains TA98 and TA100 and occasionally in the other tester strains, TA1535, TA1537, and TA1538. No detectable mutagenicity was observed with TA1535. Of the plasmid-containing strains TA98 and TA100, the former was usually the most responsive, although TA100 sometimes showed equal or higher responses. Addition of S-9 usually, but not always, decreased the mutagenicity.

It was previously reported (Löfroth, 1979) that many urban particulate samples showed a higher mutagenic activity in the assay involving anaerobic incubation than in the regular aerobic assay. This has been further confirmed, indicating that in contrast to the motor vehicle exhaust samples, urban particulate samples may contain nitro compounds that give this type of mutagenic response.

Seasonal variation. The mutagenic activity of the collected urban particulate matter varied seasonally. Although there were differences between consecutive days (see below), an average mutagenicity could be calculated for periods during which 24-h samples were collected. A number of such periods, consisting of two to four weekdays, were studied. Results to date are presented in Figure 1, which gives the average mutagenic activity in TA98 in the absence of S-9 as a function of the time of year when the samples were collected. The mutagenicity was higher during the winter than during the summer. Many factors--several meteorological parameters as well as emissions of various compounds--may have contributed to the level of mutagenic activity. Thus, it is not certain that the higher mutagenic activity during the winter months was due to heating of buildings.

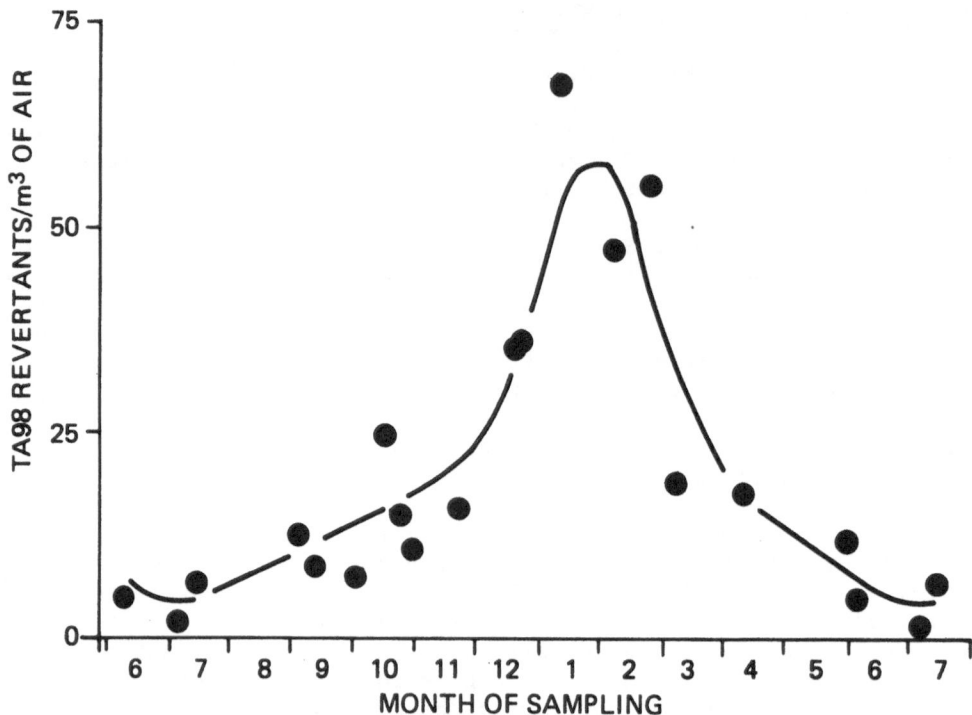

Figure 1. The mutagenicity of extracts of particulate matter
collected during weekdays above the rooftops in the
inner city of Stockholm. Each point represents the
average of two to four consecutive 24-h samples.

Inner city and suburban sites. Mutagenic activity was
somewhat higher in particulate matter from the inner city of
Stockholm than in that collected simultaneously at the suburban
site (Table 1 reports mutagenicity in the absence of S-9). The
seasonal variation observed for the inner city sampling site
(Figure 1) was also found for the suburban site (Table 1). The
suburban site was located in an area where residential heating is
generally electric.

Daily variations. The mutagenic activity varied within the
same day and between consecutive days. Mutagenic activities in the
nighttime hours, when motor vehicle traffic was low, were compared
with activities in other periods, in order to assess the
contribution from stationary sources such as residential heating.
The mutagenicity of particulate matter collected between 10 PM and
6 AM was investigated for three different periods, during which

Table 1. Mutagenic Activity of Extracts of Particulate Matter Collected Above the Rooftops in the Inner City of Stockholm and Simultaneously in a Suburban Area 22 km NNW of Stockholm.

Sampling Period	TA98 Revertants/m^3 of Air[a]	
	Inner City	Suburban
Dec. 20-22, 1978	36	27
Feb. 19-23, 1979	55	30
Apr. 9-12, 1979	18	12
May 28-June 1, 1979	12	7
July 2-6, 1979	2.2	0.9
Sept. 3-7, 1979	13	4.4
Oct. 15-19, 1979	25	13
Dec. 17-21, 1979	35	22
Feb. 4-8, 1980	48	28

[a]Average mutagenicity for two to four consecutive days.

time complete 24-h samples were also collected and assayed. The results from one of these periods are given in Table 2. Nighttime samples were appreciably less mutagenic than the 24-h samples. Similar results were obtained for the other periods: the average nighttime and 24-h samples, were respectively, 5 and 25 revertants/m^3 for Oct. 15 through 19, 1979, and 17 and 48 revertants/m^3 for Feb. 4 through 8, 1980.

These results indicate that emissions during the night, such as those associated with residential heating, were not a major direct source of the mutagenicity present in 24-h samples. However, it cannot be ruled out that some type of enhanced formation of mutagenic components occurred where motor vehicle exhaust was mixed and interacted with emissions from stationary sources.

The variations between consecutive days shown in Table 2 were typical for most of the investigated periods. Changes in meteorological parameters were probably a major reason for these variations. Although the acquired meteorological data have not yet been fully evaluated and compared with the mutagenicity results, a gross comparison revealed that wind speed may have been an important factor in the day-to-day variations in mutagenic activity.

Size distribution. Organic components associated with urban
particulate matter for the most part are adsorbed to small,
respirable particles (Van Vaeck et al., 1979). Talcott and Harger
(1979) have reported that mutagenic activity was associated
primarily with particles less than about 2 µm in samples collected
in southern California during long-term sampling.

The results for two complete sampling periods of the present
project are given in Table 3. It is evident that the smaller the
particles, the greater was the mutagenic activity. Also, the
mutagenic activity of most fractions decreased after addition of
the mammalian metabolic activation system; the exceptions were
impactor stage 5 and, in one case, impactor stage 4. Particulate
matter collected on these two stages was intensely black, in
contrast to the greyish black color of the other fractions.

Samples collected simultaneously without size fractionation
had a mutagenicity greater than the sum of the mutagenic activities
of the fractionated samples. Reconstitution of the fractionated
samples gave about the same mutagenicity as that expected from the
sum of the fractions. The lesser mutagenic activity in size-
fractionated samples, compared with unfractionated samples, cannot
be explained. However, it may have been due to loss of particulate
matter to other surfaces of the impactors (Cheng and Yeh, 1979).
It is also conceivable that, if mutagens form as artifacts during
sampling (Pitts et al., 1978), such reactions may be decreased by

Table 2. Mutagenicity of Extracts of Particulate Matter
Collected at Night and Over 24 Hours

Sampling Date (December 1979)	TA98 Revertants/m³ of Air		
	Inner City		Suburban
	8 h Night[a]	24 h	24 h
17–18	13	48	33
18–19	9	23	12
19–20	12	31	17
20–21	12	36	24
Average	11	35	22

[a]10 PM to 6 AM.

Table 3. Mutagenic Activity of Extracts of Size-fractionated
Particulate Matter Collected in the Inner City of Stockholm
Compared with the Mutagenic Activity of Samples Collected
Simultaneously Without Size Fractionation[a]

| | | TA98 Revertants/m^3 of Air | | | |
| | | Oct. 22–26, 1979 | | Nov. 19–23, 1979 | |
Sample	Particle Size (μm)	−S-9	+S-9 (50 μl/plate)	−S-9	+S-9 (50 μl/plate)
Impactor stage 1	7.2	0.3	0.1	0.2	0.2
2	3.0	0.5	0.4	0.3	0.2
3	1.5	0.5	0.5	0.4	0.3
4	1.0	0.7	1.2	0.6	0.5
5	0.5	1.4	2.5	1.0	1.4
Back-up filters		6.2	4.5[b]	5.4	2.7[b]
Sum of impactor stages 1–5 and back-up filters		10	9	8	5
Reconstituted sample: impactor stages 1–5 and back-up filters		–	–	7	7[b]
Filters without impactors		15	9[b]	16	9[b]

[a]Particle sizes are those given by the manufacturer at 50% collection efficiency for spherical particles with unit mass density.
[b]Approximate figures; assays in the presence of S-9 often give
nonlinear dose-response curves, making it difficult to arrive at a
single response.

collecting precursors on the impactor substrates instead of on the
filter through which polluted air is flowing during the entire
sampling period.

Chromatographic separation. Extracts of urban airborne
particulate matter having sufficiently high mutagenic activity were
separated by high performance liquid chromatography (HPLC), and the
eluates tested for mutagenicity. The system separates standard

polycyclic aromatic hydrocarbons and nitroarenes in narrow peaks
(1 ml, 1 or 2 fractions). The mutagenic activity in extracts of
particulate matter eluted over a wide range, with no major single
peaks (as shown in Figure 2). Most of the mutagenic activity was
found in the range where di- to tetracyclic compounds elute. The
results suggest that the mutagenicity of extracts of urban airborne
particulate matter is caused by many compounds and that it may be
unprofitable to look for single compounds as major airborne
mutagens.

Mutagenic Response in Nitroreductase-deficient Tester Strains

Salmonella strains deficient in nitroreductase activity were
developed in conjunction with mutagenicity studies of nitrofuran
derivatives (Rosenkranz and Speck, 1976). Nitrofuran derivatives
that are mutagenic in the original strains are less mutagenic in
the nitroreductase-deficient strains.

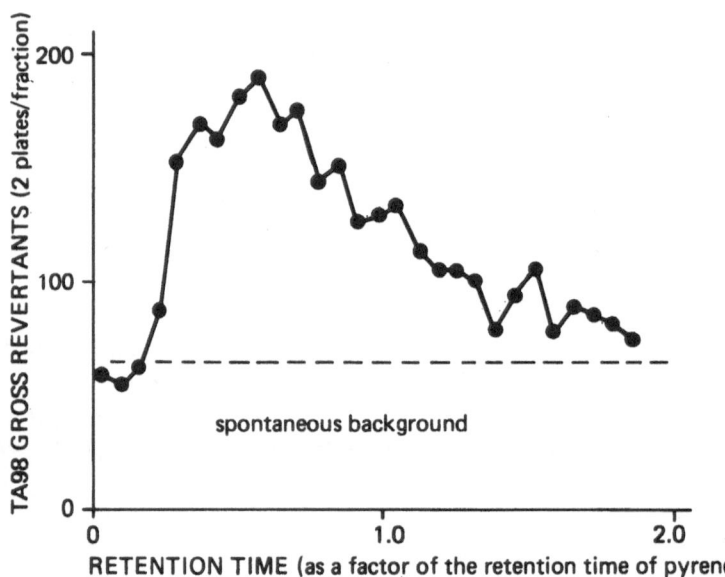

Figure 2.   Mutagenic response of fractions from a reverse-phase
            HPLC separation of an extract of airborne particulate
            matter. The sample (50 µl, containing activity
            corresponding to about 1600 revertants) was applied to
            a Spherisorb S5-ODS column (220 mm x 4.6 mm i.d.) with
            methanol-water (8:2) as elutant. Fractions of 0.67 ml
            (20 s) were collected and assayed in TA98 in the absence
            of S-9, using 0.2 and 0.4 ml.

In the present study, nitroreductase-deficient strains TA98 NR and TA100 NR were used to investigate the behavior of some nitroarenes, motor vehicle exhaust samples, and extracts of airborne urban particulate matter. Typical mutagenic nitroarenes were less mutagenic in the TA98 NR strain than in the original strain (Figure 3); but whereas the mutagenic response of the dicyclic 2-nitronaphthalene almost disappeared, the response decreased only slightly for the tetracyclic 1-nitropyrene. Strains TA100 and TA100 NR behaved similarly.

The mutagenic activity of the motor vehicle exhaust samples was more or less the same in TA98 and TA98 NR (Figures 4 and 5 and Table 4). In contrast, several extracts of urban airborne particulate matter were less mutagenic in the nitroreductase-deficient strain (Figure 6 and Table 4). Preliminary studies comparing the mutagenic activity of these samples in TA100 and TA100 NR gave similar results: the motor vehicle samples produced about the same response in both strains, whereas airborne particulate matter was less mutagenic in TA100 NR than TA100. These results indicate that extracts of urban airborne particulate matter contained certain types of nitro compounds (specifically, those that were metabolized to ultimate mutagens by the bacterial nitroreductase system that had been abolished in the deficient strains). The motor vehicle exhaust samples did not seem to contain detectable amounts of these compounds.

Thus, comparisons between the regular assay and the assay involving anaerobic incubation and between the regular and the nitroreductase-deficient strains both imply that extracts of filter-collected urban airborne particulate matter contain nitro compounds. Due to the uncertainty about artifactual formation of nitro compounds during sampling (Pitts et al., 1978), it is, however, premature to state that airborne particulate matter contains nitro compounds prior to collection. Wang et al. (1980), after comparing the mutagenic response in TA98 and a nitroreductase-deficient strain, suggested that nitroaromatic compounds are present in airborne particulate matter.

Environmental Nitro Compounds

The simultaneous presence of organic compounds and nitrogen oxides in combustion emissions and polluted air requires extensive studies on the possible presence and formation of mutagenic or carcinogenic nitro and nitroso compounds (Pitts et al., 1978; Ehrenberg et al., 1980).

Aliphatic nitro compounds should not be discounted as mutagens. In a series of tested nitroalkanes, the common industrial solvent 2-nitropropane was found to be mutagenic in the

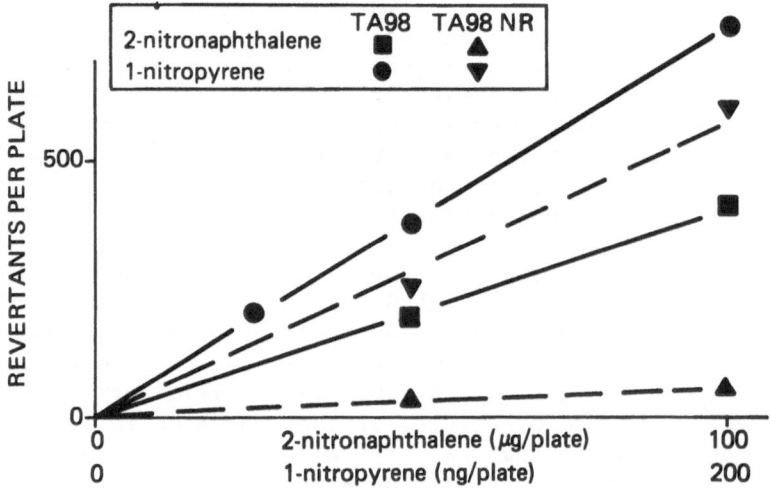

Figure 3.   Mutagenicity of 2-nitronaphthalene and 1-nitropyrene
in TA98 and TA98 NR in the absence of S-9.   The
spontaneous mutation rate has been subtracted.

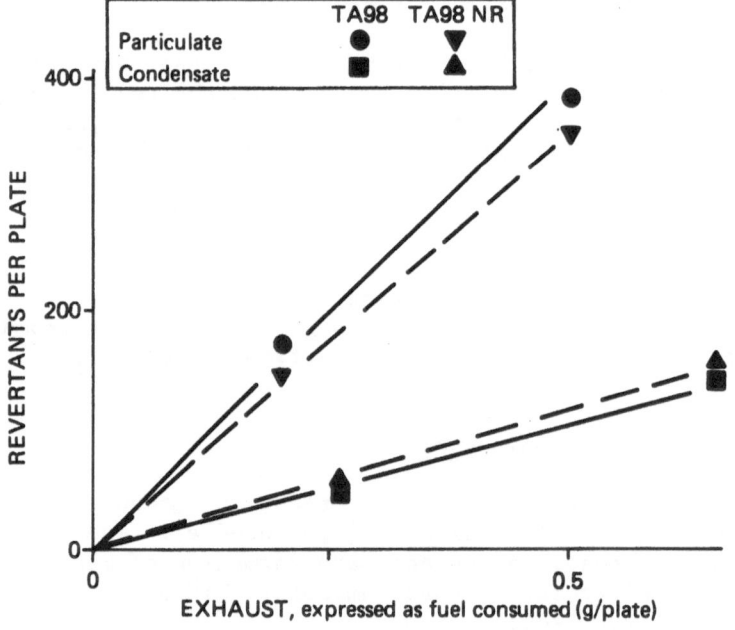

Figure 4.   Mutagenicity of diesel exhaust in TA98 and TA98 NR in
the absence of S-9.

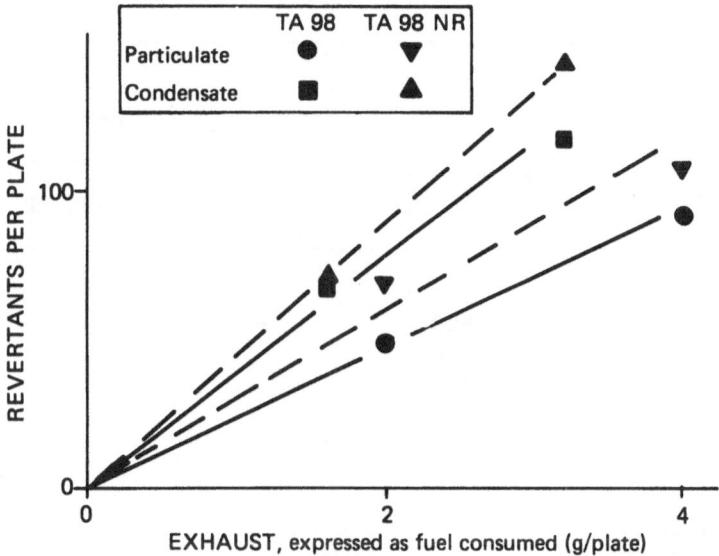

Figure 5.   Mutagenicity of gasoline exhaust in TA98 and TA98 NR
in the absence of S-9.

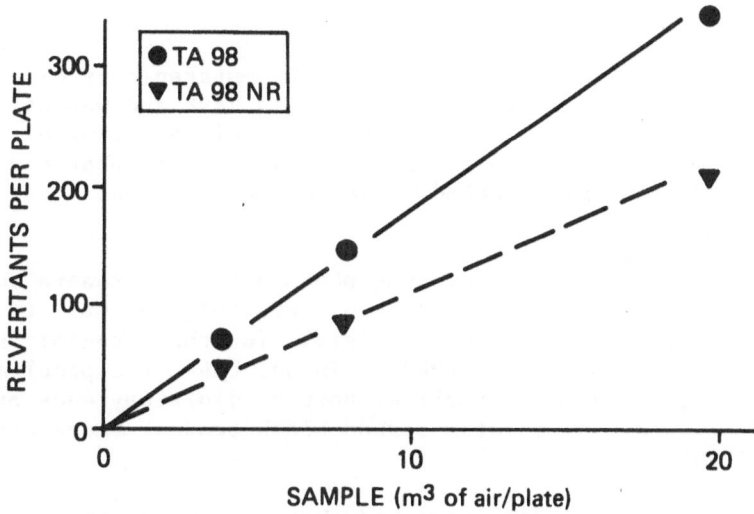

Figure 6.   Mutagenicity of urban airborne particulate matter in
TA98 and TA98 NR in the absence of S-9.

Table 4. Absolute and Relative Mutagenic Response of Motor Vehicle
       Exhaust Samples and Extracts of Urban Airborne Particulate
       Matter in TA98, TA98 NR, and TA100 in the Absence of S-9

| | TA98 | | Percent of Response in TA98 | |
Sample	Revertants/ g Fuel Used	Revertants/ m^3 of Air	TA98 NR	TA100
Diesel Exhaust				
Particulate matter	800	–	90	>100
Condensate	210	–	110	>100
Gasoline Exhaust				
Particulate matter	24	–	125	>100
Condensate	39	–	115	>100
Urban Airborne Particulate Matter				
4 days: March, 1979 (composite sample)	–	19	70	<100
4 days: Nov., 1979 (composite sample)	–	18	60	<100
4 days: Dec., 1979 (average)	–	35	60	<100

Salmonella assay (unpublished data). The mutagenicity of
2-nitropropane was higher in TA100 than in TA98 and was not related
to the mutagenicity of nitrite. The mutagenic behavior of
2-nitropropane suggests that higher homologs of aliphatic nitro
compounds may have contributed to the mutagenicity of motor vehicle
exhaust samples.

Recent studies show that some photocopies and toners contain
mutagenic compounds, and it has been suggested that tri- to
pentacyclic nitroarenes were responsible for the mutagenicity in
some cases (Löfroth et al., 1980). In one case, the problem was
traced to the presence of small amounts of dinitropyrenes present
as impurities in a particular carbon black product (Rosenkranz
et al., 1980).

Hughes, et al. (1979) reported that pyrene adsorbed to various
particulate substrates is nitrated in the presence of nitrogen
dioxide and nitric acid and that mononitropyrene can be further

nitrated to dinitropyrenes. Nitropyrenes are strongly mutagenic in the <u>Salmonella</u> system and gave the following responses in TA98:

1-nitropyrene	4 revertants/ng
1,3-dinitropyrene	300 "
1,6-dinitropyrene	450 "
1,8-dinitropyrene	750 "
1,3,6,8-tetranitropyrene	60 "

Except for 1-nitropyrene, none of the nitropyrenes showed an increased mutagenic response with anaerobic incubation. In addition, 1,6- and 1,8-dinitropyrene gave about the same mutagenic response in the nitroreductase-deficient strain TA98 NR as in TA98. Nitropyrenes and particularly dinitropyrenes do not act in the same manner as several other nitroaromatic compounds. Thus, dinitropyrenes or similarly behaving mutagenic nitroarenes can be present in motor vehicle exhaust samples and extracts of airborne particulate matter without being detected by the currently used modifications of the <u>Salmonella</u> test.

Model calculation

The investigated urban area--Stockholm and neighboring communities--has an area of approximately 3650 km^2, of which the inner city of Stockholm occupies about 50 km^2. The average fuel consumption for transportation is estimated at 0.7 metric ton/year for the whole area; 0.1 metric ton/year is used in the inner city. Approximately 10% of the fuel is diesel. If the mutagenic activity of vehicle emissions is assigned the values 200 revertants/g gasoline consumed and 2000 revertants/g diesel fuel consumed (Lofroth, 1979), the mixed use then corresponds to an emission equivalent to 400 revertants/g transportation fuel consumed. The average emission per unit area and unit time can then be calculated to be about 100 revertants/m^2/h in the inner city and 8 revertants/m^2/h in the outlying areas.

If the specific emission from an area is known, the resulting concentration in the space above the area can be approximately calculated by a box model:

$$C = \frac{E \cdot x}{L \cdot v} = \frac{1}{L \cdot v} \cdot \sum_n E_n \cdot x_n$$

where C = concentration; E = emission/(area)(time); x = distance along the emission area; L = mixing height; and v = wind speed, $\neq 0$.

If it is assumed that concentrations are monitored in the center of circular areas, the distance over the inner city is about

4 km and over the outlying areas, about 30 km. The mixing height and the wind speed may be assigned values of 200 m and 3 m/s, respectively. Using these figures, the average mutagenic activity can be calculated as 0.3 revertants/m^3 of air. Actual measurements of mutagenic activity reported in the present study were appreciably higher, e.g., 20 to 50 revertants/m^3 during the winter months.

The difference between calculated and measured average mutagenic activities seemed to be too large to be due solely to the use of a simple model. Several other explanations are more or less probable:

1) The actual mutagenic activity of motor vehicle emissions was 10 to 100 times higher than that from the tested passenger cars. Such levels would not conform with reported data (Löfroth, 1979; Ohnishi et al., 1980).

2) The emissions from DI heavy-duty diesel engines were not considered separately. However, preliminary studies of exhaust samples from such an engine indicate that the mutagenic activity is of the same order of magnitude as that from IDI diesel engines (Rehnberg and Löfroth, unpublished data).

3) There are other major sources of mutagenic components. Analyses of nighttime-collected urban particulate matter indicate that residential heating, etc., was not a major direct source of the mutagenic activity. Currently, most of the stationary energy production in the Stockholm area is from oil combustion. Residential heating is to a large extent provided by larger district plants, hopefully having low emissions of organic compounds.

4) Mutagenic components in motor vehicle exhaust samples are not the same as those in extracts of urban particulate matter. This explanation is supported by the reported differences in the mutagenic characteristics of these samples. These differences and the increased mutagenic activity could be caused by transformations during the residence time in the atmosphere or by artifactual transformations during sampling.

CONCLUSIONS

Several differences in the mutagenic characteristics and in the level of mutagenic activity have been found between motor vehicle exhaust samples and urban particulate matter collected above the rooftops. If major sampling artifacts are absent, the differences imply that transformations occur after emission, changing the composition of mutagenic compounds. It may be that either emissions or actual ambient samples are best suited for analyses.

Much information is still lacking, including evaluations of the adequacy of various sampling methods, mutagenicity studies of samples collected at street level, and mutagenicity studies in conjunction with measurements of other pollutants, including nitrogen oxides. Mutagenicity studies in the Salmonella system can only suggest human health implications, which will ultimately require further evaluation.

ACKNOWLEDGMENTS

The present project is supported by grants from the National Swedish Environment Protection Board and the Swedish Natural Science Research Council.

REFERENCES

Ames, B.N., J. McCann, and E. Yamasaki. 1975. Methods for detecting carcinogens and mutagens with the Salmonella/ mammalian-microsome mutagenicity test. Mutation Res. 31:347-364.

Cheng, Y-S., and H-C. Yeh. 1979. Particle bounce in cascade impactors. Environ. Sci. Technol. 13:1392-1396.

Ehrenberg, L., S. Hussain, M. Noor Saleh, and U. Lundqvist. 1980. Nitrous esters - A genetic hazard from nitrogen oxides (NO_x)? Hereditas 92:127-130.

Hughes, M.M., D.F.S. Natusch, D.R. Taylor, and M.V. Zeller. 1979. Chemical transformations of particulate polycyclic organic matter. Presented at the 4th International Symposium on Polynuclear Aromatic Hydrocarbons, Columbus, Ohio.

Löfroth, G. 1978. Mutagenicity assay of combustion emissions. Chemosphere 7:791-798.

Löfroth, G. 1979. Salmonella/microsome mutagenicity assays of exhaust from diesel and gasoline powered motor vehicles. Presented at the International Symposium on Health Effects of Diesel Engine Emissions, Cincinnati, Ohio.

Löfroth, G., E. Hefner, I. Alfheim, and M. Møller. 1980. Mutagenic activity in photo copies. Science 209:1037-1039.

Ohnishi, Y., K. Kachi, K. Sato, I. Tahara, H. Takeyoshi, and H. Tokiwa. 1980. Detection of mutagenic activity in automobile exhaust. Mutation Res. 77:229–240.

Pitts Jr., J.N., D. Grosjean, T.M. Mischke, V.F. Simmon, and D. Poole. 1977. Mutagenic activity of airborne particulate organic pollutants. Toxicol. Lett. 1:65–70.

Pitts Jr., J.N., K.A. Van Cauwenberghe, D. Grosjean, J.P. Schmid, D.R. Fitz, W.L. Belser Jr., G.B. Knudson, and P.M. Hynds. 1978. Atmospheric reactions of polycyclic aromatic hydrocarbons: Facile formation of mutagenic nitro derivatives. Science 202:515–519.

Rosenkranz, H.S., and W.T. Speck. 1976. Activation of nitrofurantoin to a mutagen by rat liver nitroreductase. Biochem. Pharmacol. 25:1555–1556.

Rosenkranz, H.S., E.C. McCoy, D.R. Sanders, M. Butler, D.K. Kiriazides, and R. Mermelstein. 1980. Nitropyrenes: isolation, identification and reduction of mutagenic impurities in a carbon black in toners. Science 209:1039–1043.

Talcott, R., and E. Wei. 1977. Airborne mutagens bioassayed in Salmonella typhimurium. J. Natl. Cancer Inst. 58:449–451.

Talcott, R., and W. Harger. 1979. Mutagenic activity of aerosol size fractions. EPA 600/3-79-032. U.S. Environmental Protection Agency: Research Triangle Park, NC.

Tokiwa, H., H. Takeyoshi, K. Takhashi, K. Kachi, and Y. Ohnishi. 1978. Detection of mutagenic activity in automobile exhaust emissions. Mutation Res. 54:259–260. (abstr.)

Van Vaeck, L., G. Broddin, and K. Van Cauwenberghe. 1979. Differences in particle size distributions of major organic pollutants in ambient aerosols in urban, rural, and seashore areas. Environ. Sci. Technol. 13:1494–1502.

Wang, C.Y., M-S. Lee, C.M. King, and P.O. Warner. 1980. Evidence for nitroaromatics as direct-acting mutagens of airborne particulates. Chemosphere 9:83–87.

MUTAGENIC EFFECTS OF ENVIRONMENTAL PARTICULATES IN THE CHO/HGPRT SYSTEM

John D. Shelburne
Department of Pathology
Duke University Medical Center and
Veterans Administration Hospital
Durham, North Carolina

G.M. Chescheir III and Neil E. Garrett
Health Effects Research Program
Northrop Services, Inc.
Research Triangle Park, North Carolina

Joellen Lewtas Huisingh and Michael D. Waters
Health Effects Research Laboratory
U.S. Environmental Protection Agency
Research Triangle Park, North Carolina

INTRODUCTION

The emission of particulate matter into the atmosphere from stationary fuel combustion and transportation-related sources is a serious environmental problem. Natusch (1978) has shown that trace elements and chemicals associated with particulate matter from coal combustion may constitute a health hazard. In an earlier study, Natusch and Wallace (1974) reported that many known or potential carcinogens are preferentially concentrated on the surface of respirable coal fly ash. According to Davison et al. (1974), greater quantities of trace elements are associated with particles of fly ash too small to be trapped effectively by conventional particulate-control devices. The potential hazard of respirable particles was brought into sharper focus through studies in which organic extracts of fly ash particles (Chrisp et al., 1978; Fisher et al., 1979) and diesel exhaust particulates (Huisingh et al., 1978) were shown to be mutagenic in Salmonella typhimurium.

Studies with mammalian cells are necessary to confirm the mutagenic effects of environmental particulate matter. The Chinese hamster ovary (CHO) cell is being evaluated as a test system for such particles. These cells, which form discrete cell colonies in culture, were shown by Wininger et al. (1978) to be a convenient system for testing environmental chemicals. In our laboratory we have shown that the CHO cell is capable of phagocytizing particulate matter. This characteristic has been exploited to evaluate the toxicity of a variety of particles (Garrett et al., 1979, 1980), including samples from coal gasification, fluidized bed combustion, and conventional coal combustion. Furthermore, we have shown that the CHO system is useful in determining the

337

toxicity of such diverse environmental agents as liquid effluents
from textile mills (Campbell et al., 1979), polychlorinated
biphenyls (PCB)(Garrett and Stack, 1980), and organic
condensates from a refuse energy recovery system. Because the CHO
system is useful in evaluating the toxicity of chemical and
particulate matter, the studies have been extended to detect
possible mutagenic effects of atmospheric particles. The CHO cell
assay, which measures mutation at the hypoxanthine-guanine
phosphoribosyl transferase (HGPRT) locus, detects 95% of the
chemicals known to have carcinogenic activity <u>in vivo</u> (Hsie et al.,
1978).

This report examines the use of the CHO/HGPRT system to assay
toxic and mutagenic effects of environmental particles. Since CHO
cells readily phagocytize whole particles, the system provides a
straightforward method for testing mutagenesis of particulate
matter without the complexities of extraction and fractionation.

METHODS

The Chinese hamster ovary cell line (CHO-K1) was obtained from
the American Type Culture Collection and maintained in medium
supplemented with fetal calf serum (screened for virus and
mycoplasma). Exponentially growing stock cultures were harvested
by washing the flasks once with phosphate-buffered saline and then
with 0.25% trypsin. Cells were plated in Ham's F-12 media
containing 10% serum and no antibiotics. The cells were usually
treated with a three-day exposure to 1 μM aminopterin to prevent
the appearance of spontaneous mutants and were subcultured twice
before use in a mutation assay.

Mutation induction at the HGPRT locus was measured by the
method of O'Neill (1977). Ham's F-12 media containing 5% dialyzed
fetal calf serum and antibiotics was used. After trypsinization,
0.5×10^6 cells were transferred to Corning 75-cm^2 flasks and
incubated at 37°C. After a 24-h attachment and growth period, the
test sample was added to the medium, and the cultures were
incubated for 20 h. The exposed cells were then washed three times
with saline, and the cells were harvested and counted.

The initial cell survival was determined in experiments in
which aliquots of the cell suspension were added to media in 25-cm^2
flasks (300 cells/flask). The flasks were incubated for seven days,
and the colonies were fixed and stained with 0.04% crystal violet.

Mutation induction was determined in cells that were
subcultured every 48 h. The flasks were washed and trypsinized,
and 1×10^6 cells were added to 75-cm^2 flasks. After eight days of
culture, cells were plated for selection and post-expression colony

survival. Cloning efficiency was determined in experiments in which 200 cells were seeded in hypoxanthine-free media in 25-cm^2 flasks. Colonies of mutant cells were obtained in 75-cm^2 flasks after seeding 2 x 10^5 cells in media without hypoxanthine and with 10 μM 6-thioguanine. The flasks for cloning efficiency and selection were incubated at 37°C for seven days, and the colonies were then fixed and stained. The number of phenotypic mutants was monitored by changing the media in the flasks after five days of incubation to media without thioguanine but containing hypoxanthine and 1 μM aminopterin. The flasks were then incubated an additional five days to develop the colonies.

In each experiment, two replicate cultures were formed for the negative and positive controls and each concentration of the test substance. From each replicate, three flasks were derived for measuring initial cell survival, five for selection, three for the post-expression survival, and three for determining mutants resistant to aminopterin.

Phagocytic activity was determined by adding particulate samples to CHO cells cultured in Lab-Tek microslides. Before particles were added, the cells were incubated for a 24-h attachment period at 37°C in a humidified atmosphere of 5% carbon dioxide in air. After a 24-h incubation with the particles, the cells were washed three times with saline, fixed with methanol, and stained with May Grunwald and then Giemsa solutions. The cells were washed with water and then with acetone-xylene solutions.

Phagocytosis was confirmed by preparing electron micrographs of cells exposed to particulate samples in 25-cm^2 flasks. The cells were preincubated 24 h at 37°C. After treatment with the particles for 20 h, the cells were fixed overnight with glutaraldehyde in Millonig's phosphate buffer. After post-fixation with osmium tetroxide, the monolayers were stained en bloc with uranyl acetate in water, dehydrated with ethanol, and embedded. Sections were examined with a transmission electron microscope.

Size analysis was performed using a Coulter Counter TA II after counting 10,000 to 20,000 particles. The instrument was calibrated using a 100-μm aperture and standard 10.05-μm polystyrene latex sphere. Filtered (0.45 μm) physiological saline was used as the electrolyte. Samples of 1 to 2 mg were added to polystyrene test tubes and vortexed for 2 min. Saline (5 ml) was then added, and the tubes were vortexed for 30 sec. Immediately before counting, each tube was vortexed for 10 sec, and 0.1 to 1.0 ml of the sample was added to 100 ml saline. The particles were sized, and population mean diameters for particles > 2 μm were determined from cumulative plots.

Data reduction and statistical analysis was performed using modified versions of programs written for the Texas Instruments TI-59 calculator (Garrett and Stack, 1980).

RESULTS AND DISCUSSION

The response of the CHO cells to known mutagenic agents was checked using N-methyl-N'-nitro-N-nitrosoguanidine (MNNG) and cis-dichlorodiamine platinum-II (Pt[NH$_3$]$_2$Cl$_2$). The platinum complex presumably causes base substitution in DNA by miscoding. MNNG, an alkylating agent, is a more potent mutagen in the CHO system. Both compounds increased the mutation frequency with increasing dose. The effect of platinum on mutation frequency in the CHO cells is shown in Figure 1A. The dose response was linear with a correlation coefficient of 0.99. A concentration of 1.5 µg/ml platinum yielded a mean value of 96 mutant colonies/10 flasks in 12 independent experiments. This dose of platinum gave a mutation frequency of $1.16 \pm 0.38 \times 10^{-4}$ and was selected as a positive reference against which to measure the mutagenic effects of environmental particles. Spontaneous mutation in the untreated (negative control) cells gave rise to a mean of 5 mutant colonies/ 10 flasks (N=23) and a mutation frequency of $5.3 \pm 3.8 \times 10^{-6}$.

The range of mutagenic responses obtained for the negative and positive control is shown in Figure 1B. A logarithmic plot normalizes the standard deviation of the data and illustrates the structure of the distributions. Other investigators have transformed CHO mutation data to correct for the nonhomogeneity of variances and non-normal distributions in test cultures and in untreated controls (Irr and Snee, 1979). Figure 1B shows that the mean mutation frequency for the positive reference was approximately 10 times the maximum frequency for the untreated control, facilitating a comparison with weakly mutagenic substances.

The size distribution of particles tested in the mutation assay is shown in Figure 2. Exhaust particles from two diesel engines had a similar particle size distribution (Figure 2, A and B). Brief sonication of one diesel sample apparently dissociated the particulate, so that a larger number of small particles was observed (Figure 2C). Fly ash from oil combustion (Figure 2D) was also a heterogeneous mixture with a size distribution similar to that of the diesel particles. The fly ash from coal combustion tested in these experiments was < 3 µm (Figure 2F) and 2 to 5 µm (Figure 2H). Iron oxide particles (Figure 2E) were similar in size distribution to the coal fly ashes and were used in these experiments as a reference particulate. Silica particles, obtained from a commercial source, were 4 to 9 µm (Figure 2G).

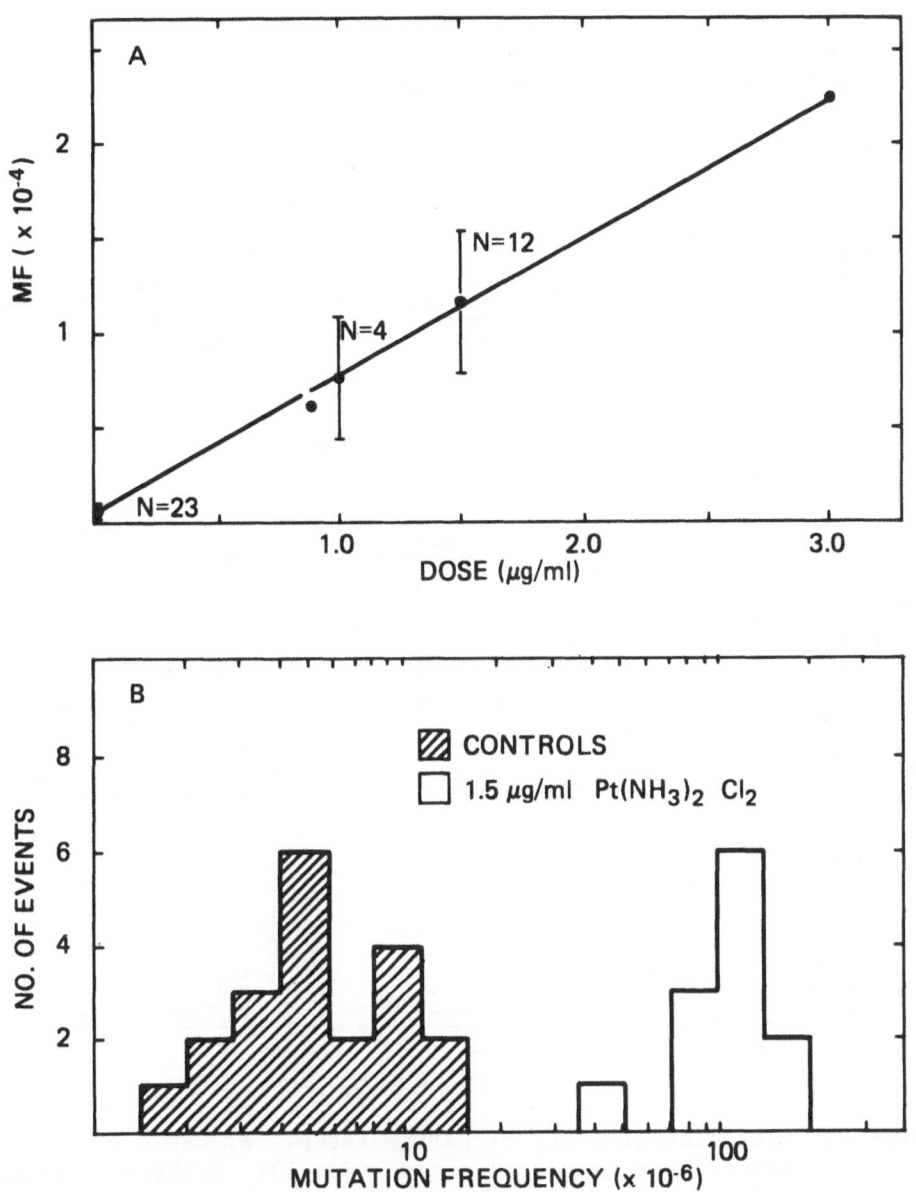

Figure 1. Mutation frequencies (MF) obtained in untreated controls
and CHO cells treated with $Pt(NH_3)_2Cl_2$. A. The effect
of $Pt(NH_3)_2Cl_2$ on mutation with increased dose (0 to 3
µg/ml). B. Histogram of the mutagenic responses
obtained in control CHO cells and cells treated with 1.5
µg/ml of $Pt(NH_3)_2Cl_2$. There were three events with
mutation frequencies equal to zero.

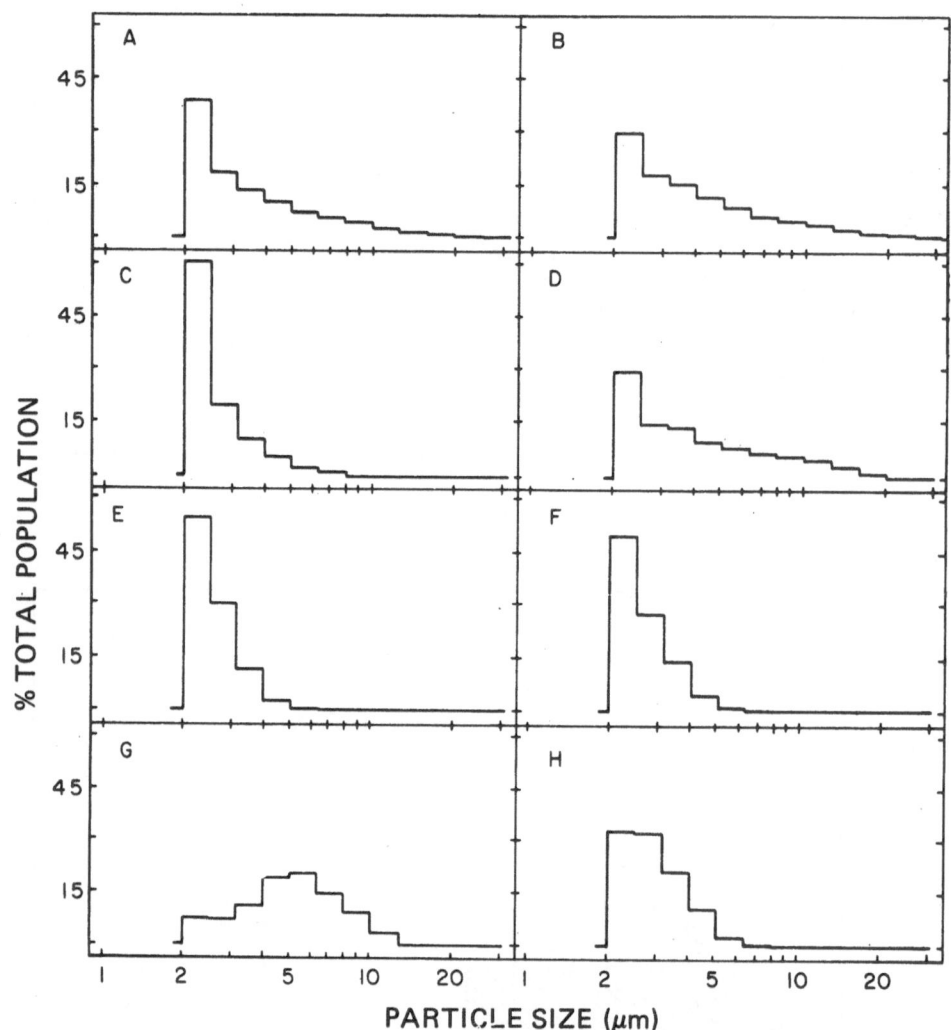

Figure 2.   Size distributions of particulates. A. Exhaust
             particles from diesel engine (No. 1). B. Diesel exhaust
             particles (No. 2). C. Diesel particulate sample No. 1
             sonicated three times for 10 sec. D. Fly ash from oil
             combustion. E. Iron oxide particles. F. Fly ash from
             coal combustion < 3 μm. G. Commercial silica particles
             4.5 to 9 μm. H. Fly ash from coal combustion size
             fractionated 2 to 5 μm.

The relative size of particles ingested by CHO cells is shown in light and electron micrographs (Figures 3 through 5). Figure 3 shows control cells with very prominent nucleoli and high ratios of nuclear material to cytoplasm; no phagocytic material is seen. Treated cells examined by phase contrast microscopy (not shown) and fixed and stained cells examined by light microscopy revealed a large number of particles in the cytoplasm of CHO cells. These particles were frequently arranged closely around the nucleus. Electron microscopy of sections of fixed and embedded cells confirmed that the particles were in the cells, often closely apposed to the nucleus. The treated cells differed from the control cells in that a membrane could often be seen around each particle or group of particles, forming a very large membrane-limited phagosome. Figure 4 shows CHO cells in which particles of fly ash were present in phagosomes. Frequently the particles either sectioned poorly or appeared to have fallen out of the section. The phagosomes of cells treated with diesel particulates contained small amorphous particles of the diesel exhaust material and empty regions suggesting fluid accumulation (Figure 5).

These experiments provided evidence that a variety of particulates of environmental concern are trapped close to the cell nucleus. The effect of these particles on mutation was investigated by studying both the particles alone and particles in combination with MNNG and $Pt(NH_3)_2Cl_2$. The latter experiments tested the possibility that the particles could facilitate passage of another mutagen to the cell nucleus, with subsequent damage to DNA. Iron oxide, silica, and fly ash particles from coal combustion did not produce a statistically significant difference in mutation frequency in cells treated simultaneously with MNNG and platinum (Figure 6, A and C). Fly ash and silica did contribute to cell toxicity (Figure 6, B and D). The toxic effect of iron oxide with mutagens was not different from that of the mutagen alone.

Because of the high baseline mutation of MNNG and $Pt(NH_3)_2Cl_2$, these experiments would detect relatively large influences on mutation, but not small changes due to the particles alone. Additional experiments tested the effects of the particles without an exogenous mutagen. As shown in Figure 7, A and B, six particles and one organic extract produced mutation in excess of the control values. These substances were toxic to the cell cultures to varying degrees (Figure 6, C and D). The model particles of coal fly ash (2 to 5 μm) produced a small increase in mutation frequency (17×10^{-6} versus 8.4×10^{-6} for controls, at a dose of 125 μg/ml). Another sample of fly ash (0 to 3 μm) from conventional coal combustion caused a four-fold increase in mutation (9.8×10^{-6} at 250 μg/ml versus 2.5×10^{-6} for controls. Similarly, fly ash from oil combustion increased the mutation frequency (29.1×10^{-6} at 75 μg/ml versus 11.8×10^{-6}; p = 0.15). Two samples of exhaust

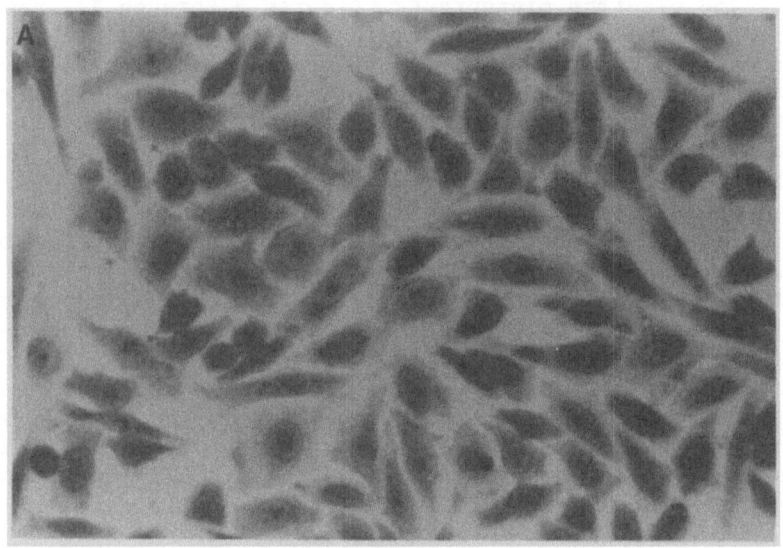

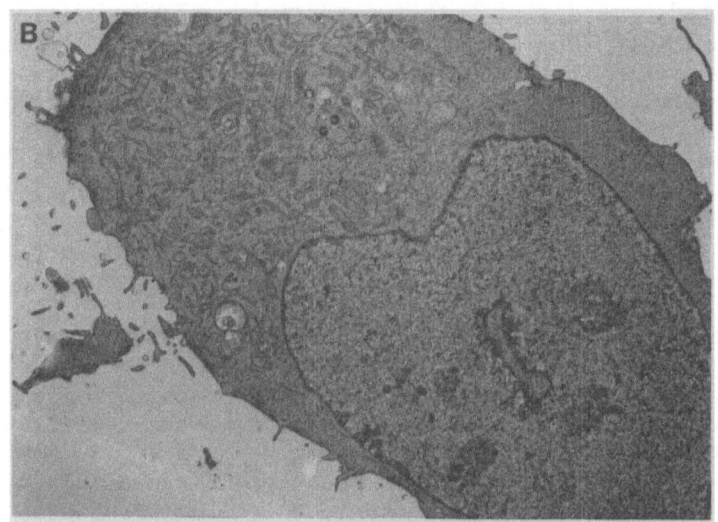

Figure 3.   Light micrograph (A) and electron micrograph (B) of
            control CHO cells.   The cells were magnified 200 times
            in the light micrograph and 5000 times in the electron
            micrograph.   The control cells exhibit prominent
            nucleoli and high nuclear-to-cytoplasm ratios.   No
            phagocytized material is seen.

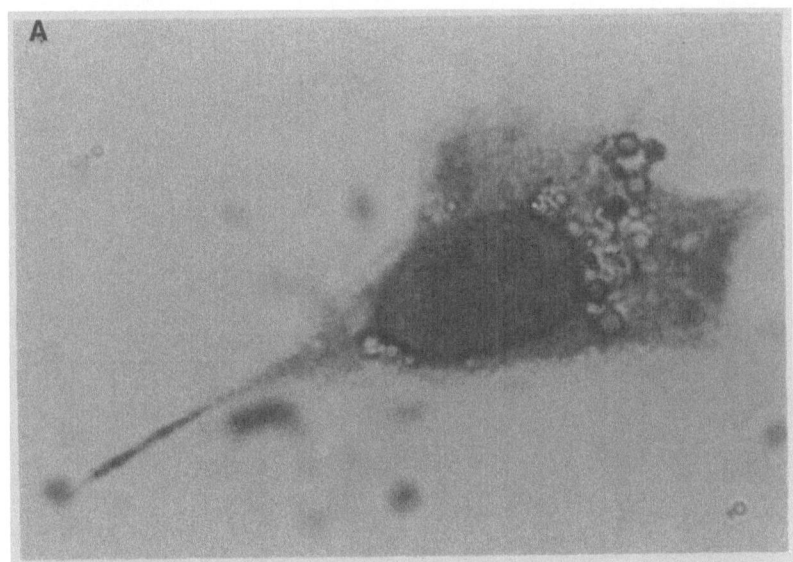

Figure 4. Light micrograph (A) and electron micrograph (B) of CHO
           cells after a 20-h exposure to coal fly ash. In A,
           cells were exposed to 200 µg/ml of fly ash, < 3 µm in
           diameter, and magnified 1000 times; in B, cells were
           exposed to 250 µg/ml of 2 to 5 µm coal fly ash and
           magnified 7100 times. Many fly ash particles are
           visible in the cytoplasm, immediately adjacent to the
           cell nucleus.

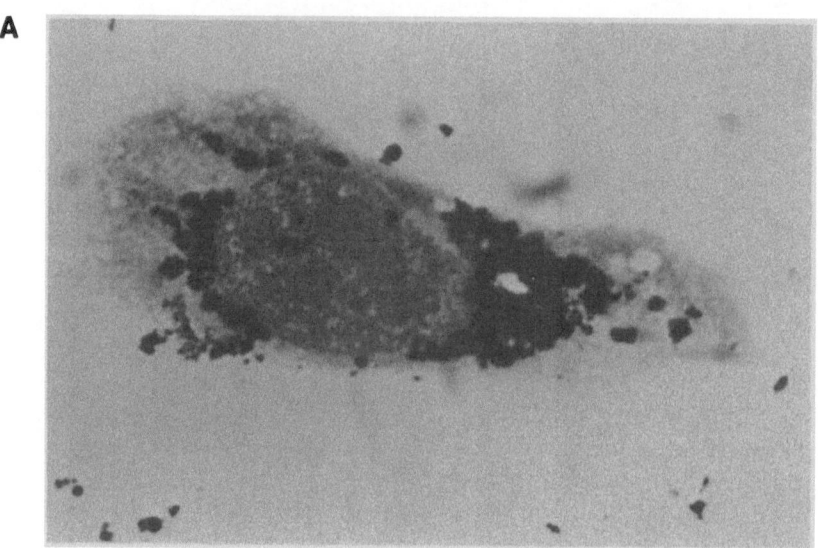

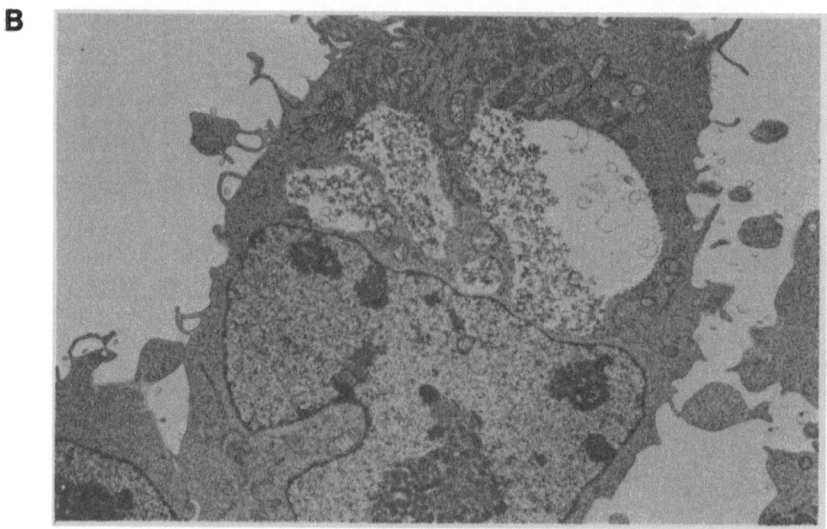

Figure 5.   Light micrograph (A) and electron micrograph (B) of CHO
cells after a 20-h treatment with sonicated diesel
(No. 1) particles at 100 µg/ml. In A, cells were
magnified 1000 times, and in B, 5000 times. Phagosomes
of the treated cells contain small amorphous particles
of the diesel exhaust material. Particles are closely
associated with the cell nucleus.

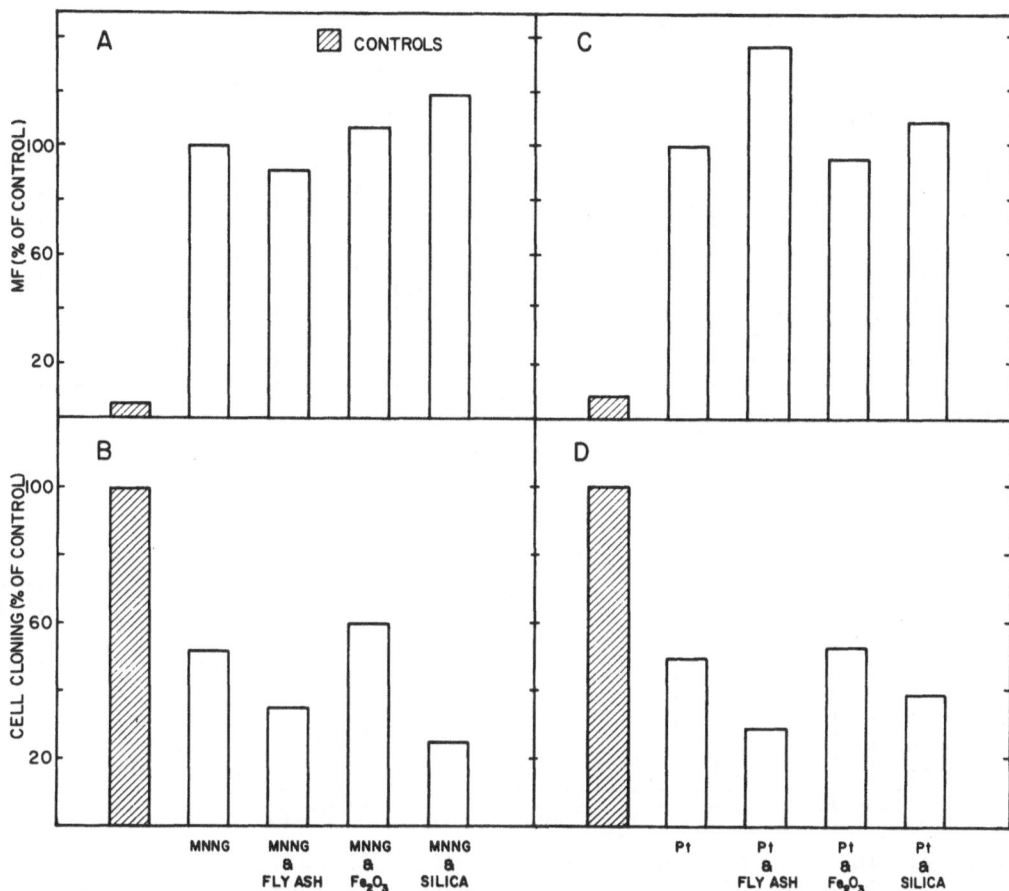

Figure 6. Mutagenicity (A, C) and toxicity (B, D) of particles in combination with MNNG and $Pt(NH_3)_2Cl_2$. Mutagenicity is expressed as a percent of the mutation frequency of MNNG or $Pt(NH_3)_2Cl_2$. CHO cells were exposed to 0 to 3 μm fly ash at 200 μg/ml, ferric oxide particles at 100 μg/ml, and silica particles at 1000 μg/ml. Toxicity was measured as the initial cell survival in the CHO cloning assay. Toxicity data are expressed as a percent of the untreated control.

particles from diesel engines produced about a three-fold increase in mutation over the control values. One sample of particles increased mutation from 2.2×10^{-6} for the control to a value of $5 \cdot \times 10^{-6}$ at 750 μg/ml. A methylene chloride extract of this diesel sample increased mutation ·to five times the control frequency

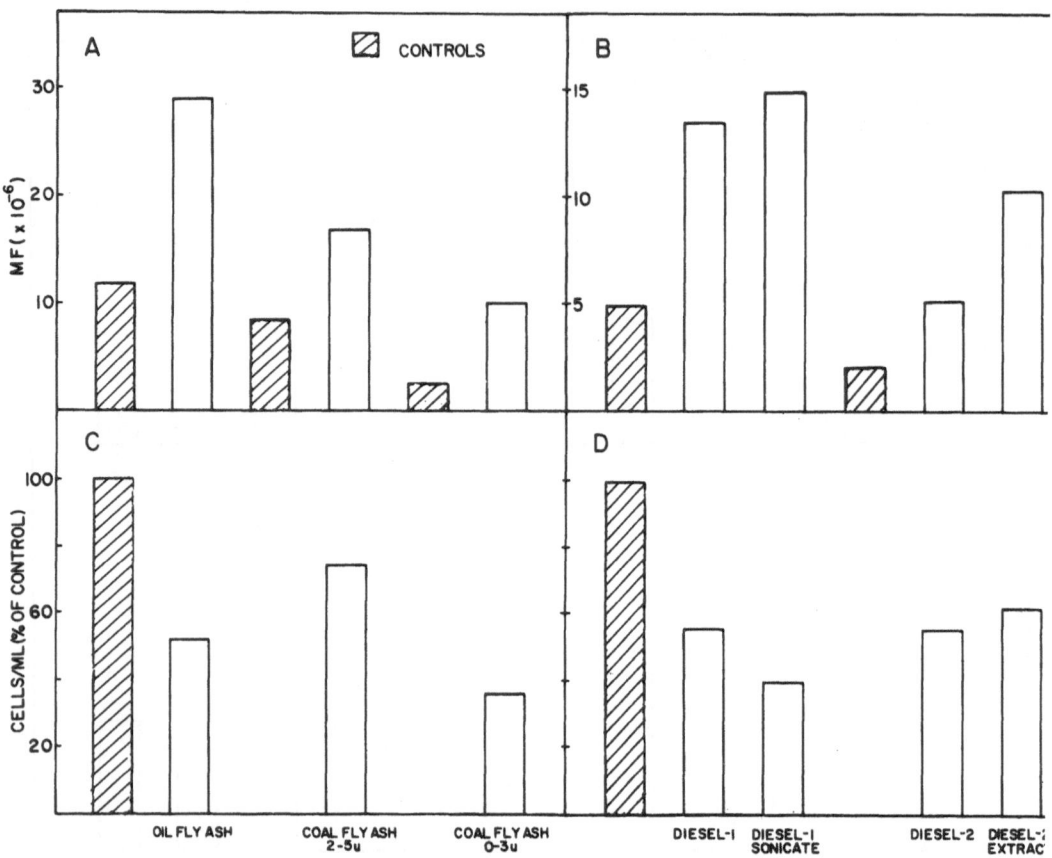

Figure 7. The effect of environmental particulates on mutation
frequency (A, B) and toxicity (C, D) in CHO cells. CHO
cells were exposed to oil fly ash at 75 μg/ml, 2 to
5 μm coal fly ash at 125 μg/ml, 0 to 3 μm coal fly ash
at 250 μg/ml, diesel exhaust particles (No. 1) at 500
μg/ml, sonicated No. 1 diesel particles at 100 μg/ml,
diesel exhaust (No. 2) at 750 μg/ml, and a methylene
chloride extract of No. 2 diesel particles at 50 μg/ml.
Toxicity or cells per milliliter is expressed as percent
of the untreated control.

(10.4 x 10^{-6} at 50 μg/ml versus 2.2 x 10^{-6} for the control;
p = 0.078). Higher concentrations of the extract were severely
toxic. Another diesel particle was evaluated with and without
sonication and mutation was also increased (13.4 x 10^{-6} at

500 µg/ml for untreated particles (p = 0.12); 14.9 x 10^{-6} at
100 µg/ml for sonicated particles (p = 0.059); the control rate was
was 4.9 x 10^{-6}).

These data are, to our knowledge, the first to demonstrate
that whole particles of fly ash from coal and oil combustion and
exhaust particles from diesel engines can cause mutation in a
mammalian cell culture system. For these experiments, it was not
necessary to extract the particles with solvents. These results
suggest that this test system might be used to assay environmental
particulates that have not been previously treated with solvents.
Such a test would more accurately reflect natural exposure to
particulate matter in the atmosphere, since extraction with organic
solvents and subsequent concentration of the extract have no
counterparts in natural exposure by inhalation.

REFERENCES

Campbell, J.A., N.E. Garrett, J.L. Huisingh, and M.D. Waters.
    1979. Cellular toxicity of liquid effluents from textile
    mills. In: Symposium Proceedings: Textile Industry
    Technology. EPA-600/2-79-104. U.S. Environmental Protection
    Agency: Research Triangle Park, NC. pp. 239-248.

Chrisp, C.E., G.L. Fisher, and J.E. Lammert. 1978. Mutagenicity
    of filtrates from respirable coal fly ash. Science 199:73-75.

Davison, R.L., D.F.S. Natusch, J.R. Wallace, and C.A. Evans. 1974.
    Trace elements in fly ash. Dependence of concentration on
    particle size. Environ. Sci. Technol. 8:1107-1113.

Fisher, G.L., C.E. Chrisp, and O.G. Raabe. 1979. Physical factors
    affecting the mutagenicity of fly ash from a coal-fired power
    plant. Science 204:879-881.

Garrett, N.E., J.A. Campbell, J.L. Huisingh, and M.D. Waters. 1979.
    The use of short term bioassay systems in the evaluation of
    environmental particulates. In: Proceedings of the Symposium
    on the Transfer and Utilization of Particulate Control
    Technology, Vol. 4. EPA-600/7-79-044d. U.S. Environmental
    Protection Agency, Research Triangle Park, NC. pp. 175-186.

Garrett, N.E., G.M. Chescheir III, N.A. Custer, J.D. Shelburne,
    J.L. Huisingh, and M.D. Waters. 1980. An evaluation of the
    cytotoxicity and mutagenicity of environmental particulates in
    the CHO/HGPRT system. In: Proceedings of the 2nd Symposium
    on the Transfer and Utilization of Particulate Control
    Technology, Vol. 4. EPA-600/9-80-039d, U.S. Environmental
    Protection Agency: Research Triangle Park, NC. pp. 524-535.

Garrett, N.E., and H.F. Stack. 1980. Cellular toxicology data analysis using a programmable calculator. Comput. Biol. Med. 10:153-167.

Hsie, A.W., J.P. O'Neill, J.R. San Sebastian, D.B. Couch, P.A. Brimer, W.N.C. Sun, J.C. Fuscoe, N.L. Forbes, R. Machanoff, J.C. Riddle, and M.H. Hsie. 1978. Quantitative mammalian cell genetic toxicology: Study of the cytotoxicity and mutagenicity of seventy individual environmental agents related to energy technologies and three subfractions of a crude synthetic oil in the CHO/HGPRT system. In: Application of Short-term Bioassays in the Fractionation and Analysis of Complex Mixtures. M.D. Waters, S. Nesnow, J.L. Huisingh, S.S. Sandhu, and L. Claxton, eds. Plenum Press: New York. pp. 292-315.

Huisingh, J., R. Bradow, R. Jungers, L. Claxton, R. Zweidinger, S. Tejada, J. Bumgarner, F. Duffield, M. Waters, V.F. Simmon, C. Hare, C. Rodriguez, and L. Snow. 1978. Application of bioassay to the characterization of diesel particle emissions. In: Application of Short-term Bioassays in the Fractionation and Analysis of Complex Environmental Mixtures. M.D. Waters, S. Nesnow, J.L. Huisingh, S.S. Sandhu, and L. Claxton, eds. Plenum Press: New York. pp. 381-418.

Irr, J.D., and R.D. Snee. 1979. Statistical evaluation of mutagenicity in the CHO/HGPRT system. In: Banbury Report 2. Mammalian Cell Mutagenesis: The Maturation of Test Systems. A.W. Hsie, J.P. O'Neill, and V.K. McElheny, eds. Cold Spring Harbor Laboratory: New York. pp. 263-275.

Natusch, D.F.S. 1978. Potentially carcinogenic species emitted to the atmosphere by fossil-fueled power plants. Environ. Hlth. Perspect. 22:79-90.

Natusch, D.F.S., and J.R. Wallace. 1974. Urban aerosol toxicity: The influence of particle size. Science 186:695-699.

O'Neill, J.P. 1977. A quantitative assay of mutation induction at the hypoxanthine-guanine phosphoribosyl transferase locus in chinese hamster ovary cells (CHO/HGPRT system): Development and definition of the system. Mutation Res. 45:91-101.

Wininger, M.T., F.A. Kulik, and W.D. Ross. 1978. In vitro clonal cytotoxicity assay using chinese hamster ovary cells (CHO-K1) for testing environmental chemicals. In Vitro 14:381.

A PRELIMINARY STUDY OF THE CLASTOGENIC EFFECTS OF DIESEL EXHAUST FUMES USING THE *TRADESCANTIA* MICRONUCLEUS BIOASSAY

Te-Hsiu Ma and Van A. Anderson
Department of Biological Sciences and
Institute for Environmental Management
Western Illinois University
Macomb, Illinois

Shahbeg S. Sandhu
Health Effects Research Laboratory
U.S. Environmental Protection Agency
Research Triangle Park, North Carolina

INTRODUCTION

Automobile engine emissions are one of the major urban air pollutants in the industrialized nations. The direct and indirect effects of these complex agents on human health are crucial problems for modern society. Epidemiological studies carried out in Europe (Barth and Blacker, 1978; Blumer et al., 1977a, b) and Japan (Shimizu et al., 1977) found that cancer mortality rates were higher in populations near heavily travelled highways than in those away from them. The increasing popularity of diesel-engine-powered vehicles, especially passenger cars, demands a better understanding of the health effects of diesel emissions.

Current information on the health effects of diesel engine exhaust fumes is limited to the results of laboratory tests with experimental animals. In an early review, Goldsmith (1964) concluded that low concentrations of diesel exhaust fumes stimulate the growth of cultured mammalian cells. Studies using mammals such as cats, rats, mice, and Guinea pigs have shown that aryl hydrocarbon hydroxylase increases in the lungs, liver, and prostate; however, no pathological change is seen in the lungs of exposed animals (see Barth and Blacker, 1978, for a review). Campbell et al. (1979), in their studies of the effects of light-duty engine diesel exhaust fumes on mice, found increased susceptibility to infection and mortality rates. Guerrero et al. (1979) found that the sister-chromatid exchange rate in golden hamsters was not altered at a dose of 20 mg/animal for 24 h, while in another study (Pereira et al., 1979) a six-month treatment gave negative results in the bone marrow micronuclei bioassay, but positive results in the sperm morphology test. Other studies have shown biochemical changes in the lungs of rats and Guinea pigs exposed to diesel exhaust fumes. At doses of 250 to 1,500 $\mu g/m^3$, 20 h/day, 5.5 days/week, for 12 to 24 weeks, prostaglandin dehydrogenase activity was reduced (Chaudart and Dutta, 1979). A 36-week exposure caused increases in lung weight and in lipids and collagen in lung tissue (Misioroski et al., 1979). In strain A mice, inhalation of diesel exhaust fumes for seven months showed no effect on lung structure (Orthoefer, 1979), and inhalation for 28 weeks did not cause changes in sperm morphology (Pereira et al., 1979b). Negative results were also obtained in the Drosophila recessive mutation test, after an 8-h exposure to diesel exhaust fumes containing 11.6 ppm of hydrocarbons (Schuler and Niemeier, 1979).

The whole-animal bioassays for diesel exhaust fume effects are time-consuming and costly. An alternative approach to this urgent problem would be using a battery of short-term bioassays similar to those proposed by Huisingh et al. (1979) for diesel exhaust particulates. The Tradescantia (spiderwort) micronucleus (Trad-MCN) is a very sensitive and quick bioassay (Ma, 1980) that is especially suitable for testing gaseous agents in ambient air (Ma et al., 1980), as well as in chambers (Ma et al., 1978; Ma, 1979). The reliability and efficiency of this bioassay were verified by tests of well-known mutagens (Ahmed and Ma, 1980; Ma, 1979; Ma and Anderson, 1979; Ma et al., 1978, 1980a) and dose-response curves established using 1,2-dibromoethane (Ma et al., 1978) and X-ray treatments (Ma et al., 1980b). Therefore, the Trad-MCN bioassay was used to determine the clastogenic effects of total diesel exhaust fumes in the present study.

MATERIALS AND METHODS

The Tradescantia paludosa clone 03 was used for all experiments. A population of this clone was cultivated in the Duke University Phytotron, Durham, NC, under optimal growing conditions in clean air. The plant cuttings bearing young inflorescences (about 5 cm long) were maintained in tap water in a plastic cup (200 ml) before, during, and after treatment. Generally, 15 to 20 cuttings constituted an experimental group, and each experiment involved three groups treated with different doses, one baseline control group (in a clean-air room), and one field control group (in the experimentation area, outside of the treatment chamber). One group exposed to 40 R of X-rays served as the positive control for all experiments.

Two pilot studies were conducted to establish an adequate dose rate and total dose for both the concentration of fumes and the duration of exposure. The final series of experiments was carried out at three different concentrations (1/200, 1/45, and 1/23

dilutions) of the total exhaust fumes. The fumes were generated by
a Nissan (1970 Datsun) diesel automobile at the Emissions
Measurements Characterization Division, Environmental Sciences
Research Laboratory, U.S. Environmental Protection Agency, Research
Triangle Park, NC. The automobile ran on a chassis dynamometer
under a simulated city driving cycle (23 min/cycle). The diluted
exhaust fumes were introduced into a specially designed Plexiglas
chamber (31-1 capacity) at a flow rate of 32 1/min. Three groups
of plant cuttings were each treated with a different dose by
exposing them to a given concentration of fumes in the treatment
chamber for one, two, or three driving cycles. Gas flow was
interrupted for 5 min between the driving cycles. The temperature
in the chamber was about 28°C, the relative humidity was in the
range of 50 to 80%, and the atmospheric pressure was 0.99 to 1.0
bar (745 to 750 mmHg). Nitrogen oxides, hydrocarbons, carbon
monoxide, and carbon dioxide in each of the dilutions of fumes were
sampled and analyzed during the treatment. The concentrations of
these gases are given in Table 1. The doses of each treatment
group were derived from a combination of dilution factor and number
of driving cycles.

There was a 30-h recovery time after the end of the treatment,
to allow the meiotic prophase I pollen mother cells to arrive at
the early tetrad stage, where the broken chromosomes appear as
micronuclei (MCN) outside of the nucleus proper. The young
inflorescences of treated and control groups were fixed in aceto-
alcohol (1:3) after the recovery period; the fixed samples were
transferred into 70% ethanol after 24 to 48 h of fixation.
Microslides of the tetrad stage of pollen mother cells were
prepared by the aceto-carmine squash method. The frequency of MCN
on each slide was expressed as the number of MCN per 100 tetrads.
An average of 350 tetrads were observed on each slide. The mean
frequency and the standard error of the mean for each experimental
group was derived from an examination of five slides. The treated
and control groups were statistically compared using the standard
error of the difference in means, with a significance level of
0.01.

RESULTS AND DISCUSSION

Since the end point of this bioassay is the frequency of MCN
in tetrads, and MCN represent the broken pieces of chromosomes, the
diesel exhaust fumes effect demonstrated in this study is
appropriately referred to as a clastogenic rather than a mutagenic
effect. The preliminary data from three series of experiments are
shown in Table 1. The only significantly positive results were for
DEF-2, T-2, and DEF-3, T-3. All lower doses (fewer driving cycles
or lower concentration of fumes or both) gave negative results.
The negative result for DEF-2, T-3 was due to an overdose, a common

Table 1. Results of Trad-MCN Test on Clastogenic Effect of Diesel Exhaust Fumes

Experimental Groups (Treatment = No. of Driving Cycles)	Concentration of Gases[a]				MCN per 100 Tetrads (mean)	Standard Error	Significance (p < 0.01)
	HC[b] (ppm)	NO (ppm)	CO (ppm)	CO$_2$ (%)			
DEF-1							
Baseline control					8.32	1.63	
Field control					5.67	0.61	
Treatment 1	3.04	1.5	0.9	0.04	5.45	0.68	-
Treatment 2	3.04	1.5	0.9	0.04	5.97	0.67	-
Treatment 3	3.04	1.5	0.9	0.04	7.75	0.78	-
DEF-2 and -3							
Baseline control					4.76	0.90	
Field control					5.04	0.70	
DEF-2							
Treatment 1	46.95	18.0	18.0	0.49	8.56	2.07	-
Treatment 2	47.36	17.0	16.0	0.49	20.81	3.13	+
Treatment 3	43.84	18.0	17.0	0.49	10.90	1.68	-
DEF-3							
Treatment 1	18.04	8.8	8.0	0.23	7.35	0.32	-
Treatment 2	20.65	10.0	9.5	0.25	7.54	1.62	-
Treatment 3	21.00	9.5	8.5	0.25	14.80	0.44	+
Positive control (40 R X-rays)					62.06	3.10	+

[a] These concentrations were obtained from the constant volume sampler before a tenfold dilution.
[b] HC = Hydrocarbons.

phenomenon in this test system. Such overdose samples usually have relatively low frequencies of MCN accompanied by abortive cells in tetrads and/or microspores.

The concentrations of fumes that induced MCN were probably comparable to some concentrations measured in a separate in situ monitoring project. Samples obtained through 1- to 6-h monitoring at public parking garages, bus stops, and truck stops (Ma et al., 1980a) sometimes gave positive results in the Trad-MCN bioassay. According to Zdrazil and Picha (1978), the particulate exhaust concentration is around 8.21 µg/m^3 at the orifice of the exhaust pipe, and around 0.205 µg/m^3 in an ordinary working garage. These figures may help to indicate the actual magnitude of exhaust pollution.

In another study (Ahmed and Ma, 1980), the Trad-MCN bioassay was used concurrently with a human lymphocyte chromosome aberration bioassay to establish the relative effectiveness of these two tests. Results of such comparative studies may be used to extrapolate to human cell systems the mutagenicity of a given agent as determined in the Trad-MCN assay.

The Tradescantia stamen hair mutation test (Schairer et al., 1978) is also capable of detecting the mutagenic effect of diesel exhaust fumes. Although no report on direct testing of diesel exhaust fumes in the laboratory is yet available, Tradescantia stamen hair in situ tests were used at interstate highway junctions and in downtown areas of several large American cities. This kind of in situ monitoring is comparable to the present laboratory study. The stamen hair mutation test, which measures the rate of somatic mutation at a particular locus, and the Trad-MCN test, which measures the frequency of chromosome damage, are an ideal combination of bioassays for better assessment of the effect of diesel exhaust fumes on living systems.

ACKNOWLEDGMENTS

The authors wish to express their deepest appreciation to Dr. Roy Zweidinger and the staff of the Emissions Measurements Characterization Division, Environmental Sciences Research Laboratory, U.S. Environmental Protection Agency, Research Triangle Park, NC, for providing diesel exhaust fumes and their dilution, characterization, and measurements; and to Dr. Henry Hellmers, Director of the Duke University Phytotron, Durham, NC, for raising the Tradescantia plants.

REFERENCES

Ahmed, I., and T.H. Ma. 1980. Chromosome breakage induced by maleic hydrazide in cultured human lymphocytes and Tradescantia pollen mother cells. Environ. Mutagen. 2:287. (abstr.).

Barth, D.S., and S.M. Blacker. 1978. The EPA program to assess the public health: Significance of diesel emissions. J. Air Pollut. Control Assoc. 28:269-771.

Blumer, M., W. Blumer, and T. Reich. 1977a. Polycyclic aromatic hydrocarbons in soil of a mountain valley: Correlation with highway traffic and cancer incidence. Environ. Sci. Technol. 11:1082-1084.

Blumer, W., M. Blumer, and T. Reich. 1977b. Carcinogenic hydrocarbons and the incidence of cancer mortality among residents near an automobile highway. Fortschro. Med. 95:1497-1498 and 1551-1552.

Campbell, K.I., E.L. George, and I.S. Washington, Jr. 1979. Enhanced susceptibility to infection in mice after exposure to diluted exhaust from light duty diesel engines. Presented at the International Symposium on Health Effects of Diesel Engine Emissions, Cincinnati, OH.

Chaudhart, A., and S. Dutta. 1979. Effect of exposure to diesel exhaust on pulmonary prostaglandin dehydrogenase (PGDH) activity. Presented at the International Symposium on Health Effects of Diesel Engine Emissions, Cincinnati, OH.

Goldsmith, G. 1964. Air pollution and health. Science 145:184-186.

Guerrero, R.R., D.E. Rounds, and J. Orthoefer. 1979. Sister chromatid exchange analysis of Syrian hamster lung cells treated in vivo with diesel exhaust particulates. Presented at the International Symposium on Health Effects of Diesel Engine Emissions, Cincinnati, OH.

Huisingh, J., S. Nesnow, R. Bradow, and M. Waters. 1979. Application of a battery of short term mutagenesis and carcinogenesis bioassays to the evaluation of soluble organics from diesel particulates. Presented at the International Symposium on Health Effects of Diesel Engine Emissions, Cincinnati, OH.

Ma, T.H. 1979. Micronuclei induced by X-rays and chemical mutagens in meiotic pollen mother cells of Tradescantia--a promising mutagen test system. Mutation Res. 64:307-313.

Ma, T.H., and V.A. Anderson. 1979. Micronuclei induced by ultraviolet light and chemical mutagens in meiotic pollen mother cells of Tradescantia. 10th Annual Meeting of the Environmental Mutagen Society, New Orleans, LA.

Ma, T.H., V.A. Anderson, and I. Ahmed. 1980a. In situ monitoring of air pollutants and screening of chemical mutagens using Tradescantia Micronucleus Bioassay. Environ. Mutagen. 20:287. (abstr.).

Ma, T.H., G.J. Kontos, Jr., and V.A. Anderson. (1980b). Stage sensitivity and dose response of meiotic chromosomes of pollen mother cells of Tradescantia to X-rays. Environ. Exp. Bot. 20:169-174.

Ma, T.H., A.H. Sparrow, L.A. Schairer, and A.F. Nauman. 1978. Effect of 1,2-dibromoethane (DBE) on meiotic chromosomes of Tradescantia. Mutation Res. 58:251-158.

Misioroski, R.L., K. Strom, and M. Schvapil. 1979. Lung biochemistry of rats chronically exposed to diesel particulate. Presented at the International Symposium on Health Effects of Diesel Engine Emissions, Cincinatti, OH.

Orthoefer, J.G. 1979. The strain A mouse as an inhalation carcinogenesis model in diesel exhaust research. Presented at the International Symposium on Health Effects of Diesel Engine Emissions, Cincinnati, OH.

Pereira, M.A., P.S. Sabharwal, C. Ross, P. Kaur, A. Choi, and T. Dixon. 1979a. In vivo studies on the mutagenic effects of inhaled diesel exhaust. Presented at the International Symposium on Health Effects of Diesel Engine Emissions, Cincinnati, OH.

Pereira, M.A., P.S. Sabharwal, and A.J. Wyrobek. 1979b. Sperm abnormality bioassay of mice exposed to diesel exhaust. Presented at the International Symposium on Health Effects of Diesel Engine Emissions, Cincinnati, OH.

Schairer, L.A., J. Van't Hof, C.C. Hayes, R.M. Burton, and F.J. de Serres. 1978. Exploratory monitoring of air pollutants for mutagenicity activity with the Tradescantia stamen hair system. Environ. Hlth. Perspect. 27:51-60.

Schuler, R.L., and R.W. Niemeier. 1979. A study of diesel
    emission on Drosophila. Presented at the International
    Symposium on Health Effects of Diesel Engine Emissions,
    Cincinnati, OH.

Shimizu, H., K. Aoki, and T. Kuroishi. 1977. An epidemiological
    study of lung cancer in relation to exhaust gas from cars.
    Lung Cancer 17:103-112.

Zdrazil, J., and F. Picha. 1978. Carcinogenic hydrocarbons from
    exhaust fumes in the working atmosphere. Cesk. Hyg.
    8:344-348.

ABILITY OF LIVER HOMOGENATES AND PROTEINS TO REDUCE THE MUTAGENIC EFFECT OF DIESEL EXHAUST PARTICULATES

Yi Y. Wang and Eddie T. Wei
School of Public Health
University of California
Berkeley, California

INTRODUCTION

An estimate by the U.S. Environmental Protection Agency (EPA) indicates that by 1985, 25% of the automobiles produced in the United States may be diesel powered, because diesel engines provide greater fuel economy than do spark-ignition gasoline engines (Santodonato et al., 1978). Automotive diesel engines may emit from 30 to 80 times more particulates than a comparable gasoline engine (Springer and Baines, 1977). Concern has been expressed about the potential health hazards of these particulates, since extracts of diesel particulates contain chemicals that are mutagenic in the Ames Salmonella typhimurium bioassay (Huisingh et al., 1978; Wei et al., 1980), a short-term test for determining the mutagenic and carcinogenic potential of chemicals (Ames and McCann, 1976).

The majority of the mutagenic activity in extracts of diesel exhausts does not require mammalian liver enzymes for activation. In fact, the addition of rodent liver homogenates (S-9) in the Ames assay mixture decreases mutagenic activity (Huisingh et al., 1978; Wei et al., 1980; Clark and Salmeen, 1980). The mutagenicity-reduction effects of S-9 are also observed when S-9 is added to mutagenic extracts of airborne particulates (Talcott and Wei, 1977) and gasoline-engine exhausts (Wang et al., 1978). Furthermore, S-9 reduces the mutagenic activity of chemicals such as sodium azide, sodium dichromate, sodium nitrite, and 5-nitro-2-furoic acid (deFlora, 1978).

Some investigators have suggested that the mutagenicity-reduction effects of S-9 are due to enzymatic degradation of the mutagen (Clark and Salmeen, 1980; deFlora, 1978); however, no experimental evidence has been brought forth to substantiate this hypothesis. We have examined here the properties and components of S-9 that account for its mutagenicity-reduction effect on diesel exhaust samples. The results indicate that the mutagenicity-reduction effect of S-9 is due to non-enzymatic binding of mutagens to liver proteins.

MATERIALS AND METHODS

The defined diesel exhaust sample used in this study was obtained from General Motors Research Laboratories (GM), Warren, MI. Details on the collection and characterization of the sample have been described by Schreck et al. (1978). The defined sample was collected on a mini baghouse filter attached to a 2.1-1 Peugeot diesel engine. The engine was operated on a water-brake dynamometer at a 69-km/m cruise condition and 9 kW (12 hp). Diesel fuel #2 was used. Particulates were weighed, mixed with an appropriate volume of dimethylsulfoxide (DMSO), sonicated for several minutes, and then vortexed immediately before bioassay. The Ames test procedures were followed without modification (Ames et al., 1975). The Salmonella strain employed was TA98, which has been shown to be the most sensitive strain for detecting diesel exhaust mutagens (Huisingh et al., 1978; Wei et al., 1980). The post-mitochondrial supernatant fraction of rat liver homogenates (S-9) was prepared from Aroclor-1254-treated rats, according to the standard methods of Ames et al. (1975). The cofactors were added to the S-9 prior to bioassay. One half milliliter of S-9 mix, containing 50 µl of S-9, was applied on each plate. All experiments have been repeated at least once, and all samples have been tested in duplicate in each experiment.

In some experiments, S-9 was separated according to the procedure of Frantz and Malling (1975). The enzymatic activity in S-9 was inactivated by heat or by omission of the NADPH-generating system. S-9 mix was heated in a boiling water bath (100°C) for five minutes. The boiled S-9 mix was cooled at room temperature and vortexed before being applied to the assay. Part of the heat-treated S-9 was filtered through a 0.45-µm Millipore filter unit to remove coagulated proteins. The filtered or unfiltered heat-treated S-9 mix was then used in the bioassay. The NADPH-generating system was omitted from the S-9 mix by replacement of the cofactor mix with an equivalent volume of sodium phosphate buffer.

Albumin was removed from the cytosol fraction by gel column chromatography. In this experiment, uninduced perfused rat liver homogenates were used. The column was prepared according to the method of Cuatrecasas (1970). Agarose gel (Sepharose 4B-CL; Pharmacia, Sweden) was activated with cyanogen bromide (approximately 250 mg/ml of settled gel) and the cyanogen-bromide-agarose was stirred overnight at 8 to 10°C in 2 vol of 0.1 M sodium carbonate, pH 9, containing 30 mg/ml ξ-aminocaproic acid. The ξ-aminocaproic-acid-agarose produced was washed, resuspended in two volumes of methanol containing 20 mg/ml L-tryptophan methyl ester and 20 mg/ml dicyclohexylcarbodiimide, and stirred for 4 h at room temperature. The gel was washed with several volumes of methanol prior to resuspension in 50 mM monobasic potassium phosphate. Cytosol fractions were passed through the column, and the filtrates collected were termed the "low-albumin cytosol."

The protein content of S-9, cytosol, and low-albumin cytosol was determined according to a method based on Bradford's Coomassie Brilliant Blue G250 dye-binding assay (Bradford, 1976) (BioRad Protein Assay Kit with a BioRad bovine plasma albumin standard). The albumin content of S-9, cytosol, and low-albumin cytosol was measured with a colorimetric method based on the formation of an intense blue albumin-bromocresol green complex (Doumas and Biggs, 1972). Glutathione was measured according to the method of Ellman (1959), which is specific for thiol groups. Reduced glutathione (Sigma) and bovine serum albumin (Metrix) were obtained from commercial sources. Positive controls in the Ames test were 2-nitrofluorene and 2-aminofluorene (Aldrich).

RESULTS AND DISCUSSION

The addition of S-9 to the incubation mixture of the Ames bioassay reduced the mutagenic activity of the GM defined diesel exhaust sample (see Table 1). The decrease in mutagenic activity was proportional to the amount of S-9 added to the incubation mixture (see Table 2). As shown in Table 3, the mutagenicity-reduction effect was still observed with heat treatment of the S-9 mix or without the NADPH-generating system. The mutagenicity-reduction effect disappeared when the heated S-9 mix was filtered to remove coagulated materials. The above results indicate that enzymatic activity was not responsible for the mutagenicity-reducing ability of S-9. It is evident that proteins, the principal denatured materials in heat-treated S-9, produced the mutagenicity-reduction effect.

Table 1. Mutagenic Activity of the GM Defined Diesel Exhaust
Particulate Sample in the Presence or Absence of S-9 Mix[a]

	TA98 Mean Net Revertants/Plate[b]	
Particulates (mg/plate)	Without S-9 Mix	With S-9 Mix[c]
Spontaneous	34	43
0.125	584	446
0.25	890	500
0.5	1138	888
1	1567	1090

[a]For details on the collection of the sample, see Schreck et al.,
1978.
[b]The mean number of spontaneous revertants was subtracted.
[c]The concentration of protein in the S-9 mix used was 2.9 mg/plate.

Table 2. Relationship Between the Amount of S-9 Added in the
Assay Mixture and the Number of Revertants Induced by the
Diesel Exhaust Mutagens[a]

S-9 (µl/0.5 ml S-9 Mix)	TA98 Mean Net Revertants/plate	Protein (mg/plate)	Spontaneous Revertants/ Plate
0	1385	0	23
5	1175	0.3	25
10	1052	0.6	25
20	882	1.2	31
50	770	2.9	33
100	702	5.8	34
150	649	8.7	35

[a]One half milligram GM defined diesel particulates was used in each
plate. The protein concentration in the S-9 was 58 mg/ml. One
half milliliter S-9 mix was applied on each plate.

Table 3.   Effects of S-9 Mix Subjected to Various Treatments on
the GM Defined Diesel Exhaust Mutagens[a]

Particulates (mg/plate)	TA98 Mean Net Revertants/Plate			
	Without S-9 Mix	With Boiled S-9 Mix	With Boiled and Filtered S-9 Mix	S-9 Mix Without Cofactors
Spontaneous	29	40	44	44
0.0625	185	126	231	15
0.125	380	197	330	103
0.25	641	392	697	395
0.5	1075	855	1130	854

[a]S-9 mix was inactivated either by heat or by omission of the
NADPH-generating cofactor system. Heat-denatured materials,
mainly coagulated proteins, were removed from the heat-treated S-9
mix by filtration. The filtrate was also applied on the assay.
The protein content of the S-9 mix was 2.9 mg/plate.

Albumin is the principal protein formed in the liver
(Rothschild et al., 1972). When albumin is removed from the
cytosol fraction by gel filtration, the detoxifying activity is
reduced by nearly 90% (see Table 4). Exogenous bovine serum
albumin, when added to the Ames bioassay, simulates the
mutagenicity-reducing activity of S-9 on diesel exhaust mutagens
(see Table 5). These results provide strong evidence that albumin
in S-9 or from exogenous sources acts as a nonspecific
mutagenicity-reduction agent for diesel exhaust mutagens.

Albumin contains thiol groups that may interact with
electrophilic mutagens (Peters and Reed, 1978; Ross, 1962). As
glutathione is present in S-9, we investigated the activity of
glutathione on the diesel exhaust mutagens. Glutathione and other
sulfhydryl compounds have been shown to reduce the mutagenicity of
certain mutagens (Rosin and Stich, 1979; Hollstein et al., 1978;
Srinivasan and Fugimori, 1979). Table 6 shows that glutathione
significantly reduced the mutagenic activity of diesel exhaust at
doses of 25 μmol/plate or more, but was ineffective at 0.8 μmol
/plate. The amount of endogenous glutathione in S-9, 0.05 μmol
/plate (in 50 μl of S-9), was too small to account for the overall
mutagenicity-reduction activity of S-9.

Table 4. The Effects of S–9 Mix, Cytosol, and Low–Albumin
Cytosol on the Mutagenic Activity of the GM Defined
Diesel Exhaust Particulate Sample[a]

	TA98 Mean Net Revertants/Plate			
	Without S–9 Mix	With S–9 Mix[b]	With Cytosol[c]	With Low–Albumin Cytosol
Particulates (mg/plate):				
Spontaneous	27	42	33	30
0.125	461	362	153	480
0.25	781	656	309	744
0.5	1197	945	760	1061
Protein (mg/plate)	0	1.9	7.5	3.7
Albumin (mg/plate)	0	1.4	5.3	0.3

[a]One half milliliter S–9 mix, containing 50 μl S–9, 0.5 ml cytosol,
or 0.5 ml low–albumin cytosol, was applied on each plate. The
amount of total protein, and more specifically, the amount of
albumin used on each plate are given in table.
[b]S–9 was prepared from perfused livers of uninduced male Sprague-
Dawley rats.
[c]Cytosol fraction was obtained from the S–9 according to the method
of Frantz and Malling (1975).

These results clearly show that the ability of liver
homogenates to reduce the mutagenic activity of diesel exhaust
samples is due not to enzymatic activity but to an interaction
between mutagens and liver albumin. The nature of the interaction
is not known. It may be adsorption of mutagens onto proteins by
van der Waal forces, a process frequently observed in the binding
of drugs to protein. Binding of mutagens to protein may reduce the
effective dose of mutagens to bacterial DNA, so that the observed
mutagenicity is decreased. Another possible explanation of the
decrease would be that protein might block the transport of the
mutagens to the DNA.

Table 5.   Mutagenic Activity of the GM Defined Diesel Exhaust
Particulate Sample in the Absence or Presence of Various
Amounts of Bovine Serum Albumin

	TA98 Mean Net Revertants/Plate			
Particulates (mg/Plate)	Without Bovine Serum Albumin	With Bovine Serum Albumin		
		2.5 mg/plate	12.5 mg/plate	20 mg/plate
Spontaneous	24	23	27	20
0.125	318	207	198	148
0.25	579	362	255	170
0.5	1024	736	534	392
1	1278	1146	859	569

Table 6.   Mutagenic Activity of the GM Defined Diesel Exhaust
Sample in the Absence or Presence of Exogenous Glutathione

	TA98 Mean Net Revertants/Plate		
Particulates (mg/plate)	Without Glutathione	With Glutathione	
		0.8 μmol/plate	25 μmol/plate
Spontaneous	21	20	20
0.125	456	435	262
0.25	712	673	472
0.375	907	835	670
0.5	1062	1055	836

The chemical identification of the direct-acting mutagens is an important objective in safety evaluation of diesel exhausts. When this objective is obtained, the molecular interactions between the mutagens, glutathione, and albumin may be revealed.

ACKNOWLEDGMENTS

This research has been supported by the Northern California Occupational Health Center. Y. Y. Wang is the recipient of the Grossman Endowment for the School of Public Health, University of California, Berkeley.

REFERENCES

Ames, B.N., and J. McCann. 1976. Carcinogens are mutagens: a simple test system. In: Screening Tests in Chemical Carcinogenesis. R. Montesano, H. Bartsch, and L. Tomatis, eds. IARC Publications V. 12, IARC, Lyon, France. pp. 493-504.

Ames, B.N., J. McCann, and E. Yamasaki. 1975. Methods for detecting carcinogens and mutagens with the Salmonella/ mammalian microsome mutagenicity test. Mutation Res. 31:347-364.

Bradford, M. 1976. A rapid and sensitive method for the quantitation of microgram quantities of protein utilizing the principle of protein-dye binding. Anal. Biochem. 72:248-254.

Clark, C.K., and I.T. Salmeen. 1980. Influence of sampling filter media on the mutagenicity of diesel soot. In: Lovelace Institute's 1978-1979 Annual Report to the Department of Energy. Lovelace Inhalation Toxicology Research Institute: Albuquerue, NM. pp. 211-216.

Cuatrecasas, P. 1970. Protein purification by affinity chromatography. J. Biol. Chem. 245:3059-3065.

deFlora, S. 1978. Metabolic deactivation of mutagens in the Salmonella-microsome test. Nature 271:455-456.

Doumas, B.T., and H.G. Biggs. 1972. Determination of serum albumin. In: Standard Methods of Clinical Chemistry, Vol. 7. G.R. Cooper, ed. Academic Press: New York. pp. 175-188.

Ellman, G.L. 1959. Tissue sulfhydryl groups. Arch. Biochem. Biophys. 82:70-77.

Frantz, C.N., and H.V. Malling. 1975. The quantitative microsomal mutagenesis assay method. Mutation Res. 31:165-380.

Hollstein, M., R. Talcott, and E. Wei. 1978. Quinoline: conversion to a mutagen by human and rodent liver. J. Natl. Cancer Inst. 60:405-410.

Huisingh, J., R. Bradow, R. Jungers, L. Claxton, R. Zweidinger, S. Tejada, J. Bumgarner, F. Duffield, M. Waters, V.F. Simmon, C. Hare, C. Rodriguez, and L. Snow. 1978. Application of bioassay to the characterization of diesel particle emissions. In: Application of Short-term Bioassays in the Fractionation and Analysis of Complex Environmental Mixtures. M.D. Waters, S.L. Nesnow, J.L. Huisingh, S.S. Sandhu, and L. Claxton, eds. Plenum Press: New York. pp. 381-418.

Peters, T., Jr., and R.G. Reed. 1978. Serum albumin: conformation and active sites. In: Albumin, Structure, Biosynthesis, Function. T. Peters and I. Sjoholm, eds. Pergamon Press: Oxford, England. pp. 11-20.

Rosin, M.P., and H.F. Stich. 1979. Assessment of the use of the Salmonella mutagenesis assay to determine the influence of antioxidants on carcinogen-induced mutagenesis. Int. J. Cancer 23:722-727.

Ross, W.C.J. 1962. Biological Alkylating Agents. Butterworths: London, England.

Rothschild, M.A., M. Oratz, and S.S. Schreiber. 1972. Albumin synthesis. New Eng. J. Med. 286:748-757.

Santodonato, J., D. Basu, and P. Howard. 1978. Health Effects Associated with Diesel Exhaust Emissions: Literature Review and Evaluation. EPA-600/1-78-063. U.S. Environmental Protection Agency: Research Triangle Park, NC. p. 9.

Schreck, R.M., J.J. McGrath, S.J. Swarin, W.E. Hering, P.J. Groblicki, and J.S. MacDonald. 1978. Characterization of Diesel Exhaust Particulate for Mutagenic Testing. GMR-2755. General Motors Research Laboratories: Warren, MI.

Springer, K.J., and T.M. Baines. 1977. Emission from Diesel Version of Production Passenger Cars. SAE paper no. 770818. Society of Automotive Engineers: Detroit, MI.

Srinivasan, B.N., and E. Fugimori. 1979. Benzo(a)pyrene-serum: albumin/cysteine interactions: fluorescence and electron spin resonance studies. Chem.-Biol. Interact. 28:1-15.

Talcott, R.E., and E.T. Wei. 1977. Airborne mutagens bioassayed
    in <u>Salmonella typhimurium</u>. J. Natl. Cancer Inst. 58:449-451.

Wang, Y.Y., S.M. Rappaport, R.F. Sawyer, R.E. Talcott, and E.T. Wei.
    1978. Direct-acting mutagens in automobile exhaust. Cancer
    Lett. 5:39-47.

Wei, E.T., Y.Y. Wang, and S.M. Rappaport. 1980. Diesel emissions
    and the Ames test: a commentary. J. Air Pollut. Control
    Assoc. 30:267-271.

SESSION 5

STATIONARY SOURCES

BIOASSAYS OF EFFLUENTS FROM STATIONARY SOURCES: AN OVERVIEW

R.G. Merrill, Jr., W.W. McFee, and N.A. Jaworski
Industrial Environmental Research Laboratory
U.S. Environmental Protection Agency
Research Triangle Park, North Carolina

Major point sources of pollutants will be with us for a long time. Ever since western civilization began to industrialize and to require high inputs of energy from combustion, pollutants have been emitted in increasing amounts. In 1977, stationary electric power sources were emitting 19.3 million tons of sulfur oxides, 7 million tons of nitrogen oxides, 3.1 million tons of suspended particles, and 0.1 million tons of volatile organic compounds into the atmosphere of the United States (U.S. EPA, 1978). These figures represent about 70% of the sulfur oxides and 40% of the nitrogen oxides emitted each year. Clearly, stationary sources of air pollutants deserve close attention on the basis of the mass of emissions alone. The upward trend in emissions from stationary sources is expected to level off or decrease in some regions of the United States after 1990 but to continue upward in others, in spite of advances in control technologies. New systems of energy conversion and fuel processing, for example, fluidized bed combustion and coal gasification, will introduce new pollution control problems. The industrial growth anticipated in synthetic fuels will increase the need for stringent analysis of the potential hazardous impact of pollutants in the coming years.

In the next few years, we, as a nation, are faced with important energy-related decisions that will have long-term effects on our quality of life. A certain trade-off between abundant, cheap energy and a pleasant, healthy environment seems inevitable. We should make these hard decisions based on the best information available.

To secure this information is the task of the Environmental Assessment Program of the U.S. Environmental Protection Agency

(EPA). We have learned a lot about the threats of pollution from
industrial air, water, and residual waste streams in the last few
years. However, we want to do all we can to avoid being surprised
by the cumulative effects of effluents from present processes or
the introduction of new materials by a shift in technology. The
nation's economic growth depends, in large measure, on industrial
expansion. To aid industry and regulating agencies in protecting
our environment, we must know the chemistry and the potential
effects of products, by-products, and wastes. We can and must
protect the environment, even while we allow industry to grow.

However, environmental assessment is no easy task. For
several years we have been systematically measuring, as precisely
as possible, the emissions of many industrial technologies,
especially energy-processing technologies. The major components of
the waste streams are generally well-characterized, but with the
increasing concern over trace-level contaminants, complex mixtures,
and their subtle, cumulative effects on the environment, we have
been faced with increasingly difficult assessment problems.

At a conference two years ago, Stephen Gage (1978) said, "The
emergence of the Environmental Assessment Program as a distinct and
important part of EPA's environmental research efforts over the
past several years is an excellent example of how the Agency's
efforts have turned from primarily research in reaction to known
environmental problems to include research which anticipates and
tries to avoid future environmental problems. This change of
emphasis also predated the rechartering of EPA's position toward
toxics."

Some of EPA's best efforts in environmental assessment are in
the area of coal utilization. It is obvious that the United States
is going to have to use much more coal if our economy is to grow.
Continued electricity generation is important, and part of the coal
will probably have to be converted to gases and liquids to help
meet our varied fuel needs. These processes introduce new
compounds and mixtures into the stationary source emission picture.
Some of them may present a serious health risk. EPA's Environmental Assessment Program activities have made considerable
progress identifying the hazards in the processes, but much remains
to be lone.

It is extremely difficult to obtain representative samples
from process streams that are at high temperature and pressure and
that often include highly reactive chemical species. Chemically
analyzing complex mixtures of chemicals is another great challenge,
even with today's powerful analytical chemistry techniques. Once
the emissions are chemically characterized, determination of the
relative degree of health or environmental hazard is the next
difficult step. This activity is particularly troublesome when

cumulative effects, such as those which might result from exposure to carcinogenic, mutagenic, or teratogenic agents, are considered. In this third area--determination of potential health or ecological hazards--bioassays add an important dimension to the process of environmental assessment.

Since 1975, EPA's Industrial Environmental Research Laboratory (IERL), in Research Triangle Park, NC, has been conducting a series of environmental assessment programs designed to

1)   systematically characterize the physical, chemical, and biological characteristics of all effluent streams from an energy conversion or industrial process;

2)   rank those streams according to hazard potential;

3)   identify control technology programs needed to reduce the hazard of those streams; and

4)   predict the effects of those streams on the environment, in conjunction with the health and ecological research laboratories of EPA's Office of Research and Development.

Examples of EPA environmental assessment programs currently underway include assessments of coal gasification, fluidized bed combustion, stationary conventional combustion, and coal cleaning and liquefication processes. In performing these environmental assessments, IERL is supporting the regulatory and enforcement offices of EPA by anticipating future control technology needs and developing the data bases needed to support development of standards.

The phased approach to source sampling, chemical analysis, and bioassay has been designed to provide the data needed to evaluate potential environmental impact (Dorsey et al., 1978). This phased approach incorporates the three levels of sampling and analysis shown in Figure 1. This scheme, which relies heavily on bioassays, is successfully meeting the environmental assessment goals listed above.

The phased approach was adopted because it offered potential cost savings over a direct approach in which all streams would be carefully sampled and completely analyzed in one pass. In each level of this approach (see Figure 2), both chemical and biological characterizations of an effluent stream are performed. The chemical characterization provides a quantitative numerical rank of a stream's potential hazard based on an engineering model for potential discharge severity. The bioassay characterization provides a direct measure of a biological response. The dual

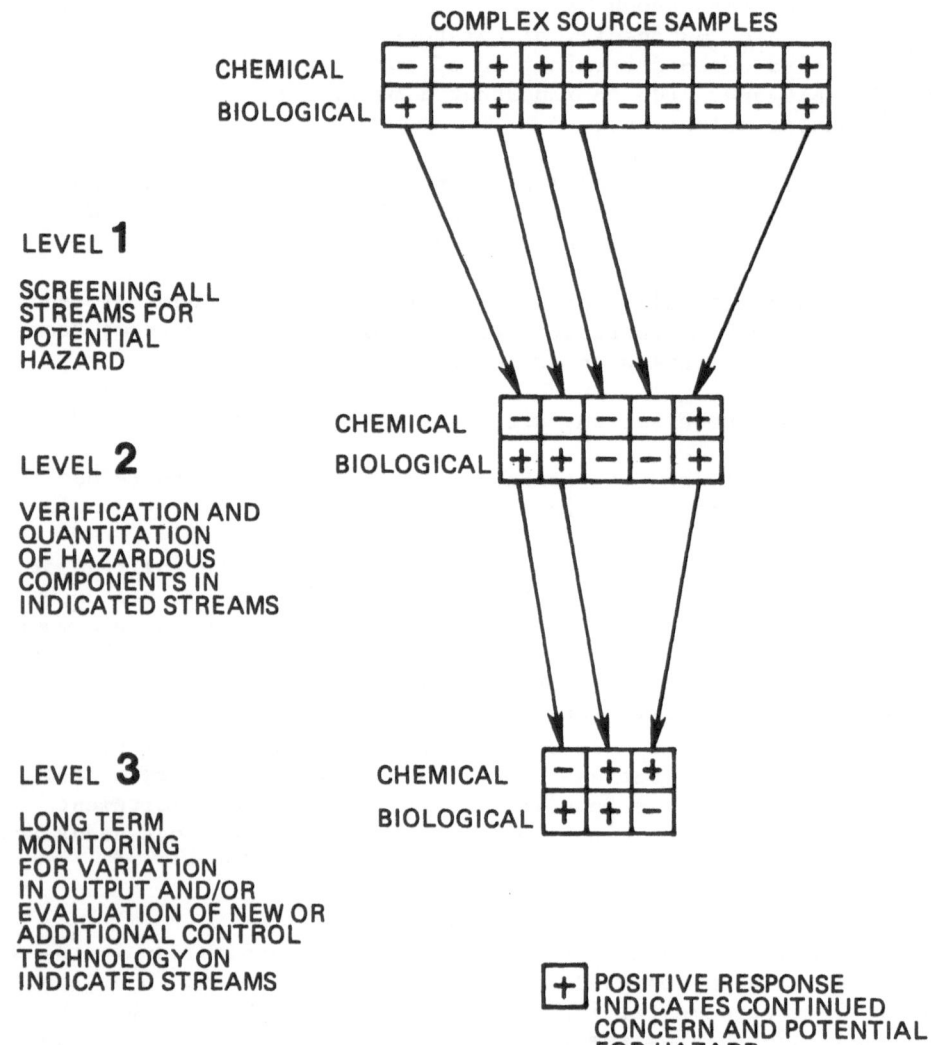

Figure 1. The phased approach to environmental assessment.

chemical and biological characterizations are designed to
complement each other.

Numerous advantages can be gained by including bioassays in
the assessment of stationary sources of pollutants. For example,
it may be possible to use organisms the same as or similiar to
those likely to be affected by the source. In bioassays, one can
frequently use the whole sample without separation or modification,

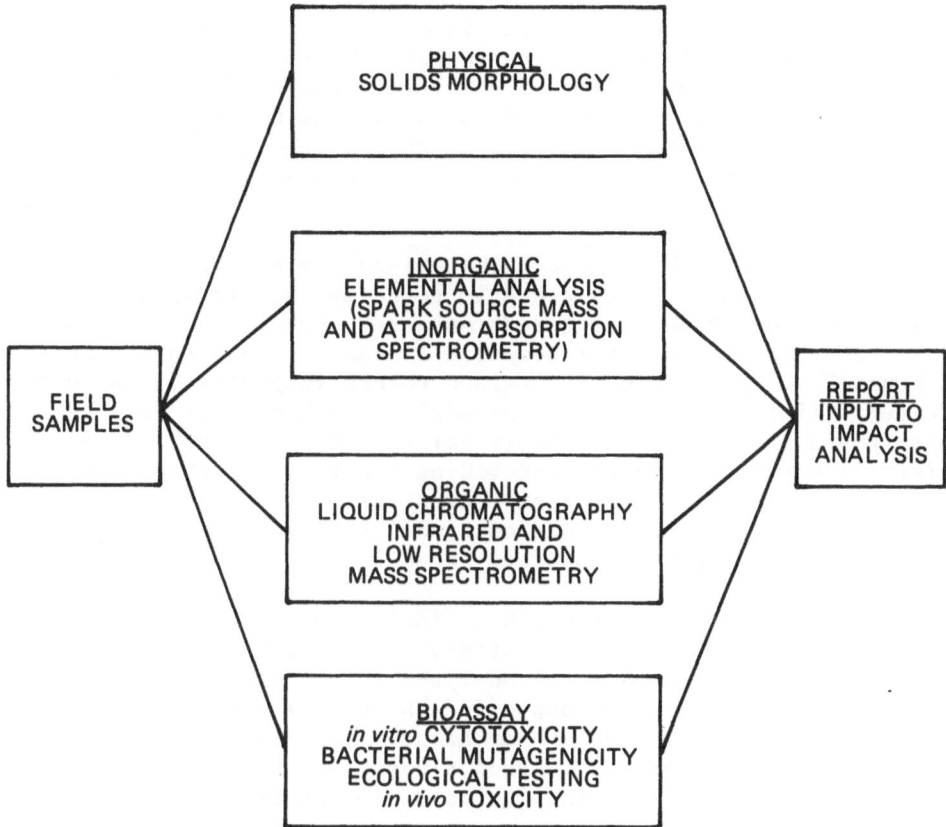

Figure 2.   Flow chart of Level 1 scheme.

thus allowing the combined synergistic or antagonistic effects to
be expressed, which might be either lost in separation or
unobserved in chemical analyses alone. We are constantly faced
with the dilemma of whether we should separate, extract, and
isolate, in an attempt to understand what is present and how it
behaves, or work with the whole sample, so that it can express the
combined properties. The use of bioassays allows both approaches.

Another advantage is that bioassay results have some built-in
interpretation that is absent in chemical analysis. A chemical
result indicating the presence of substance X is meaningful only
insofar as the potency of X in causing health or environmental
damage is known. In contrast, when all the minnows die in an
aquatic test, we don't necessarily know why, but it is obvious that
this material contains one or more substances likely to be
dangerous in aquatic systems.

One disadvantage is apparent in the example above. Bioassays do not always satisfy our desire to know the specific components. Another disadvantage is that of complexity and costs. It is more difficult to set up a laboratory to do routine bioassays at low costs. Bioassays often cannot be started on short notice and thus may require several weeks or months to complete, creating delays and higher costs.

Even though we have argued that bioassays have some built-in interpretative value, they are also sometimes difficult to evaluate. Some often-asked questions are: "Do the results suggest a similar response in other organisms? Can the results be applied to humans? Does mutagenicity relate to carcinogenicity? So what if fruit flies die or don't reproduce?" The bioassays used in the first phase don't provide unequivocal answers. However, they are effective screening devices, providing a basis for subsequent, more definitive tests. All tests must eventually be related to biologic effects. Bioassays, like chemical results, are easier to interpret when supported by other bioassays or analyses.

The best situation we can hope for in environmental assessment is to have a combination of a battery of bioassays and the most complete chemical and physical analyses we can afford. The results of chemistry and bioassays support each other and provide a much more complete picture than either alone. Bioassays should complement chemical assays in such a way that the task of assessing potential environmental impact is possible and practical.

The presentations in this session deal with advances in the application of short-term bioassays to complex mixtures from stationary energy-related sources. In particular, the application of in vitro and in vivo bioassays to fly ash, fluidized bed combustion effluents, and coal-liquefication material are discussed.

It is important that we look not only at present-day processes, but also at technologies of the future, so that when they are commercially applied, we are not surprised by their effects. It is a distorted perspective to consider only today's problems while allowing tomorrow's to get ahead of us and possibly out of hand. It should be our goal to help avoid any potential environmental problems that may be associated with new combustion processes, synthetic fuel production, or energy-conversion processes.

Bioassays of complex mixtures have already contributed to our understanding of processes and hazards. Engineers, chemists, and biologists have worked together in developing new techniques and in interpreting the results. A series of pilot studies on selected stationary sources have taught us a lot about the combined use of

chemical and biological tests in environmental assessment.
Experience in these research and development programs has allowed
refinements in the bioassays to provide cost-efficient, reliable
assessment of environmental impact from a wide variety of source
and sample types. It should be reiterated that health, ecological,
and industrial laboratories are working together on the collection
and interpretation of results. The people involved have a right to
be proud of their accomplishments, but we all realize the
inadequacies of our techniques and the need to improve them. The
work reported at this conference is contributing to our
margin of safety and flexibility in future energy production.

REFERENCES

Dorsey, J.A., L.D. Johnson, and R.G. Merrill. 1978. A phased
approach for characterization of multimedia discharges from
processes. In: Monitoring Toxic Substances. D. Schuetzle,
ed. ACS Symposium Series, No. 94. American Chemical Society:
Washington, DC. pp. 29-48.

Gage, S.J. 1978. Keynote address. In: Symposium Proceedings:
Process Measurements for Environmental Assessment, Atlanta,
GA. E.A. Burns, ed. EPA 600/7-78-168. U.S. Environmental
Protection Agency: Research Triangle Park, N.C. pp. 1-3.

U.S. Environmental Protection Agency. 1978. National Air Quality,
Monitoring and Emissions Trends Report, 1977.
EPA-450/2-78-052. U.S. Environmental Protection Agency:
Research Triangle Park, NC.

COAL FLY ASH AS A MODEL COMPLEX MIXTURE FOR SHORT-TERM BIOASSAY

Gerald L. Fisher, Clarence E. Chrisp, and Floyd D. Wilson
Battelle Columbus Laboratories
Columbus, Ohio

INTRODUCTION

Combustion of coal for electric power generation has increased markedly throughout the past decade and is expected to continue to increase throughout the remainder of this century. Coal combustion produces a variety of biologically active inorganic and organic compounds. Of major concern is the release of oxides of carbon, nitrogen, and sulfur; biologically active trace elements: siliceous primary particulate matter; and organic compounds. Most of the primary particulate matter produced during coal combustion is coal fly ash. Interaction of fly ash particles with inorganic and organic compounds formed during the combustion process produces a unique complex mixture. This mixture may serve as a model for other mixtures resulting from interaction of relatively inert carrier particles with biologically active metals and organic compounds; the bulk of atmospheric particulate matter may serve as a matrix for subsequent interaction of airborne vapors and condensable gases. In this report, we review our studies that indicate the complexity of the physical and chemical properties of coal fly ash and attempt to relate such properties to the biological activity of coal fly ash.

SAMPLE COLLECTION

The fly ash samples described below were collected from a western U.S. power plant burning low-sulfur (0.5%), high-ash (20%) coal (McFarland et al., 1977a). Unless otherwise indicated, the samples were collected and size-fractioned in situ at 95°C downstream of the plant's electrostatic precipitator (ESP). On

occasion, ESP-collected ash was also studied. Four fractions of
stack-collected material with volume median diameters (VMD) of 20,
6.3, 3.2, and 2.2 μm and geometric standard deviations of
approximately 1.8 were obtained. Material was size-classified
using a specially thermostatically controlled (95°C) system
containing two cyclones in series followed by a centripeter with 25
parallel jets. Cyclone-separated material was deposited in a
collection hopper, while centripeter particles were collected on
fabric filters and removed for hopper deposition by cleaning with
reverse air jets operating at one-minute intervals. Thus, in
contrast to standard filter collection techniques, collected fly
ash samples were not continually exposed to the reactive gases in
the flue stream. Such an exposure may lead to changes in the
chemical composition and biological activity of collected
polynuclear aromatic hydrocarbons (Chrisp and Fisher, in press).

     Our samples were collected continuously over a 30-day period.
Thus, variations in coal composition, combustion conditions, and
other parameters of plant operation that may affect chemical
composition should be reflected in these samples. In this regard,
Kubitschek and Kirchner (1980) have demonstrated the important
effects on mutagenic activity of combustion conditions during
start-up and shut-down of a bench-model fluidized-bed combustion
system.

PHYSICAL AND CHEMICAL CHARACTERIZATION

Microscopic Studies

     In previous studies (Fisher, 1979; Fisher et al., 1976;
1978b), we have described the physical and morphological properties
of coal fly ash and have generally found it to be an extremely
heterogeneous, complex mixture with a variety of morphological
forms. Morphology generally depends both on matrix composition and
on exposure conditions during combustion. Upon heating,
aluminosilicate inclusions in the coal initially become rounded and
then through degassing, become vesicular (Fisher, 1979). Further
heating results in the formation of solid spheres, hollow spheres
(cenosphere), or sphere-within-sphere structures (plerosphere).
Crystals form somewhat later in morphogenesis, with internal
(quench) crystals forming rapidly during the transition from liquid
to solid phase (Fisher et al., 1976). Such quench crystals have
been identified as mullite by Gibbon (1979). While internal
crystal formation occurs in milliseconds, surface crystal formation
appears to be a much slower process, taking days or months.
Surface crystals apparently form through sulfuric acid interaction
with metals found on fly ash surfaces.

Previous analysis of individual fly ash particles demonstrated extreme matrix heterogeneity among morphologically similar particles (Pawley and Fisher, 1977). More recently, we have completed a detailed comparison of light-microscopic morphology and individual-particle elemental composition using scanning electron microscopic (SEM) X-ray analysis. Comparative light and electron microscopy is a powerful tool for assessing physical and chemical properties of coal fly ash (Fisher et al., in press). We have demonstrated the presence of relatively pure mineral phases, although the bulk of the ash is amorphous and appears to have the composition of the clay minerals generally associated with coal. Pure quartz, calcium phosphate, titanium dioxide, calcium oxide, iron oxide, and alumina have been observed (T.L. Hayes, E.E. Lai, B.A. Prentice, and G.L. Fisher; University of California at Berkeley; unpublished data; 1979). The degree of pigmentation of spherical particles depends on their iron content. Statistical cluster analysis has confirmed the usefulness of the light-microscopic morphological classifications (Fisher et al., in press). In particular, within each of the specific morphological classes defined by light microscopy, the individual particles generally have very similar elemental compositions. However, the most morphologically amorphous particles tend to be most diverse in elemental composition.

We have also recently completed X-ray diffraction analysis of the crystalline phases within coal fly ash (Hansen et al., MS). These studies show the predominant crystalline species in coal fly ash to be quartz, mullite, and magnetic iron oxides. The highest concentration of crystalline material is found in the coarsest fly ash fractions; we found quartz concentrations of 4%, mullite 8%, and iron oxide 0.6%. The crystalline mineral concentrations decrease with decreasing particle size; in the finest fly ash fraction, quartz was 1.3%, mullite 4%, and magnetic oxides 0.03%. We hypothesize that quartz intrusions within the coal itself produce the silica, whereas mullite and magnetite are formed during the combustion and cooling processes. Mullite is usually associated with a quench crystal phase that occurs during rapid cooling and hence is generally encapsulated within the aluminosilicate matrix of the clay mineral. This does not appear to be the case for quartz or magnetic iron oxides.

As described in our earlier work, surface crystal may form (Fisher et al., 1976; 1978b) as the result of chemical interaction or formation of sulfuric acid on fly ash surfaces, possibly through leaching of minerals and heavy metals from within the fly ash by the surface-associated sulfuric acid. Electron microprobe analysis of the larger crystals in fly ash has identified only calcium and sulfur, thus the majority of crystals appear to be calcium sulfite, present as either gypsum or anhydrite. We postulate, however, that sulfate crystal formation may also increase the biological availability of refractory metal oxides through conversion to the more soluble metal sulfates.

We found (as have many other investigators: see Coles et al., 1979; Ondov et al., 1977) that the concentration of the volatile (at coal-combustion temperatures) trace elements or their oxides are highest in the finest fly ash particles. The elements most enriched in the finest fly ash particles are cadmium, zinc, selenium, arsenic, antimony, tungsten, molybdenum, gallium, lead, vanadium, fluorine, and sulfur. However, the relatively refractory elements uranium, chromium, barium, copper, beryllium, and manganese also are enhanced in the finest fly ash fractions. While it appears that vapor-phase condensation enhances volatile trace elements in fine fly ash particles (Natusch et al., 1974), other mechanisms also may be important in this phenomenon. Filtration studies with neutron-activated coal fly ash indicate that some elements are highly enriched in particles of the size range from 0.2 to 0.4 μm VMD (Fisher et al., 1979). In particular, the elements antimony, arsenic, tungsten, uranium, and chromium have a disproportionately high concentration in particles less than 0.4 μm. It has been hypothesized that homogeneous nucleation and subsequent coagulation of primary particles or the condensation of reaction products enhance submicron particles. Relatively high concentrations of biologically active trace elements in submicron particles may present special technological problems in particulate abatement.

We have compared the distribution of elements in fly ash associated with the aluminosilicate matrix with that of elements either in separate mineral phases or associated with the particulate surface (Hansen and Fisher, 1980). Preferential dissolution with either hydrochloric or hydrofluoric acid indicated that more than 70% of the titanium, sodium, potassium, magnesium, hafnium, thorium, and iron is associated with the aluminosilicate matrix. On the other hand, more than 70% of the volatile elements arsenic, selenium, molybdenum, zinc, cadmium, tungsten, vanadium, uranium, and antimony are associated with the particulate surface. It also appears that the larger portion of the calcium, scandium, strontium, lanthanum, and the rare earth elements is associated with a separate mineral phase, possibly an apatite phase, which has a particle size distribution similar to that of the aluminosilicate phase. These findings, then, allow computation of the probable biological availability of trace elements in coal fly ash. In agreement with these observations, we have found relatively high solubilities for molybdenum, calcium, selenium, barium, arsenic, tungsten, zinc, and antimony in coal fly ash treated with a tris-hydrochloric acid buffer at pH 7.4 (Fisher et al., 1979b).

IMMUNOTOXICOLOGY STUDIES

Although many sensitive in vitro bioassays exist for
mutagenesis, few are available for screening environmental
toxicants for their potential effects on host cellular defense
factors. The development of assays for monitoring immunity effects
is hampered by the complexity of the cellular and humoral factors
involved in the host reaction to neoplasia. We have begun to
develop a variety of methods for the study of environmental factors
affecting cellular immunities. A major effort has been made toward
developing assays that reflect inhibition of macrophage function.

In Vitro Studies

Using SEM X-ray analysis to investigate particles contained in
macrophages, we find that the matrix composition of particles
contained within phagocytes may vary dramatically (Hayes et al.,
1978; Pawley and Fisher, 1977). Furthermore, we hypothesize that
variation in the chemical composition of phagocytized particles may
also represent variation in the toxicological potential of such
particles (Hayes et al., 1978; 1980). We have proposed a model
for the exposure of individual lung cells to the foreign elements
in fly ash. Segregation of elements in specific particles of fly
ash allows much higher exposure levels within individual cells than
are predicted by a model based on uniform distribution of elemental
concentrations among all particles. Furthermore, we have
demonstrated that comparative microscopic techniques can be used to
correlate the viability of individual cells with the elemental
composition of phagocytized particles. Light-microscopic analysis
of macrophages deposited on SEM finder grids and treated with
trypan blue dye indicates the viability of the cells; subsequent
electron-microscopic analysis of the same cells indicates the
elemental composition of the phagocytized particles. Studies are
now under way to compare the toxicities of the various fly ash
compositions.

To determine whether trace elements in fly ash can alter the
function or viability of macrophages, we have calculated the
elemental content of macrophages that have phagocytized a few fly
ash particles (Hayes et al., 1980). Our calculations indicate
that the concentrations of many biologically active trace elements
generally are increased by one, two, or even three orders of
magnitude. These calculations, then, indicate that the trace
elements in fly ash potentially could produce cellular damage.
Further studies are necessary to evaluate the biological
availability of trace elements in the fly ash. We have found that
many trace elements in fly ash inhibit either lectin-induced or
mixed-lymphocyte-induced lymphocyte blastogenesis (Shifrine et al.,
in press). Chromium, lead, vanadium, and copper appear to be

effective inhibitors in the lymphocyte-stimulation assays, at
concentrations similar to those found in coal fly ash.
Interestingly, both lymphocytes and macrophages appear to be
extremely sensitive to vanadium.

In vitro exposure of macrophages to particulate matter of
similar size distributions allows us to compare the toxicity of fly
ash, silica, and glass beads (Fisher et al., 1977; Whaley et al.,
1977). Silica was chosen as a positive control because it is a
well-documented macrophage toxicant. Glass beads (aluminosilicate
particles) were chosen as a negative control because they are
apparently inert. The effects of in vivo exposure to these
particles have been studied using both rat and mouse pulmonary
macrophages (Fisher et al., 1977). Particulate exposure was
performed at a 40:1 particle:cell ratio, comparable to test
particle combinations used in the phagocytic assay (Fisher et al.,
1978a). We find that the phagocytic activity of macrophages
increases with time in incubation media (Fisher et al., 1977). The
degree and rate of enhancement of control phagocytosis depends on
the species derivation of macrophages; rat macrophages show the
effect after two hours in culture, while murine macrophages take
two days in culture. Exposure to fly ash delays the increase in
phagocytic rates. The lag can be seen after two and four hours in
culture for rat macrophages and after one and two days for murine
macrophages. Interestingly, although silica exposure did not delay
the increase in phagocytosis, the final phagocytic capability
(after seven days) was markedly below that of controls. Further
studies are necessary to determine the nature of the enhanced
phagocytosis associated with in vitro culture and to evaluate the
significance of the lag in this effect caused by fly ash.

We have also developed techniques for evaluating the
proliferative capacity of lavaged cells from the lung (Boorman et
al., 1979a, b). Clonogenic technique was adapted from previously
described techniques for quantitation of bone marrow granulocyte-
monocyte progenitors (Wilson et al., 1974). The basic culture
system involves plating lung-lavaged macrophages into a semisolid
methyl cellulose medium. Colonies tend to be smaller and slower to
grow than those of hematopoietic cells. Preliminary studies
indicate that all particle types (i.e., fly ash, silica, or glass
beads) may affect the ability of these lavaged cells to divide
(Whaley et al., 1977). Agents that tend to stimulate phagocytosis
appear to decrease the cells' proliferative capacity. Similarly,
agents that inhibit phagocytosis tend to stimulate proliferation.
These observations suggest that particle-induced phagocytic
functions may preclude differentiation and subsequent division of
progenitors.

We have also demonstrated that cloning techniques are useful
in studying the in vitro dose-response characteristics of

lymphohematopoietic progenitors exposed to trace elements in semisolid culture systems (Wilson et al., 1980). Because of the enhanced concentrations of zinc and selenium in fine fly ash particles and because of the known biological activity of these elements, we evaluated the responses of murine spleen B-lymphocyte progenitors and bone marrow granulocyte-monocyte progenitors to selenite and zinc exposures. At physiological concentrations, both elements significantly suppressed cell proliferation from splenic B-lymphocytes, but not from granulocyte-monocyte progenitors. The results demonstrate the feasibility of using lymphohematopoietic cloning techniques as sensitive short-term bioassays to determine the effects of fossil fuel combustion products on cellular pathways involved in hematopoiesis and immunological processes. We are presently studying the effects of in vivo exposure on the progenitor cell function.

In Vivo Studies

We have also performed in vivo inhalation studies with mice acutely exposed to fly ash and silica aerosols and with rats chemically exposed to fly ash alone. As part of this effort, we have developed techniques for generating well-dispersed fly ash aerosols. We use a Wright dustfeed mechanism for fly ash deagglomeration and aerosolization, a cyclone for separating larger particles, and a krypton-85 discharger for reducing particulate charge to Boltzmann equilibrium (Raabe et al., 1979). In an attempt to improve aerosolization procedures, we compared the efficacy of a fluidized-bed generator to that of the Wright dustfeed system (McFarland et al., 1977b). The results of the study indicate that aerosols produced by the fluidized-bed generator are relatively unstable over time and that deagglomeration is markedly less effective than with the Wright dustfeed mechanism. Aerosols produced by the Wright dustfeeder had smaller aerodynamic size and broader size distributions than those produced by the fluidized bed, with or without use of the Wright dustfeed as a feed mechanism. For these reasons, we have continued to use the Wright dustfeed rather than the fluidized bed for generation of stable deagglomerated aerosols of fly ash, as well as other nonhygroscopic particles.

For acute inhalation studies, we have used exposure via the nose only for up to two hours in small chambers. Chronic inhalation studies have been performed in immersion chambers for periods of up to 20 h per day for 180 days (Raabe et al., 1979). Particle size distributions are continuously monitored using a light-scattering particle counter. We also obtain size-distribution and aerodynamic data using SEM analysis of point-to-plane ESP samples or cascade impactor samples. Total mass is measured periodically in filter samples. For acute inhalation

studies, we have used the finest fly ash fractions of the size-
classified stack-collected fly ash. However, because chronic
inhalation studies require relatively large masses of material, we
have employed size-classified material collected from the hopper of
the power plant's ESP.

    In acute inhalation studies, using a stack-collected fly ash,
mice were sacrificed 2, 6, and 15 days after exposure, and
macrophage function, pulmonary pathology, and progenitor cell
kinetics were evaluated (Fisher and Wilson, 1980). Macrophage
functional assays indicated a depression in phagocytic capacity of
fly-ash-exposed mice (compared with controls) at 6 and 15 days
after exposure. Similar depressed phagocytic activity was observed
in silica-exposed animals. Progenitor-cell assays indicated an
initial depression in pulmonary alveolar macrophage colonies at 2
days after exposure and a marked elevation at 15 days after
exposure. In contrast to the changes observed in pulmonary
macrophage progenitors, macrophage precursors in bone marrow and
spleen were not significantly affected. These data suggest that
the elevated progenitor cell activity is due to recruitment of
progenitors from the lung itself (i.e., local production). The
increased proliferation of progenitor cells two weeks after acute
exposure in vivo contrasts with the continued depression observed
in in vitro studies. Similarly, preliminary studies with
intraperitoneal exposure to zinc indicate that lymphohematopoietic
progenitors are less sensitive to in vivo exposure than are cells
exposed in vitro.

    Further evidence to support the recruitment hypothesis is
provided by observing changes in the number of particles within
phagocytes (Fisher and Wilson, 1980). For the 2-, 6-, and 15-day
observation periods, the number of particles continually decreased
within the cells, suggesting enhanced production of phagocytic
cells in the lung. On the other hand, we have not observed similar
effects with long-term low-level chronic exposure to fly-ash
aerosols derived from the power plant's ESP. It is not clear
whether the difference in biological response reflects the
difference in concentration (200 vs. 2 mg/m^3), the difference in
species (rat vs. mouse), or perhaps, most importantly, the
difference in fly ash source (stack vs. ESP hopper-collected ash).

    We have also developed techniques for quantifying fly ash
deposited in rat lungs during the chronic inhalation studies
(Fisher et al., 1980). Because it is difficult to separate
particulate matter from lung tissue, we chose to evaluate elemental
analysis as a measure of fly ash in the lung. Selection of the
appropriate element was based on the following criteria:

1) the elemental analysis should be specific, and detection limits should be appropriate for a sensitive indicator of of lung burden;

2) the levels of the element in the tissue should be low and relatively constant;

3) the dissolution of the element should parallel the particulate mass dissolution; and

4) the element should be uniformly distributed throughout the size range of the fly ash under study.

Only aluminum and silicon met these criteria; aluminum was chosen because its analysis is more sensitive and less troublesome than that of silicon. The lung content of fly ash calculated from the aluminum analysis was in quantitative agreement with calculations based on available deposition and clearance data.

We have evaluated the mutagenic properties of coal fly ash extracts using the Ames Salmonella assay system (Ames et al., 1975: Chrisp et al., 1978; Fisher et al., 1979a). Our studies indicate that in keeping with a model of surface deposition, the finest fly ash fractions are indeed the most mutagenic (Fisher, 1980). However, the 3.2-μm fraction of fly ash is more mutagenic than the 2.2-μm fraction. At first we assumed that this was due to antimutagens in the fly ash, and we evaluated the possible role of selenium, as the selenite, or fluorine, as the fluoride. These elements were chosen because of their relatively higher concentrations in the finest fly ash fraction than in the 3.2-μm fraction, and because selenium, as the selenite, is an antimutagen for acetylaminofluorene and its derivatives. Fluoride is generally recognized as an enzyme inhibitor. Adding these elements to extracts of the 3.2-μm fraction, however, did not alter its mutagenic activity. Thus, we do not have experimental support for the hypothesis that antimutagens are present in the finest fly ash fraction.

Natusch (Colorado State University, personal communication, 1980) suggests that the difference in mutagenic activity may be due to differences in chemical absorption of mutagens by the two fly ash fractions. Further evidence for the chemisorption of mutagens on fly ash surfaces is the photostability of these compounds. We have irradiated fly ash samples with ultraviolet light, sunlight, and X-irradiation, with no decrease in mutagenic activity. However, heating the fly ash to temperatures of 200 to 250°C results in the loss of approximately half of the mutagenic activity, while heating above 300°C results in complete loss of detectable mutagenic activity (Fisher et al., 1979a). The biphasic nature of the loss in mutagenic activity with heating indicates the

presence of at least two mutagens or classes of mutagens. Most
recently, we have demonstrated that the loss of mutagenicity is due
to decomposition of surface-associated materials, as opposed to
volatilization (Hansen et al., MS), further supporting the
suggestion that mutagens are on fly ash surfaces.

Fly ash collected by the power plant's ESP does not appear to
be mutagenic in the Ames test (Fisher et al., 1979a). Furthermore,
even when classified by size and sampled in a size distribution
equivalent to that of our finest stack-collected fraction, ESP-
collected materials are still not mutagenic. Our studies indicate
that temperatures of approximately 100°C may be critical for
absorption of mutagens on fly ash surfaces. Indeed, the
calculations of Natusch and Tomkins (1978) indicate that the
deposition of vapor-phase polynuclear aromatic hydrocarbons on fly
ash particles is extremely sensitive to temperature changes around
100°C.

We have compared the efficiencies of mutagen extraction by a
variety of solvents. Normal saline is a very poor extractant of
mutagenic activity, whereas horse serum and serum from other
species are fairly efficient extractants of fly ash mutagens
(Chrisp et al., 1978). We found that a mutagen-serum protein
complex forms that can be isolated from the serum, although it is
available to the bacterial genome. Further studies indicated that
albumen alone is nearly as efficient as the total serum in
extracting mutagens from fly ash (C.E. Chrisp and G.L. Fisher,
unpublished data). Direct extraction of fly ash with
dimethylsulfoxide and sonication results in the highest detectable
levels of mutagens.

We have not identified the chemical composition of the fly ash
mutagens, although recent studies using acidic, neutral, and basic
aqueous fly ash extracts further indicate that mutagens in fly ash
are not inorganic. Acidic aqueous fractions were not mutagenic,
whereas basic aqueous fractions contained approximately half of the
mutagens extractable with dimethylsulfoxide. These results
indicated that a significant portion of the mutagenic activity in
coal fly ash could be accounted for by the presence of weak organic
acids (Hansen et al., 1980).

To evaluate the carcinogenic potential of coal fly ash, we
have modified the tracheal implant system described by Griesemer et
al. (1974). For these studies, we have packaged fly ash in 0.2-μm
Nuclepore filters. This method allows for the slow release of
mutagens to the sensitive tracheal epithelial cells and minimizes
the risks of local tissue damage and toxicity. We have used the
Ames bacterial mutagenesis system to monitor the release of
mutagens from fly ash in tracheal implants (Chrisp and Fisher,

1980). Studies are now in progress to evaluate the carcinogenic potential of the fly ash using the tracheal implant system.

CONCLUSIONS

In conclusion, our results demonstrate the extreme complexity of coal fly ash in terms of matrix composition, morphological appearance, and surface trace element and organic chemical composition. Assays are being developed to measure the potential immunotoxicity of fly ash. Acute inhalation studies have demonstrated that coal fly ash may be as toxic as quartz to the pulmonary alveolar macrophage. Further in vivo studies comparing the cytotoxicity of fly ash and ∝-quartz are required to substantiate this hypothesis. The feasibility of applying sophisticated cloning techniques to the evaluation of potential lymphohematopoetic effects from complex mixtures has been demonstrated. Mutagens in coal fly ash appear to be absorbed to fly ash surfaces and hence may exist in the environment for relatively long periods of time. Techniques are now being developed to evaluate the carcinogenic potential of coal fly ash through a combination of bacterial mutagenesis assays and tracheal implant carcinogenesis assays. Detailed chemical, morphological, and toxicological analyses indicate the usefulness of coal fly ash as a model complex mixture.

ACKNOWLEDGMENTS

This work was supported by the U.S. Department of Energy through the Laboratory for Energy-Related Health Research, University of California, Davis, CA. This manuscript was also presented at the Second Symposium on Process Measurements for Environmental Assessment, February 1980, Atlanta, GA.

REFERENCES

Ames, B.N., J. McCann, and E. Yamasaki. 1975. Methods for detecting carcinogens and mutagens with the Salmonella/mammalian-microsome mutagenicity test. Mutation Res. 31:347-360.

Boorman, G.A., L.W. Schwartz, and F.D. Wilson. 1979a. Formation of macrophage colonies in vitro by free lung cells obtained from rats. J. Reticuloendothel. Soc. 26:855-866.

Boorman, G.A., L.W. Schwartz, and F.D. Wilson. 1979b. In vitro macrophage colony formation by free lung cells during pulmonary injury. J. Reticuloendothel. Soc. 26:867-872.

Chrisp, C.E., and G.L. Fisher. 1980. Mutagenesis of coal fly ash
    linked to tracheal graft assay for carcinogenesis. Presented
    at the 11th Annual Meeting of the Environmental Mutagenesis
    Society.

Chrisp, C.E., and G.L. Fisher. (in press). Mutagenicity of
    airborne particles. Mutation Res.

Chrisp, C.E., G.L. Fisher, and J. Lammert. 1978. Mutagenicity of
    respirable coal fly ash. Science 199:73-75.

Coles, D.G., R.C. Ragaini, J.M. Ondov, G.L. Fisher, D. Silberman,
    and B.A. Prentice. 1979. Chemical studies of stack fly ash
    from a coal-fired power plant. Environ. Sci. Technol.
    13:455-459.

Fisher, G.L. 1980. Size-related chemical and physical properties
    of power plant fly ash. In: Aerosol Generation and Exposure
    Facilities. K. Willeke, ed. Ann Arbor Science: Ann Arbor,
    MI. pp. 203-214.

Fisher, G.L. 1979. The morphogenesis of coal fly ash. In:
    Proceedings of the Symposium on the Transfer and Utilization
    of Particulate Control Technology, IV, Denver, CO.
    pp. 433-440.

Fisher, G.L., D.P.Y. Chang, and M. Brummer. 1976. Fly ash
    collected from electrostatic precipitators: Microcrystalline
    structures and the mystery of the spheres. Science
    192:553-557.

Fisher, G.L., C.E. Chrisp, and O.G. Raabe. 1979a. Physical
    factors affecting the mutagenicity of fly ash from a
    coal-fired power plant. Science 205:879-881.

Fisher, G.L., K.L. McNeill, C.B. Whaley, and J. Fong. 1978a.
    Attachment and phagocytosis studies with murine pulmonary
    alveolar macrophages. J. Reticuloendothel. Soc. 24:243-252.

Fisher, G.L., K.L. McNeill, C.B. Whaley, and J. Fong. 1977.
    Functional studies of lavaged pulmonary alveolar macrophages
    exposed to particles in vitro. In: Radiobiology Laboratory
    Annual Report. UCD 472-124. University of California:
    Davis, CA. p. 50.

Fisher, G.L., and D.F.S. Natusch. 1979. Size-dependence of the
    physical and chemical properties of fly ash. In: Analytical
    Methods for Coal and Coal Products, III. C. Karr, ed.
    Academic Press. pp. 489-541.

Fisher, G.L., B.A. Prentice, T.L. Hayes, and C.E. Lai. (in press). Comparative analysis of coal fly ash by light and electron microscopy. In: Air. American Institute of Chemical Engineers.

Fisher, G.L., B.A. Prentice, D. Silberman, J.M. Ondov, A.H. Biermann, R.C. Ragaini, and A.R. McFarland. 1978b. Physical and morphological studies of size-classified coal fly ash. Environ. Sci. Technol. 12:447-451.

Fisher, G.L., D. Silberman, B.A. Prentice, R.E. Heft, and J.M. Ondov. 1979b. Filtration studies with neutron-activated coal fly ash. Environ. Sci. Technol. 13:689-693.

Fisher, G.L., D. Silberman, and O.G. Raabe. (1980). Chemical characterization of coal fly ash in rodent inhalation studies. Environ. Res. 22:298-306.

Fisher, G.L., and F.D. Wilson. 1980. The effects of coal fly ash and silica inhalation on macrophage function and progenitors. J. Reticuloendothel. Soc. 27:513-524.

Gibbon, D.L. 1979. Microcharacterization of fly ash and analogs: the role of SEM and TEM. In: Scanning Electron Microscopy, Vol. 1. O. Johari, ed. Scanning Electron Microscopy, Inc.: Chicago. pp. 501-510.

Griesemer, R.A., J. Kendrick, and P. Nettesheim. 1974. Tracheal grafts. In: Experimental Lung Cancer. E. Karbe and J.F. Park, eds. Springer-Verlag: Berlin. pp. 539-547.

Hansen, L.D., and G.L. Fisher. 1980. Elemental distribution in coal fly ash particles. Environ. Sci. Technol. 14:1111-1117.

Hansen, L.D., G.L. Fisher, C.E. Chrisp, and D.J. Eatough. 1980. Chemical properties of bacterial mutagens in stack-collected coal fly ash. Presented at the Fifth International Symposium on Polynuclear Aromatic Hydrocarbons, Columbus, OH.

Hansen, L.D., D. Silberman, and G.L. Fisher. (MS). Crystalline components of stack-collected, size-fractionated coal fly ash.

Hayes, T.L., J.B. Pawley, and G.L. Fisher. 1978. The effect of chemical variability of individual fly ash particles on cell exposure. In: Scanning Electron Microscopy. Scanning Electron Microscopy, Inc., AMF: O'Hare, IL. pp. 239-244.

Hayes, T.L., J.B. Pawley, G.L. Fisher, and M. Goldman. 1980.
A toxicological model of fly ash exposure to lung cells.
Environ. Res. 22:499-509.

Kubitschek, H.E., and F.R. Kirchner. 1980. Biological monitoring
of fluidized bed combustion effluents. Presented at the
U.S. Environmental Protection Agency Second Symposium on the
Application of Short-term Bioassays in the Fractionation and
Analysis of Complex Environmental Mixtures, Williamsburg, VA.

McFarland, A.R., R.W. Bertch, G.L. Fisher, and B.A. Prentice.
1977a. A fractionator for size-classification of aerosolized
solid particulate matter. Environ. Sci. Technol. 11:781-784.

McFarland, A.R., C. Ortiz, G.L. Fisher, and O.G. Raabe. 1977b.
Aerosolization of bulk powders utilizing fluidized bed and
Wright dust feed techniques. In: Radiobiology Laboratory
Annual Report. UCD 472-124. University of California:
Davis, CA. pp. 10-13.

Natusch, D.F.S., and B.A. Tomkins. 1978. Theoretical
consideration of the adsorption of polynuclear aromatic
hydrocarbon vapor onto fly ash in a coal-fired power plant.
In: Carcinogenesis 3. P.W. Jones and R.I. Freudenthal, eds.
Raven Press: New York. pp. 145-154.

Natusch, D.F.S., J.R. Wallace, and C.A. Evans. 1974. Toxic trace
elements: preferential concentration in respirable particles.
Science 183:202-204.

Ondov, J.M., R.C. Ragaini, R.E. Heft, G.L. Fisher, D. Silberman,
and B.A. Prentice. 1977. Interlaboratory comparison of
neutron activation and atomic absorption analyses of
size-classified stack fly ash. In: Proceedings of the Eighth
Symposium on Methods and Standards for Environmental
Measurement Materials Research. National Bureau of Standards:
Gaithersburg, MD. pp. 565-572.

Pawley, J.B., and G.L. Fisher. 1977. Using simultaneous three
color x-ray mapping and digital-scan-stop for rapid elemental
characterization of coal combustion by-products. J. Microsc.
110:87-101.

Raabe, O.G., K.D. McFarland, and B.K. Tarkington. 1979.
Generation of respirable aerosols of powerplant fly ash for
inhalation studies with experimental animals. Environ. Sci.
Technol. 13:836-840.

Shifrine, M., G.L. Fisher, and N.J. Taylor. (in press). Effect of trace elements in coal fly ash on lymphocyte blastogenesis. J. Environ. Pathol. Toxicol.

Whaley, C.B., F.D. Wilson, G.L. Fisher, M. Shifrine, and K.L. McNeill. 1977. Growth of canine alveolar macrophage colonies. In: Radiobiology Laboratory Annual Report. UCD 472-124. University of California: Davis, CA. pp. 47-49.

Wilson, F.D., G.L. Fisher, and B.A. Concoby. 1980. Studies on in vitro dose-response characteristics of trace elements (Zn, Se) on lymphohematopoietic progenitors using semisolid culture systems. In: Pulmonary Toxicology of Respirable Particles. C.L. Sanders, F.T. Cross, G.E. Dagle, and J.A. Mahaffey, eds. CONF-791002. U.S. Department of Energy: Washington, DC.

Wilson, F.D., L. O'Grady, C. McNeill, and S.L. Munn. 1974. The formation of bone marrow derived fibroblastic plaques in vitro, preliminary results contrasting these populations to CFU-C. Exp. Hematol. 3:353-354.

POSSIBLE EFFECTS OF COLLECTION METHODS AND SAMPLE PREPARATION ON LEVEL 1 HEALTH EFFECTS TESTING OF COMPLEX MIXTURES

D.J. Brusick
Department of Genetics and Cell Biology
Litton Bionetics, Inc.
Kensington, Maryland

INTRODUCTION

The Level 1 Environmental Assessment Program of the U.S. Environmental Protection Agency (EPA) Industrial Environmental Research Laboratory (IERL) has two main goals: first, to detect potentially hazardous emissions from stationary sources, and second, to develop a data base that will permit a relative ranking of industrial streams with respect to their potential biohazard. The level of bioassay applied does not permit either qualitative or quantitative risk assessment. However, since both chemical and biological effects are evaluated, the level of data integration and coordination is increased, reducing the chance that a potentially hazardous stream will go undetected.

During the initial phases of the IERL Environmental Assessment Program, considerable time was given to methods of sample collection, preparation, and analysis for chemistry assessment (Lentzen et al., 1978). The recent introduction of bioassays requires a similar appraisal of the functions of sample collection, storage, and pretest handling as they relate to the specific health effects and ecological tests proposed for Level 1 biological assessment.

Because of the need to rank streams according to potential biohazard, the initial approach taken by IERL was to evaluate samples in the specific bioassays in a state as similar as possible to that found at the time of sampling. However, recent results reported in the scientific literature on analysis of complex environmental mixtures have shown that pretest processing of samples (concentration of liquids, extraction of particulate) often

results in enhanced biological activity (Klekowski and Levin, 1979;
Kubitschek and Venta, 1979; Löfroth, 1978; Pitts et al., 1977;
Teraniski et al., 1978). In this report, several of the sampling
and pretest procedures are reviewed to illustrate how the
application of these techniques to Level 1 Environmental Assessment
will affect the test responses and ultimately the goals of this
program.

METHODOLOGY

The IERL Level 1 Environmental Assessment Program evaluates
source emissions according to the following five parameters:

1) rate of release into the environment,

2) distribution between physical states,

3) chemical composition,

4) detection of potential specific health effects, and

5) detection of potential impact on the ecosystem.

A schematic of the above parameters is given in Figure 1.

The initial Level 1 methods manual for biological testing
contained protocols for five health effects tests and eight
biological tests (Duke et al., 1977). The types of data obtained
from these tests were varied and not amenable to interpretation by
anyone other than a biologist. Clearly, a method of developing
uniform data was needed. Out of a review of Level 1 bioassays and
data from several pilot studies, a system of uniform data analysis
and formatting was recommended (Brusick, 1980). The review study
also recommended elimination or substitution of certain test
procedures. Table 1 identifies the current status of
recommendations for Level 1 bioassays. No procedures have yet
been approved for the Soil Microorganism Toxicity Assay; test
substitutions or protocol modifications are being considered for
some of the other tests as well.

The data of three pilot studies have been evaluated following
the recently proposed data-formulating procedures (Brusick, 1980).
The results were recorded as high (H), moderate (M), low (L), or
nondetectable (ND) on a summary sheet such as in Figure 2. Table 2
defines the limits of each category.

The studies on Coal Gasification, Fluidized Bed Combustion,
and Textile Plant Liquid Effluents received very little pretest
sampling or sample history evaluation. Most of the samples were

TYPICAL STEPS INVOLVED IN ENVIRONMENTAL ASSESSMENT

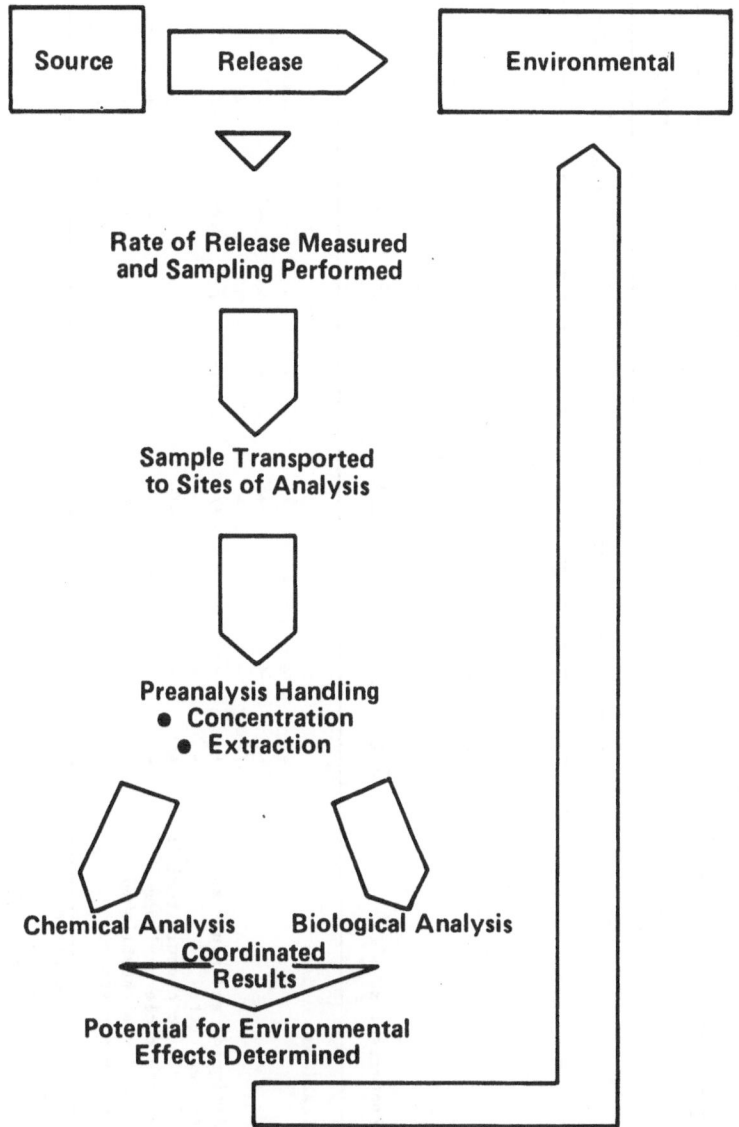

Figure 1.   An overview of the steps involved in Level 1
environmental assessment.

Table 1. Level Bioassay Current Status Of Recommendation

| | Health Effects Bioassays | | | | Ecological Assays | | | | | |
| | | | | | Freshwater/Marine Aquatic | | | Terrestrial | | |
Characteristics	Bacteria Mutagenesis	CHO Clonal Toxicity	RAM	In Vivo Rodent Toxicity	Fish	Invertebrate	Algae	Insect Toxicity	Soil Microorganism Toxicity	Plant Toxicity
Test sample applications										
solids	+	+	+	+	+	+	+	+a	-	-
liquids	+/-	+	+	+	+	+	-	+	+	+
gases	+/-	+	-	-	+/-	-	-	+	+	+
mixtures (solid/liquid) and slurries	+	+	+	+/-	-	+/-	+/-	+/-	+/-	-
filter or sorbent extracts	+	+	+	+/-	-	-	-	+/-	-	-
Test results applicable to chronic toxicity	+	-	-	+/-	-	-	-	+/-	-	-
Bioassay includes life cycle analysis	-	-	-	-	-	-	-	+	-	-
Quality control procedures established	+	+	+	+	+	-	-	-	-	-
Test method in development state	-	-	-	-b	+	-	+/-	+/-	+c	+

a Depends on solubility of materials.
b Protocol changed from rat to mouse.
c No test currently proposed for Level 1.

BIOASSAY SUMMARY TABLE

Technical Directive or Project No. _____

Contract No. _____

Sample Identification	Health Effects Tests					Ecological Effects Tests									Notes
						Aquatic						Terrestrial			
						Fresh Water			Marine						
	AMES Salmonella	RAM Cytotoxicity	CHO Cytotoxicity	Rodent Toxicity	Fish	Invertebrate	Algal	Fish	Invertebrate	Algal	Plant Stress Ethylene	Root Elongation	Insect Toxicity		
_____															
_____															
_____															
_____															
_____															
_____															
_____															
_____															
_____															
_____															
_____															
_____															
_____															
_____															
_____															
_____															
_____															
_____															
_____															
_____															
_____															
_____															

ND = No Detectable Toxicity
L = Low Toxicity
M = Moderate Toxicity
H = High Toxicity

LBI-0468 R 11/80

Figure 2. Example of bioassay summary sheet.

Table 2. Definition of Effectiveness Categories

Assay	Activity Measured	Units	MAD[a]	Range of Concentration or Dosage			
				High	Moderate	Low	Not Detectable
Ames test	Mutagenesis[b]	mg/plate	5	<0.05	0.05-0.5	0.5-5	ND at >5
		µl/plate	50	<0.5	0.5-5.0	5-60	ND at >50
RAM/WI-38 and CHO toxicity	Lethality EC50[c]	mg/ml	1	<0.01	0.01-0.1	0.1-1	ND at >1
		µl/ml	600	<6.0	6.0-60	60-600	ND at >600
		µl/ml	20[d]				ND at >20
Rodent toxicity	Lethality LD50[e]	gm/kg	10	<0.1	0.1-1	1 -10	ND at >10
Aquatic Tests							
Algae	Growth inhibition EC50	gm/l	1	>0.01	0.01-0.1	0.1-1	ND at >1
		%	100	<20	20-75	75-100	ND at >100
Fish	Lethality LC50[f]	gm/l	1	<0.01	0.01-0.1	0.1-1	ND at >1
		%	100	<20	20-75	75-100	ND at >100
Invertebrate	Lethality LC50	gm/l	1	<0.01	0.01-0.1	0.1-1	ND at >1
		%	100	<20	20-75	75-100	ND at >100

[a] Maximum applicable dose (technical limitations).
[b] Negative response at 5 mg/plate or at level of toxicity is given as ND: positive response requires calculation of minimum effective concentration (MEC) to produce a positive mutagenic response; H, M, and L designations are made from MEC values of positive agents.
[c] Calculated concentration expected to produce effect in 50% of population.
[d] Volumes used for solvent exchange samples (this maximum keeps DMSO below level of toxicity).
[e] Calculated dosage expected to kill 50% of population.
[f] Calculated concentration expected to kill 50% of population.

placed directly into the bioassays; only a few filters or sorbent collectors were extracted or sonicated with solvents. A severity assessment was conducted both on chemicals detected and biological responses; a comparison of the analyses indicated generally complementary results (Sexton, 1979). In a few instances, toxicity was not predicted by the chemical analysis. Pretest processing could have a significant impact on such data. With the increased use of Level 1 testing procedures, the questions of whether to use a sample directly or to reduce it to its active constituents or to concentrate a dilute sample become more relevant.

Important components in the final data analysis and interpretation of Level 1 bioassays are

1) sampling methods (selecting a representative sample),

2) storage (maintaining proper composition and preventing degradation),

3) shipping (same as for storage), and

4) pretest handling (altering chemical composition, physical state, or preferential extraction and concentration of potential toxicants and mutagens).

Sampling methods, storage, and shipping can be grouped into a single parameter called sample history (i.e., documentation of sample handling until receipt at the testing laboratory). Pretest handling encompasses the pretest sample processing techniques conducted at the testing facilities. Table 3 describes the possible effects of each technique.

Sample History

Both the final interpretations of the test results and the ranking of emission sources according to potential hazard can be affected by sample history. Representative sampling is particularly decisive.

The environmental fate of the toxic substances in emissions, for example, profoundly influences the final assessment. An emission with extremely low volume and release rate may represent a negligible health or ecological hazard even if it is highly toxic. However, if the emission were large (e.g., of fly ash from power plants), significant human and ecosystem exposure would result (Fisher et al., 1979; Kubitschek and Venta, 1979). If a potential hazard is identified as a function of its release rate (volume per unit time), toxicity, and environmental fate, a quantified assessment may be possible. Thus, if the bioaccumulation of any

Table 3. Effects of Pretest Handling Methods

Pretest Handling	Method Employed	Possible Effect
Particulate extraction	(a) Organic solvent/sonication (b) Soxhlet/organic solvent	Preferential releases to toxic and mutagenic organic materials from bound state. If these organics are not released under normal environmental or physiological conditions, they can skew ranking scheme.
Liquid concentration	Flow through XAD-2 resin column followed by soxhlet extraction from resin and solvent exchange	Concentrates organics and permits inorganics to pass through. Some of the inorganics may be important toxicants. Again, preferential concentration of organics might increase detection of mutagens and skew ranking scheme. Concentration of chemicals may introduce artifacts by altering chemical: chemical dynamics not encountered in dilute solutions.
Extraction from collection filter	Sonication in organic solvent	Preferential release of toxic and mutagenic agents preferentially attached to filter material. Degradation and alteration of chemicals is known to occur when bound to filter. Sonication of filter may release small particles that would be toxic in RAM assay.
Solvent exchange and solubilization	Addition of sample to methylene chloride, acetone, benzene, DMSO, etc.	Preferential release of soluble compounds to the target organism. Nonrepresentative composition of original sample by extraction of organics.
Particulate sizing and grinding	(a) Filter collection (b) Grinding and sizing with mill and wire screen	Development of a physical state more amenable to cellular phagocytosis or animal absorption. This could skew ranking by producing abnormally high levels of toxicity.

single component is very large, the potential hazard could be significant. However, only toxicity is being recorded at present (see Figure 2). This factor is not being related to the type of emission release rate or to the volume of emission sample. Attempts should be made to document the history of each sample.

Pretest Handling

Several laboratories in the United States (particularly IERL of EPA in Research Triangle Park, NC, and Oak Ridge National Laboratory in Oak Ridge, TN) have initiated programs to analyze complex mixtures for mutagenic activity (Epler, 1979; Huisingh et al., 1978). The results of the Ames Salmonella assay serve as a preliminary indicator for further fractionation tests to determine the biologically active components (pure substances).

The Ames test procedures specify how to collect the desired sample (particulate, gas, or liquid) on a filter or solid sorbent; extract the organics from the filter or sorbent (by sonication or Soxhlet methods) into an organic solvent; exchange solvents to dimethylsulfoxide (DMSO) or evaporate to dryness and resuspend in DMSO; and conduct the bioassay on the concentrated extract. Chemical fractionation and further testing may eventually lead to specific associations between chemicals or chemical classes with biological activity (Huisingh et al., 1978). Table 3 describes some of the pretest sample processing techniques currently used and indicates their possible effect on the interpretation of Level 1 bioassay data.

The Level 1 toxicity assessments listed in the Bioassay Summary (Figure 2) include several types of biological systems and phylogenetic levels. If a significant amount of pretest processing is anticipated, all the Level 1 tests, and not just the Ames or one or two selected tests, should be evaluated. Otherwise, the data balance will be upset and the ranking of the test site might be biased by an abnormally toxic response from an extract or concentrate in a single assay, erroneously exaggerating the potential hazard. This precaution is not meant to preclude pretest processing. However, as with the sample history factor (release rate), the pretest sample processing must be factored into the final assignment of a toxicity value and its contribution to the potential hazard. A mechanism needs to be developed to normalize the data obtained from preprocessed samples. The simplest approach might be to divide the actual test response by a concentration factor.

RECOMMENDATIONS

The following recommendations would facilitate the
implementation of sample history documentation and pretest sample
processing and enhance the reliability of the Level 1 bioassays:

1) Documentation of sample collection, storage, shipping, and
   pretest processing should be available for all test
   samples. Examples of forms for this purpose are shown in
   Figures 3 and 4.

2) Pretest processing should be factored into the final
   toxicity designations of H, M, L, and ND. Specific
   methods need to be developed to normalize data obtained
   from samples modified prior to evaluation.

3) Emissions from a given site should be applied uniformly to
   the spectrum of Level 1 bioassays so as not to bias the
   final interpretation. Data related to environmental fate
   should be included.

4) Discharge-severity calculations used to factor chemical
   and physical information should be expanded to include
   biological response and fate. Specific methods need to be
   developed.

CONCLUSIONS

Level 1 environmental assessment bioassays should permit
accurate ranking of emissions from stationary-site sources with
respect to their potential hazard. Pretest processing should be
kept to a minimum and applied uniformly across all Level 1
bioassays. The ranking must ensure that the potential hazard will
be derived from emissions in the state in which they were released
into the environment.

The Level 1 Environmental Assessment Program should include
the environmental fate and emission release rate along with the
chemical and bioassay toxicity determinations. This approach would
result in a second level determination called a severity potential
hazard. A general scheme has been proposed to develop this
assessment and is given in Figure 5. If a potential hazard can be
calculated with reasonable accuracy, the usefulness of Level 1
results will be greatly enhanced.

Sample Collection Form No._____

LEVEL 1 SAMPLE COLLECTION FORM

I.   SAMPLE HISTORY

    A.   Company Name _____

    B.   Sampling Manager _____

    C.   Contract No. _____

    D.   Sampling Date _____

    E.   Emission Source _____

    F.   Approximate Rate of Emission (Volume/Time) _____

    G.   Proportion of Emission Sampled (Volume Captured) _____

II. SAMPLE TYPE

    A.   Sample No. _____

    B.   Name _____

    C.   Sample Description _____
        _____
        _____

III. HANDLING CONDITIONS

    A.   Storage

Container	Temperature	Light
☐ Amber Glass Bottle	☐ Ambient	☐ Keep in Dark
☐ Polyethylene Bottle	☐ Refrigerate (0 to 4°C)	
☐ Coated Bag or Bottle	☐ Freeze (-20°C)	

    B.   Approximate Time of Storage Before Shipping _____

IV. SHIPPING HISTORY

    A.   Biological Contractor _____

    B.   Address:

    C.   Carrier _____

    D.   Date Shipped _____ By _____

    E.   Special Packaging _____

Two copies of this form must accompany each sample.

Figure 3.   Example of Level 1 sample collection form.

Sample Processing Form No._____

LEVEL 1 SAMPLE PROCESSING FORM

I.   SAMPLE IDENTIFICATION

    A.   Sample Collection Form No. _____
    B.   Contract No. _____
    C.   Project Officer _____
    D.   Sample No. _____

II. SAMPLE TYPE AND PROCESSING REQUIREMENTS

Basic Type	Subtype	Processing

☐ Solid	☐ Solid Granular
	☐ Slurry (>50% Solids)
	☐ Particulates from
	   Filter
	☐ Filter/Unit Particulates
		☐ Grind to <5µ size
		☐ Extract Particulates with
		   Organic Solvent
		☐ Remove Particulate
		   from Filter
		☐ Prepare Water Leachate

☐ Liquids	☐ Suspensions (<50% Solids) ☐ Effluent ☐ Leachate ☐ Extract ☐ Condensate	☐ Concentrate with XAD-2 ☐ Solvent Exchange ☐ Evaporate to Dryness
☐ Gas	☐ Pressure Collection ☐ Vacuum Collection	

III. BIOASSAYS REQUESTED

☐ Ames, Salmonella   ☐ Freshwater Fish Toxicity
☐ RAM Toxicity   ☐ Freshwater Invertebrate
☐ CHO Clonal Toxicity   ☐ Algal Test
☐ Rodent Quantal Toxicity ☐ Insect Toxicity
                          ☐ Plant Test
                          ☐ Soil Test

Figure 4.   Example of Level 1 sample processing form.

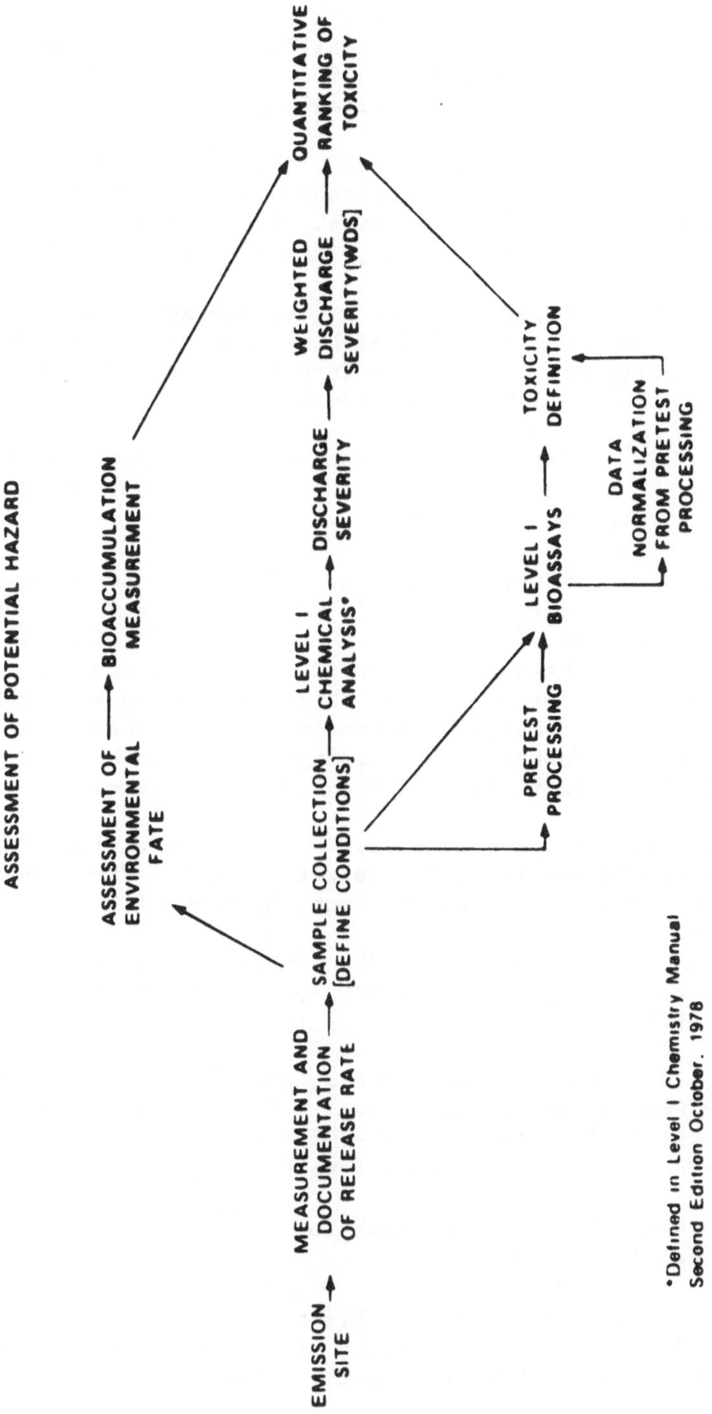

Figure 5. Proposed scheme for a second stage evaluation of Level 1 results (defined in Lentzen et al., 1978).

REFERENCES

Brusick, D.J. 1980. Level 1 Biological Testing Assessment and
     Data Formatting. EPA-600/7-80-079. U.S. Environmental
     Protection Agency: Research Triangle Park, NC.

Duke, K.M., M.E. Davis, and A.J. Dennis. 1977. IERL-RTP
     Procedures Manual: Level 1 Environmental Assessment
     Biological Tests for Pilot Studies. EPA-600/7-77-043. U.S.
     Environmental Protection Agency: Research Triangle Park, NC.

Epler, J.L. 1979. Mutagenicity testing of energy-related
     compounds. In: Energy and Health Proceedings. N.E. Breslow
     and A.S. Whittemore, eds. Siam Institute for Mathematics and
     Sociology: Philadelphia. pp. 17-36.

Fisher, G.L., C.E. Chrisp, and O.G. Raabe. 1979. Physical factors
     affecting the mutagenicity of fly ash from a coal-fired power
     plant. Science 204:879-881.

Huisingh, J., R. Bradow, R. Jungers, L. Claxton, R. Zweidinger, S.
     Tejada, J. Bumgarner, F. Duffield, M. Waters, V.F. Simmon, C.
     Hare, C. Rodriguez, and L. Snow. 1978. Application of
     bioassay to the characterization of diesel particulate
     emissions. In: Application of Short-term Bioassays in the
     Fractionation and Analysis of Complex Environmental Mixtures.
     M.D. Waters, S. Nesnow, J.L. Huisingh, S.S. Sandhu, and L.
     Claxton, eds. Plenum Press: New York. pp. 381-418.

Klekowski, E., and D.E. Levin. 1979. Mutagens in a river heavily
     polluted with paper recycling wastes: Results from field and
     laboratory mutagen assays. Environ. Mutagen. 1:209-219.

Kubitschek, H.E., and L. Venta. 1979. Mutagenicity of coal fly
     ash from electric power plant precipitators. Environ.
     Mutagen. 1:79-82.

Lentzen, D.E., D.E. Wagoner, E.D. Estes, and W.F. Gutknecht. 1978.
     IERL-RTP Procedures Manual: Level 1 Environmental Assessment
     (Second Edition). EPA-600/7-78-201. U.S. Environmental
     Protection Agency: Research Triangle Park, NC.

Löfroth, G. 1978. Mutagenicity assay of combustion emissions.
     Chemosphere 10:791-798.

Pitts, J.N., Jr., D. Grosjean, T.M. Mischke, V. Simmon, and D.
     Poole. 1977. Mutagenic activity of airborne particulate
     organic pollutants. Toxicol. Lett. 1:65-70.

Sexton, N.G. 1979. Biological Screening of Complex Samples from Industrial/Energy Processes. EPA-600/8-79-021. U.S. Environmental Protection Agency: Research Triangle Park, NC.

Teranisiki, K., K. Hamada, and H. Watanabe. 1978. Mutagenicity in Salmonella typhimurium mutants of the benzene-soluble organic matter derived from airborne particulate matter and its five fractions. Mutation Res. 56:273-280.

BIOLOGICAL MONITORING OF FLUIDIZED BED COAL COMBUSTION OPERATIONS I. INCREASED MUTAGENICITY DURING PERIODS OF INCOMPLETE COMBUSTION

H.E. Kubitschek, D.M. Williams, and F.R. Kirchner
Division of Biological and Medical Research
Argonne National Laboratory
Argonne, Illinois

INTRODUCTION

The Ames <u>Salmonella</u> microsome assay (Ames and Yamasaki, 1975) is used extensively to screen environmental pollutants because of its rapidity, its relatively low cost, and its sensitivity in detecting carcinogens (McCann et al., 1975). We applied the Ames assay to the study of mutagenic particulates produced by an experimental process-development fluidized bed combustor (FBC) at Argonne National Laboratory. This FBC also was used for mouse inhalation toxicology experiments (Kirchner et al., 1980; this study is reported in this volume as Biological Monitoring of Fluidized Bed Coal Combustion Operations II.).

The original plan was to determine the relative mutagenicities of the different effluents produced by the FBC and to measure the mutagenicity of the particulate effluent during the mouse exposures. However, our earliest measurements (Kubitschek and Haugen, 1980) indicated that the mutagenicity was much greater than that for a larger FBC (Clark et al., 1978). This excessive mutagenicity in the Argonne FBC was traced to operating conditions: fly ash deposited on the final filter during start-up periods was up to 60 times as mutagenic as that produced during steady operation (Kubitschek and Williams, 1980). This finding, and the variability in the mutagenicity of the fly ash produced during a 1000-h (40-day) run, made it apparent that the Ames assay might be used to examine the effects of process conditions on levels of mutagenicity.

411

MATERIALS AND METHOD

The Argonne FBC, operated by the Chemical Engineering Division, is a process-development, 6-in.- (15.2-cm-) diameter combustor maintained at atmospheric pressure. This FBC burned high-sulfur (5.5%) bituminous coal (Sewickly) and used as the sorbent either of two calcitic limestones (Greer or Grove) treated with sodium carbonate. The combustor operated at a power level of 21 kW, with coal feed controlled by the concentration of oxygen in the off-gas stream. Oxygen was normally maintained at 3%, and sulfur dioxide at 700 ppm. The operating temperature was 850°C.

Particulates were collected in cyclones and also on a porous metal filter, located downstream in the off-gas line. This filter had an efficiency greater than 98% for particles 6 μm in diameter or larger. Particle cutoff diameters for the first and second cyclones were approximately 8 and 5 μm, respectively. An important design feature of the FBC was the availability of twin off-gas clean-up trains downstream from the combustion chamber, each containing a primary cyclone, secondary cyclone, and porous metal filter. The effluent was vented to the exhaust system. With this equipment, off-gas trains were used alternately to collect filter particulate samples, and equipment could be cleaned to reduce cross-contamination between the samples.

Initially, particulate samples (fly ash) were collected after runs of a single day's duration (8 to 10 h). These samples were extracted for 45 min at 37°C with dimethylsulfoxide, which yielded greater observable mutagenicity than did any of the other organic solvents tested (Kubitschek and Haugen, 1980). Metabolic activation did not increase observable mutagenicity (Kubitschek and Haugen, 1980), which agreed with similar observations by Fisher et al. (1979). In all of the experiments described below, mutagenic activity was determined with the Ames Salmonella assay using strain TA98 without microsomal enzyme activation.

RESULTS

Mutagenicity During Start-up and Steady Operation

In a series of tests of fly ash from single daily runs of the Argonne FBC, we observed mutagenicity levels in excess of 1000 revertants/mg (Kubitschek and Haugen, 1980; Kubitschek and Williams, 1980). These values were much greater than those obtained by Clark et al. (1978) for fly ash samples from the 18-in.- (45.7-cm-) diameter FBC at the Morgantown Technology Center. Clark et al. found that when mutagenicity was detectable, activities were less than 3 revertants/mg.

We first considered the possibility that these very different mutagenicities might be due to the different temperatures at which the samples were collected. Natusch and Tompkins (1978) predicted that the adsorption of mutagenic polycyclic aromatic hydrocarbons would increase greatly as tempertures were decreased below 150°C, and the collection temperature of the Argonne filter (70°C) was known to be somewhat less than that for the Morgantown samples (estimated to be in the range of 70° to 90°C). However, when the Argonne FBC was operated for longer periods, somewhat lower sample activities were observed, indicating that sample activities might depend on operating conditions.

To distinguish between these possible explanations, fly ash samples were collected from the first and second cyclones and the filter both during start-up periods and during steady operation. The observed mutagenicities (Kubitschek and Williams, 1980) are shown in Table 1. An inverse relationship between mutagenicity and particulate collection temperature can be seen, in agreement with the predictions of Natusch and Tompkins (1978). However, mutagenicity levels varied much more widely when operating conditions were changed, and fly ash mutagenicity during start-up of operations was as much as 60-fold greater than that observed during later steady operation. Clearly, the great bulk (> 98%) of the difference between our earlier determinations and those of Clark et al. (1978) can be assigned to excessive mutagenicity that was deposited in our samples during start-up and shut-down of operations.

Table 1. Particulate Effluent Mutagenicity During Start-up and Steady FBC Operation[a,b]

		Operation		
		---	---	---
Site	Temperature	Start-up	Steady	Ratio
Primary cyclone	150°C	3 ± 2	0 ± 1	–
Secondary cyclone	95°C	470 ± 25	8 ± 1	59
Filter	70°C	1400 ± 20	22 ± 2	64

[a]From Kubitschek and Williams, 1980, by permission of the publisher.
[b]Values are averages ± standard errors for net numbers of TA98 His[+] revertants/mg ash extracted.

While the data shown in Table 1 are those for the experiment
with the greatest difference between start-up and steady operating
conditions, the same qualitative results were obtained in each of
three experiments. These results suggest that excessive
mutagenicity might be produced during periods of incomplete
combustion, as would be expected during start-up of combustion
(Kubitschek and Williams, 1980).

Mutagenicity During Departures from Steady Operation

Particulate mutagenicity was monitored for samples collected
daily during a 1000-h exposure of mice to gaseous and particulate
effluents from the FBC. During this period, the specific
mutagenicity of the fly ash varied widely (Figure 1) with unusually
high levels of mutagenic activity produced during six transient
periods. Later, by examining the record of operations, we found
that each of these peaks of mutagenic activity (Table 2) was
associated with a departure from steady combustor operation, due
either to start-up of operations or to mechanical breakdown
(Kubitschek et al., 1980).

Examination of the gas concentration records indicated that
carbon monoxide (CO) concentrations also increased during the same
periods of high fly ash mutagenicity. Figure 1 shows the sum of
the peak values for CO concentrations in excess of 1000 ppm during
each 12-h period of the run. This level was chosen because
mutagenicity did not appear to be closely correlated with the CO
peaks of smaller magnitude, which became more frequent as CO
concentrations approached the background level of approximately 500
ppm. Good correlation between this peak CO concentration and
sample mutagenicity is evident in Figure 1; the correlation
coefficient is 0.72.

Correlations in time and in intensity for the individual
transients supported the correlation between mutagenicity and peak
CO concentrations. The mean time of occurrence of mutagenicity in
each peak and the mean time of peak CO production for the
corresponding periods are shown in Figure 2 and Table 2. The
coefficient of correlation between the mean times of occurrence was
0.9998. This close correlation is especially noteworthy
considering that the average duration of these six transient
periods was more than three days, while the average deviation
between corresponding CO and mutagenicity peaks was less than 5 h.

Good correlation also was observed between the magnitude of CO
peak production and mutagenicity (shown in Figure 3 and Table 2);
the correlation coefficient was 0.81. The average net number of
revertants per milligram produced per CO value at the standard
level of 1000 ppm CO was 16.4 (standard error = 2.1). Thus, both

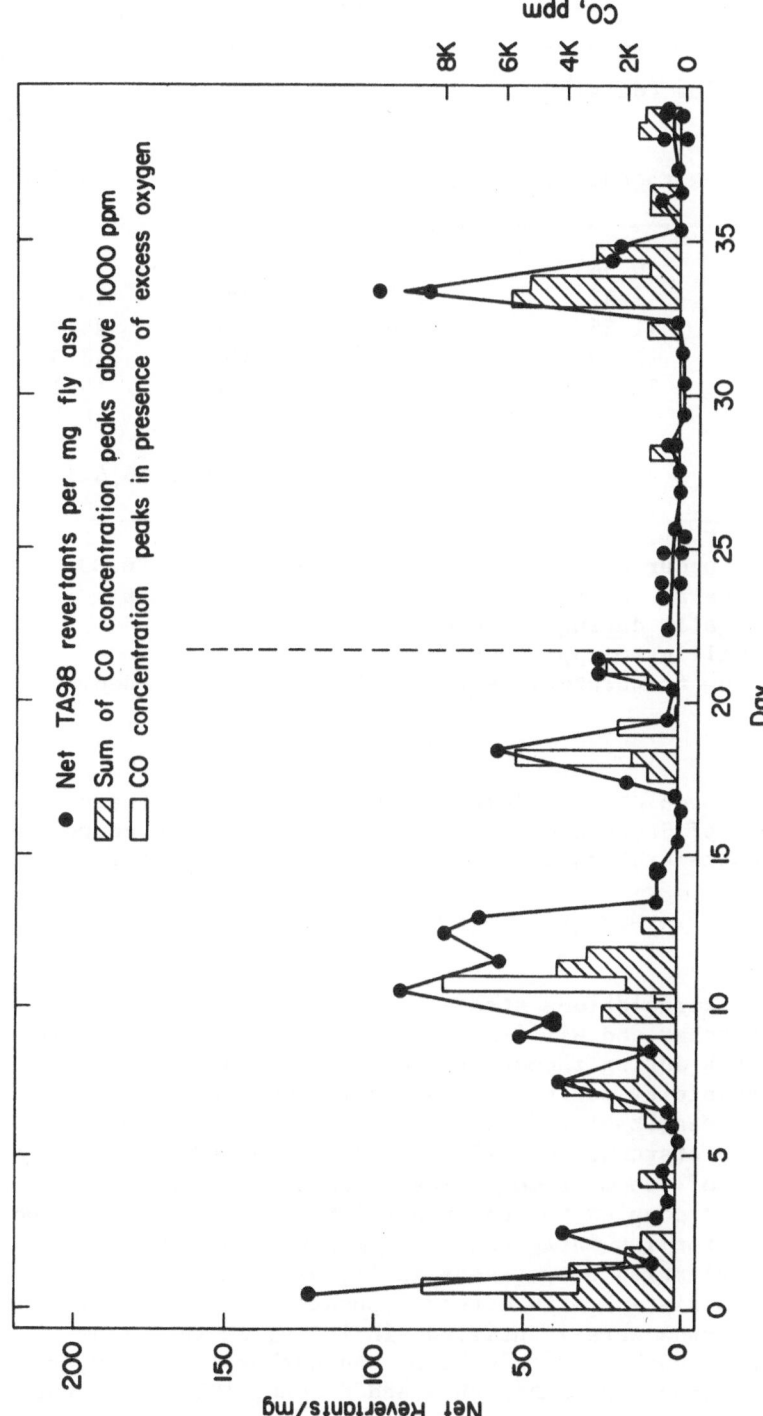

Figure 1. Fly ash mutagenicity and peak CO concentrations during a 40-day period of FBC operation (Kubitschek et al., 1980). The bar graph shows the sum of the CO concentration peaks for each half-day period. The letter T (day 10) indicates a large decrease in operating temperature. The dashed line indicates a period of about 10 days during which the FBC was shut down for repairs.

Table 2. Mutagenicity and CO Production During Departures
from Steady FBC Operation[a]

Period (days)	Mean Time (days)[b]		CO (ppm x 10^{-3})	Mutagenicity (rev/mg)	Ratio[c]
	CO Peaks	Mutagenicity			
0-2	0.55	0.58	21.2	274	12.9
6-8	7.42	7.67	11.7	126	10.8
9-13[d]	11.60	11.35	18.4	458	24.9
17-19	18.50	18.09	8.4	154	18.3
20-21	21.10	20.97	3.7	56	15.1
32-35	33.80	33.63	14.4	237	16.5

Mean: 16.4 ± 2.0

[a]From Kubitschek et al., 1980.
[b]The mean time of occurrence for each transient increase in CO
production (CO peaks) and in mutagenicity is given in days
numbered sequentially during the tested periods.
[c]Revertants per milligram: ppm CO x 10^{-3}.
[d]Values for the low temperature period on day 10 are excluded.

the time of occurrence and the degree of mutagenicity of these
transient periods of departure from steady operation were correlated
with increased concentration of CO in the combustion chamber.

CONCLUSIONS

The observed correlations strongly support our earlier
suggestion (Kubitschek and Williams, 1980) that excessive effluent
mutagenicity occurs during incomplete coal combustion. The
increase in mutagenicity during periods of incomplete combustion
might have been a result of the evolution of adsorbed hydrocarbons
(especially during start-up) or, alternatively, the result of only
partial oxidation of the macromolecules of which coal is composed,
with the release of complex hydrocarbon moieties that are mutagens
or are capable of forming mutagens. Preliminary results for
chemical characterization of mutagens in fly ash samples from the
Argonne FBC are consistent with either hypothesis: several classes
of mutagenic compounds were identified, including 2- to 6-ring
aromatics, phenols, carbonyls, alcohols, and carboxylic acids, and
other very polar compounds (Kubitschek and Haugen, 1980), and the

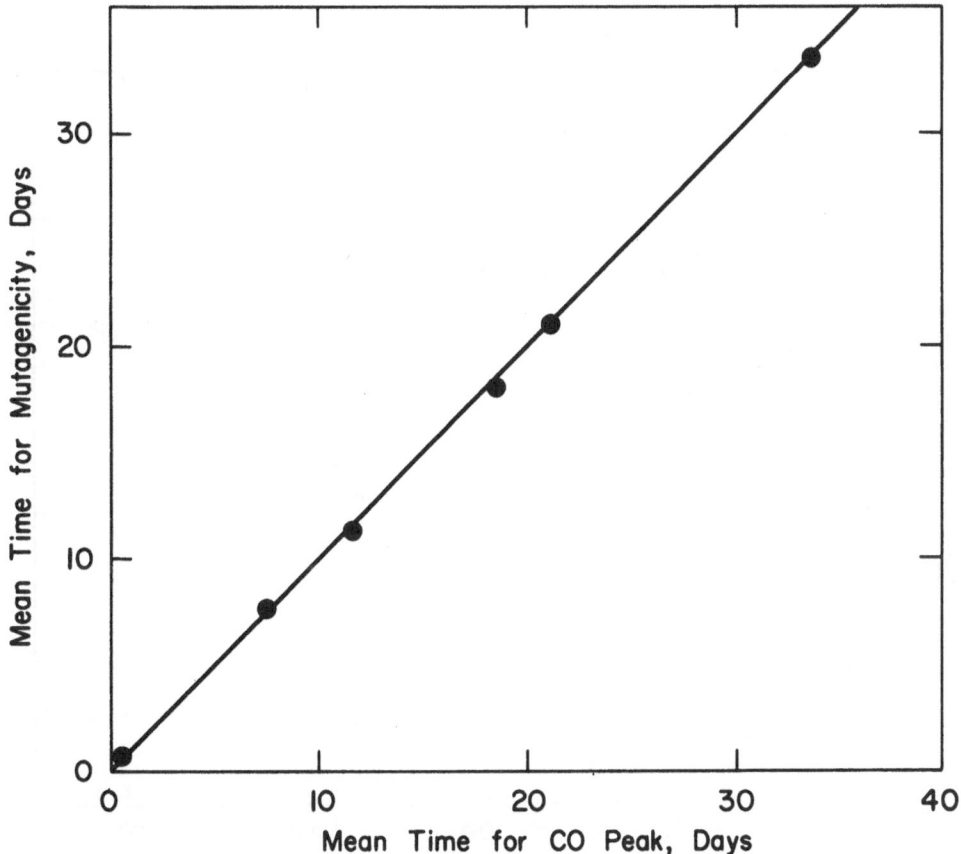

Figure 2.   Correlation between mean times of occurrence of
            increased CO production and mutagenicity during
            transient departures from steady FBC operation (redrawn
            from figure in Kubitschek et al., 1980).

amounts of these mutagens decreased during steady operation of the
combustor (Kubitschek and Williams, 1980).

These results could influence estimates and comparisons of the
biological risk from the various coal technologies. If
mutagenicity increases with incomplete combustion, then any
estimate of biological hazards or comparison among different coal
technologies should take into account the mutagenic activities
associated with both steady and nonsteady operating conditions. If,
as in our experiments, mutagen production occurs primarily during
start-up periods or other transient departures from steady

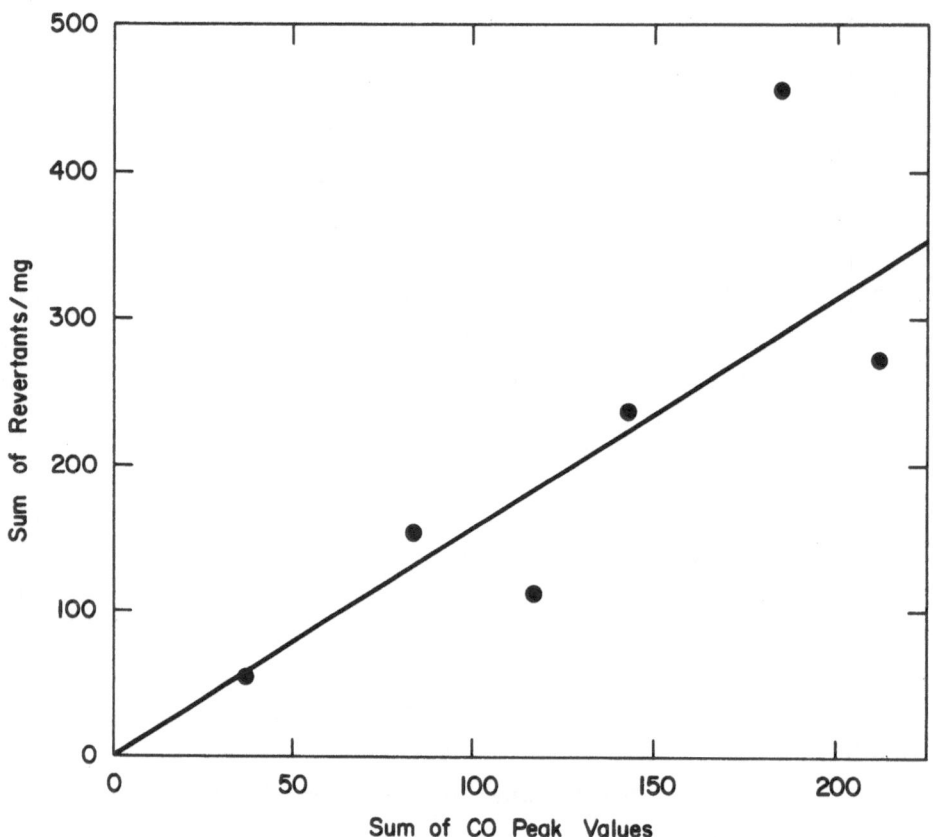

Figure 3.   Correlation between specific mutagenicity and the sum of
            the CO concentration peak values (redrawn from figure in
            Kubitschek et al., 1980).

operation, then biological risk would be determined mainly by the
frequency and duration of those operational transients.

ACKNOWLEDGMENTS

These studies could not have been carried out without the generous
assistance of K.M. Myles and G.W. Smith, of the Chemical
Engineering Division, during the course of the run. We thank them
for their help and for making the records of operation available to

us. We also thank V.A. Pahnke, of our division, for collecting samples. This work was supported by the U.S. Department of Energy under contract No. W-31-109-ENG-38.

REFERENCES

Ames, B.N., J. McCann, and E. Yamasaki. 1975. Methods for detecting carcinogens and mutagens with the Salmonella/ mammalian-microsome mutagenicity test. Mutation Res. 31:347-364.

Clark, C.R., R.L. Hanson, and A. Sanchez. 1978. Mutagenicity of effluents associated with the fluidized bed combustion of coal. In: Inhalation Toxicology Research Annual Report 1977-1978. Lovelace Biomedical and Environmental Research Institute.

Fisher, G.L., C.E. Chrisp, and O.G. Raabe. 1979. Physical factors affecting the mutagenicity of fly ash from a coal-fired power plant. Science 204:879-881.

Kirchner, F.R., D.M. Buchholz, V.A. Pahnke, and C.A. Reilly, Jr. 1980. Biological monitoring of fluidized bed combustion operations II. Mammalian responses following exposure to the gaseous effluents. Presented at the U.S. Environmental Protection Agency Second Symposium on the Application of Short-term Bioassays in the Fractionation and Analysis of Complex Environmental Mixtures, Williamsburg, VA.

Kubitschek, H.E., and D.A. Haugen. 1980. Biological activity of effluents from fluidized bed combustion of high-sulfur coal. In: Health Implications of New Energy Technologies. W.N. Rom and W.E. Archer, eds. Ann Arbor Science Publishers: Ann Arbor, MI. pp. 381-394.

Kubitschek, H.E., and D.M. Williams. 1980. Mutagenicity of fly ash from a fluidized bed combustor during start-up and steady operating conditions. Mutation Res. 77:287-291.

Kubitschek, H.E., D.M. Williams, and F.R. Kirchner. 1980. Correlation between particulate effluent mutagenicity and increased carbon monoxide concentration in a fluidized bed coal combustor. Mutation Res. 74:329-333.

McCann, J., E. Choi, E. Yamasaki, and B.N. Ames. 1975. Detection of carcinogens as mutagens in the Salmonella/microsome test: Assay of 300 chemicals. Proc. Nat. Acad. Sci. USA 72:5135-5139.

Natusch, D.F.S., and B.A. Tomkins. 1978. Theoretical
    consideration of the adsorption of polynuclear aromatic
    hydrocarbon vapor onto fly ash in a coal-fired power plant.
    In: Carcinogenesis, Vol. 3: Polynuclear Aromatic
    Hydrocarbons. P.W. Jones and R.I. Freudenthal, eds. Raven
    Press: New York. pp. 145-303.

BIOLOGICAL MONITORING OF FLUIDIZED BED COAL COMBUSTION OPERATIONS II. MAMMALIAN RESPONSES FOLLOWING EXPOSURE TO GASEOUS EFFLUENTS

F.R. Kirchner, D.M. Buchholz, V.A. Pahnke, and
C.A. Reilly, Jr.
Division of Biological and Medical Research
Argonne National Laboratory
Argonne, Illinois

INTRODUCTION

Several coal combustion technologies, including fluidized bed combustion, have potential for meeting government-imposed environmental pollution guidelines. Fluidized bed combustion enables the combustion of high sulfur coal while limiting the emission of sulfur dioxide.

In a previous study (Kirchner et al., 1980), detrimental biological effects were observed in animals exposed to whole (gaseous and particulate) effluents derived from the atmospheric pressure fluidized bed coal combustor (FBC) operated by the Chemical Engineering Division at Argonne National Laboratory (ANL), Argonne, IL. In the study presented here, the animals were exposed only to gaseous FBC components in order to determine which effluent components are responsible for these observed biological effects. Three testing systems were applied.

To assess pulmonary alveolar macrophage (PAM) function, one of the organism's first lines of defense against airborne pollutants, Brennan et al. (1980) modified the assay of neutrophil function first described by Tan et al. (1971). The assay differentiates between defective phagocytosis and impaired intracellular killing. Effective PAM function is required to eliminate inhaled micro-organisms, particulate pollutants, and other inhaled particles.

Increased numbers of PAMs were observed in the lungs of animals exposed only to FBC fly ash (Brennan et al., 1980), taxing the lung macrophage progenitor stem cell compartment. To assess the systemic toxic effects following acute exposure to FBC

421

gaseous effluent, the Till and McCulloch spleen colony assay (1961) was used. The lungs and other tissues thought to be at risk were also examined histopathologically during and after the exposures.

METHODOLOGY

The process-development-scale atmospheric pressure FBC described by Kirchner et al. (1980) was used to expose mice and rats to gaseous effluents of fluidized bed combustion. The effluent was diluted in two stages by an effluent dilution system. The effluents were taken from a point downstream of the final filter particle cleanup system, to eliminate the particulate component of the effluents (Figure 1). Following a 20-fold dilution, the effluents were delivered to the top of an Atmospheric Effects Simulator (AES)(Figure 2) that exposed the diluted effluent to simulated sunlight for 15 min. The animals were thus exposed to diluted gaseous combustion effluents only. The levels of carbon monoxide, nitrogen monoxide, sulfur dioxide, nitrogen oxides, and total vapor-phase hydrocarbons were monitored instrumentally in the exposure chambers throughout the experiments.

PURPOSE

Two experiments, each with a 500-h (~ 21-day) continuous exposure period, were conducted. In Experiment I, 154 male B6CF$_1$/Anl mice, each 120 days old, and 8 male Fisher F-344 rats, each 90 days old, were exposed in chambers designed and built in this laboratory (Kirchner et al., 1980; Figure 3). At the same time, similar control animals (66 mice and 8 rats) were placed in identical chambers through which only HEPA-filtered room air was passed. In Experiment II, previously exposed animals (88 mice and 8 rats) were re-exposed for a second 500-h period. In addition, previously untreated animals (88 mice and 8 rats) were exposed for 500 h. The two control groups in Experiment II were also exposed to HEPA-filtered room air: one group consisted of 44 control mice from Experiment I, and the other 44 untreated mice. The animals were maintained on a cycle of 12 h light and 12 h darkness. They had food (Wayne Lab Blox, Allied Mills) and water ad libitum. The animals' condition was checked daily.

The animals were exposed to gaseous effluents obtained only during steady-state operation conditions. Steady state is defined by the following conditions: 1) a period at least 12 h after initiation of coal feed; 2) sulfur dioxide concentrations in the undiluted off-gas at a constant 700 ± 50 ppm; 3) a bed temperature of 850 ± 5°C; and 4) carbon monoxide, nitrogen monoxide, and total vapor-phase hydrocarbon concentrations stabilized.

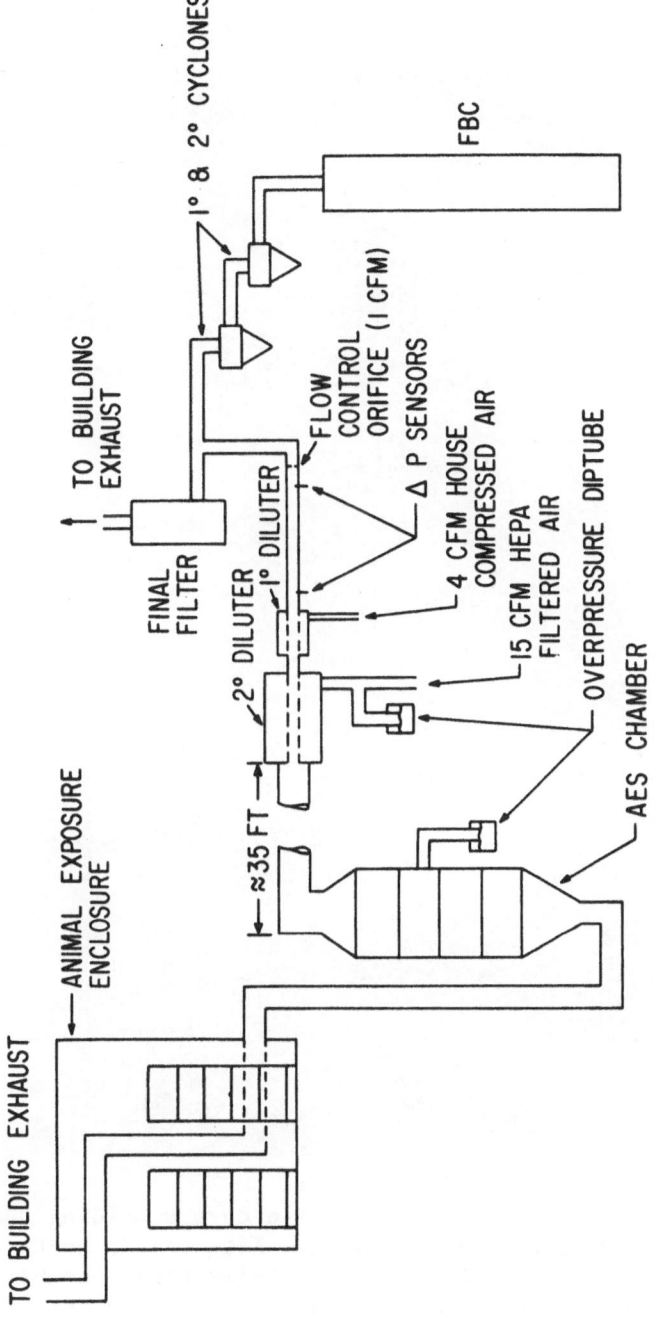

Figure 1.   Schematic diagram of the FBC and the associated effluent delivery system, the AES (reproduced from Kirchner et al., 1980).

Figure 2.   AES with one of the protective covers holding the
             aluminum reflectors removed.   Note sample ports on far
             side of the chamber used for collecting gaseous and
             particulate samples from the AES (reproduced from
             Kirchner et al., 1980).

Figure 3.   Animal exposure system shown with doors to aluminum enclosure removed (reproduced from Kirchner et al., 1980).

To verify that the animals were exposed only to the gaseous
component of the effluent, the concentration of particulate in the
diluted off-gas was monitored with optical and electrical aerosol
analyzers. For the smallest-diameter particles (0.01 to 1.0 μm)
the Thermo Systems, Inc., electrical aerosol analyzer was used.
For particles in the range of 0.3 to 3.0 μm or greater, the Royco
forward-light-scattering particle counter was used. These analyses
also indicated the efficiency of the particle cleanup system.

After 250, 500, and 1000 h of exposure, two randomly selected
animals from each group were sacrificed by cervical dislocation.
Tissues from the lung, liver, kidney, spleen, and heart were fixed,
stained with hematoxylin and eosin, and examined histologically.
Four to eight days after termination of exposures, the PAM assays
(from mice and rats), femoral bone marrow spleen colony assays
(mice only), and histopathological examinations (mice only) were
performed.

RESULTS

The levels of particulate effluent measured by the electronic
particle counters were less than 10% of those observed during
exposures in which the effluent was taken from a point just before
the final filter of the particle cleanup system. No particles
larger than 1.0 μm were detected.

The combustion-gas concentrations during the two experiments,
given in Table 1, are similar to those previously reported by
Kirchner et al. (1980). The only toxicant above the human
threshold limit value (TLV)(ACGIH, 1977) was sulfur dioxide (TLV =
5 ppm)(animals exposed to 22 ± 5.1 ppm). The high value for
sulfur dioxide resulted from the 20-fold dilution factor; the
concentration would have been below the TLV following atmospheric
dilution. (Note: TLVs are based on time-weighted average
concentrations to which workers can be exposed for a normal 8-h
workday or 40-h workweek. The mice, however, were exposed
continuously during the 500-h exposures.)

In Experiment I, three experimental mice died (1.5% of the
number entering the experiment); the deaths occurred at days 6, 11,
and 13 of exposure. In Experiment II, four effluent-exposed mice
died, two from the 500-h exposed group (one at day 12 and one at
day 16) and two from the 1000-h exposed group (both on day 16).
These mice represented 2.3% of the animals entering the experiment
from each group. No moribund animals were observed and no rats or
control animals died during either experiment.

Table 1. Concentration of Gaseous Effluents in the
Atmospheric Effects Simulator During a 500-Hour Exposure[a]

Gas	Concentration (ppm)[b]	TLV (ppm)[c]
Carbon monoxide	17 ± 11	50
Nitrogen monoxide	19.0 ± 6.3	25
Nitrogen oxides	1.1 ± 0.1	5
Sulfur dioxide	22.0 ± 5.1	5
Total hydrocarbons (vapor phase)	2.1 ± 0.75	--

[a]Averages of readings taken every 6 h during the 500-h exposure
± 1 standard error).
[b]Determined by instrumental gas analyzers.
[c]Threshold limit values for chemical substances in workroom air
adopted by the American Conference of Governmental Industrial
Hygienists (1977).

Four to eight days after the termination of the experiments,
PAMs were taken from both rats and mice. No impairment was found
in their ability to engulf and kill a challenge dose of
Staphylococcus aureus (Table 2).

The spleen colony assay showed that a 500-h exposure to the
effluent increased spleen colony formation 167% over control
values. The sizes of the spleen colonies from mice exposed 500 h
were generally larger than those from the control mice. In mice
exposed for 1000 h, however, the number of spleen-colony-forming
units (CFUs) was comparable to the control value (Table 3).

Histological changes in the lungs of exposed mice were limited
to a slight accumulation of fly ash in the macrophages in mice
sacrificed after 500 and 1000 h of exposure. Fly ash was not
present in the control mice or the animals exposed for 250 h.
These histological changes were mild compared with all levels and
times of exposure for previous exposures to the whole effluent
(Kirchner et al., 1980). No evidence of epithelial hyperplasia
or obvious proliferation of macrophages was seen. The gross and
histological appearances of all other organs were normal.

Three days after termination of Experiment II, the 500- and
1000-h exposed mice had lost 10% of their initial body weights,
while the control mice had insignificant weight loss (< 1%). The
rats, too, had moderate weight loss (500-h exposed, 9%; 1000-h

Table 2. Pulmonary Alveolar Macrophage Function
Four to Eight Days After In Vivo Exposures to Effluents[a]

| | | | Staphylococcus aureus | |
| | | Duration of | Phagocytized | Phagocytized but |
Experiment	Treatment	Exposure (h)	& Killed (%)	Not Killed (%)
I. Mice	control	500	98.7	0.2
	exposed	500	99.3	0.0
Rats	control	500	95.1	1.5
	exposed	500	93.8	1.8
II. Mice	control	500	99.9	0.0
	exposed	500	98.2	0.1
	exposed	1000	99.9	0.0
Rats	control	500	93.8	1.5
	exposed	500	97.6	0.3
	exposed	1000	98.1	0.7

[a]Pooled alveolar macrophages from 12 mice/group; results are
expressed as the percent of the total challenge dose of bacteria.

Table 3. Hemopoietic Colony-forming Units (CFUs) in Mice
Four to Eight Days After Termination of Exposures to FBC Effluents[a]

Treatment	Duration of Exposure (h)	$CFUs/10^5$ [b] Nucleated Cells
Control	--	46.5 ± 3.65
Exposed	500	78.0 ± 4.16
	1000	41.6 ± 2.50

[a]For each determination, pooled femoral bone marrow cells (2.0 x
10^4/recipient) from two donors were injected into 15 irradiated
recipient mice.
[b]Mean number of spleen colonies per mouse from 15 mice (± 1
standard error).

exposed, 13%), with a negligible (< 2%) weight loss in the
controls.

CONCLUSIONS

In the previous study (Kirchner et al., 1980), exposure of
the rodents to whole (gaseous and particulate) FBC effluent for
500 h significantly impaired the PAMs' ability to phagocytize and
kill a challenge dose of <u>Staphylococcus aureus</u>. However, after an
exposure of 1000 h, the exposed animals' macrophages appeared to
accommodate the exposure conditions and were again able to engulf
and kill the bacteria at the same level as those of the control
animals. In the present studies with FBC effluent gas only, the
bacterial cell killing was unimpaired in either the 500- or 1000-h
exposure groups, indicating that the temporary reduction in
macrophage function was probably due solely to the high level of
particulate matter.

The spleen colony assay results indicated that exposure to the
gaseous effluent for 500 h acted as a strong proliferation and
differentiation stimulus to the pluripotential stem cells. The
numerous colonies represent large numbers of proliferating
pluripotent stem cells; the large size of the nodules indicates the
rapidity of the proliferative response of these cells to the
stimulus (McCulloch, 1970). Prolonged exposures (1000 h) resulted
in CFUs similar to those of unexposed mice in both colony size and
numbers of colonies. These results differed from those observed in
the previous work with particulate effluents (Kirchner et al.,
1980), in which there appeared to be a cumulative toxic effect
(i.e., decrease in CFUs) in the longer exposures, with little, if
any, recovery or adaptation to the effluent exposures. The
observed decrease in CFUs from 500 to 1000 h of exposure may have
been due to a depletion of the stem cell compartment resulting from
the strong persistent stimulus (500-h exposure) for stem cell
proliferation (Till, 1976). The initial demand for rapid
proliferation may have reduced the pluripotent stem cell pool,
perhaps simultaneously decreasing the committed stem cell pool.
The systemic stress on the hematopoietic stem cell compartment may
ultimately impair one or more of the functional end cells. This
possible alteration of hemopoiesis is currently under
investigation.

ACKNOWLEDGMENTS

The authors would like to thank V. Ann Ludeman for her
technical assistance and Dr. T.E. Fritz for conducting the
histopathologic diagnosis. This work was supported by the U.S.
Department of Energy under contract no. W-31-109-ENG-38.

REFERENCES

ACGIH, American Conference of Governmental and Industrial
    Hygienists. 1977. Threshold Limit Values for Chemical
    Substances in Workroom Air Adopted by ACGIH for 1977.

Brennan, P.C., F.R. Kirchner, J.O. Hutchens, and W.P. Norris.
    1980. The effect of reaerosolized fly ash from an atmospheric
    fluidized bed combustor on murine alveolar macrophages. In:
    Pulmonary Toxicology of Respirable Particles. C.L. Sanders,
    F.T. Cross, G.E. Dagle, and J.T. Mahaffey, eds. CONF-791002.
    U.S. Department of Energy: Washington, DC. pp. 279-288.

Kirchner, F.R., J.O. Hutchens, P.C. Brennan, D.A. Haugen, H.E.
    Kubitschek, D.M. Buchholz, R. Kumar, K.M. Myles, and W.P.
    Norris. 1980. Mammalian responses following exposure
    to the total diluted effluent from fluidized bed combustion of
    coal. In: Pulmonary Toxicology of Respirable Particles.
    C.L. Sanders, F.T. Cross, G.E. Dagle, and J.A. Mahaffey, eds.
    CONF-791002. U.S. Department of Energy: Washington, DC.
    pp. 29-46.

McCulloch, E.A. 1970. Control of hematopoiesis at the cellular
    level. In: Regulation of Hematopoiesis, Volume 1. A.S.
    Gordon, ed. Appleton-Century-Crofts: New York. pp. 133-
    159.

Tan, J.S., C. Watanahunakorn, and J.P. Phair. 1971. A modified
    assay of neutrophil function: use of lysostaphin to
    differentiate defective phagocytosis from impaired
    intracellular killing. J. Lab. Clin. Med. 78:316-322.

Till, J.E. 1976. Regulation of hemopoietic stem cells. In: Stem
    Cells of Renewing Cell Populations. A.B. Cairnie, P.K. Lala,
    and D.G. Osmond, eds. Academic Press: New York. pp. 143-155.

Till, J.E., and E.A. McCulloch. 1961. A direct measurement of the
    radiation sensitivity of normal mouse bone marrow cells.
    Radiat. Res. 14:213-222.

IN VITRO AND *IN VIVO* EVALUATION OF POTENTIAL TOXICITY OF INDUSTRIAL PARTICLES

Catherine Aranyi and Jeannie Bradof
Illinois Institute of Technology Research Institute
Chicago, Illinois

Donald E. Gardner and Joellen Lewtas Huisingh
Health Effects Research Laboratory
U.S. Environmental Protection Agency
Research Triangle Park, North Carolina

INTRODUCTION

Alveolar macrophages (AM) protect the lungs by phagocytosis and digestion of inhaled irritant particles and infectious agents. Reduced activity of the AM system can impair the lung's defensive capacity and increase susceptibility to respiratory disease. Since resistance to infection may be lowered by exposure to an inhalation hazard, changes in the major functional characteristics of AM can be used to monitor environmental stresses in the intact animal. In addition, since these cells can be obtained easily by tracheo-bronchial lavage and maintained in culture, they are frequently used in in vitro cellular toxicology to assess the potential inhalation hazard of various substances.

The advantages of in vitro screening assays in terms of cost and time efficiency are well known. The rabbit alveolar macrophage (RAM) test, a rapid, efficient in vitro assay, has been used extensively by the U.S. Environmental Protection Agency (EPA) and in our laboratories to compare the cytotoxicity of a variety of soluble compounds and particulate materials (Aranyi et al., 1977, 1979; Mahar, 1976; Waters et al., 1974a, b; 1975a, b; 1978). This system is capable of rapid screening and toxicity ranking of test materials and thereby identifies not only the potentially hazardous but also the inert agents. Based on the in vitro results, the number of samples to be studied further in vivo can be reduced considerably.

The purpose of these studies was to determine whether in vitro exposure of AM to various complex industrial particles produced the same relative toxicity ranking as inhalation exposure to aerosols of these particles in vivo. Our main objective was to establish the correlation between the effects of in vitro and in vivo exposures and to determine whether inhalation hazards could be predicted on the basis of in vitro screening assays.

MATERIALS AND METHODS

Particulate Stationary Source Samples

The particles were collected as baghouse samples, from electrostatic precipitators (ESP) or by cyclone sampling train from various stationary point sources, including four coal-fired power plants (three conventional and one fluidized-bed process), a steel foundry, and an aluminum and a copper smelter. Because of the large sample requirement for the aerosol inhalation exposure studies, it was not possible to provide sufficient amounts of samples collected by special emission source samplers located beyond the normal in-plant control devices. Thus, the samples reported here were not true emission or effluent samples and were not necessarily similar in composition or toxicity to emission samples.

Foundry, smelter, and selected coal combustion samples (fly ash nos. 1, 2, and 3) were provided and inorganic analysis of these particulate materials was performed by EPA Industrial Environmental Research Laboratory (IERL) at Research Triangle Park, N.C. The conventional power plant coal fly ashes (nos. 1, 3, and 4) originated from high-sulfur-containing Eastern coals, and nos. 1 and 3 were collected as baghouse samples. Fly ash no. 4, a < 3.3-μm aerodynamic size fraction of an ESP hopper fly ash, has been studied in depth by Griest and Guerin (1979) and contains arsenic, barium, cobalt, chromium, copper, lead, strontium, and zinc as some of the more prevalent trace metals, as well as traces of polycyclic aromatic hydrocarbons. The no. 2 fly ash was collected at 840°C (1550°F) from the second cyclone of a calcium-oxide-fluidized-bed coal combustion process in an experimental demonstration plant. Because of the high collection temperature, residual organic compounds were < 0.1%. Major inorganic components identified by spark source mass spectroscopy were calcium, magnesium, iron, silicon, and sulfur. The steel foundry particles were collected as a baghouse sample. No extractable organics were found, and the major inorganic constituents were iron, silicon, magnesium, and zinc. The copper smelter dust was collected by ESP at 200°C (400°F). No organic components have been identified, but trace metals such as lead, arsenic, copper, iron, antimony, and zinc were found in high concentrations. The aluminum smelter sample, collected as a baghouse dust, contained 3 mg/g of extractable organic compounds,

which were mostly fused aromatics. No information on the inorganic
constituents was available.

     All collected samples were air-classified, and only the
particles in the size-fraction of < 3 μm were used in the
experiments. Since our previous studies (Aranyi et al., 1979)
demonstrated that the smallest particles had the most deleterious
effects on AM in vitro, we wanted to explore the in vivo
correlation for this size range in particular.

In Vitro Methods

     The in vitro experimental procedures have been described
previously in more detail (Aranyi et al., 1979). Briefly, AM
obtained from rabbits by tracheobronchial lavage were centrifuged
and washed in Hanks' Balanced Salt Solution (HBSS), and total and
differential cell counts and viability were determined. AM
suspensions and separate particle suspensions at twice the
projected exposure concentrations were prepared in Medium 199/HBSS
supplemented with serum and antibiotics, and equal volumes of the
two suspensions were mixed. The final concentration of AM in the
test suspensions was maintained constant at 10^6 AM/ml; the
concentration of the particles was increased stepwise to 1000 μg/ml
to attain a dose-related change in viability from approximately 20
to 90%. The test suspensions were incubated in wells of disposable
plastic cluster dishes placed on a rocker platform for 20 h at 37°C
in a humidified 4% CO_2 atmosphere. Immediately after incubation,
percent viability was determined by dye exclusion. The test
suspensions were subsequently washed, centrifuged, and resuspended
in HBSS before total cellular protein and adenosine triphosphate
(ATP) levels were monitored. An aliquot was treated with sodium
deoxycholate, the resulting cell lysate was centrifuged at 10,000 x
g, and the supernatant was used for the Lowry protein assay. A
second aliquot was used for ATP determination; after extraction
from the cells with dimethylsulfoxide (DMSO), ATP was determined
through the luciferin-luciferase reaction in a Dupont 760
Luminescence Biometer.

In Vivo Methods

     Inhalation exposures. All inhalation exposure facilities are
located in rooms maintained under negative pressure relative to
outside areas. The animal exposure chambers as well as the aerosol
generation and dilution systems are housed in second chamber
enclosures (safety cabinet-glove box) that permit safe handling of
the animals and maintenance and monitoring of the experimental
environment.

Animals were exposed to the experimental environment in Plexiglas chambers of various sizes (87 to 476 liters) that can hold up to 240 mice in individual compartments of wire cages. The compressed air supplied to the exposure chambers for dilution or dissemination of the test agents was passed through appropriate filter systems to dry the air and remove all traces of oil and particulates. Ambient temperature and humidity were maintained throughout the exposures by providing an adequate flow of conditioned air.

Aerosols of the coal fly ash and the copper smelter dust were generated with a Wright Dust Feeder. Mass concentration was monitored optically, using a Phoenix JM7000 Aerosol Smoke and Dust Photometer, and gravimetrically, by weighing particles collected on membrane filters on an analytical microbalance and measuring the air volumes sampled by a gas meter for the corresponding time intervals. Aerosol particle size distribution was monitored by an ASAS-300A Active Scattering Aerosol Spectrometer (Particle Measuring Systems, Inc.).

The infectious aerosol was generated with a Model 841 DeVilbiss nebulizer using Streptococcus pyogenes (Lancefield Group C), grown in Todd Hewitt Broth from stock cultures obtained from colonies isolated from the hearts of infected mice. For the bactericidal activity assay, aerosols of ^{35}S-labeled K. pneumoniae were disseminated with a Retec X-70 disposable nebulizer. Aerosol particle size produced by both nebulizers was between 1 and 5 µm MMD.

Radiolabeled K. pneumoniae were cultured in modified Anderson's medium in which the sulfate requirement of the bacteria was provided by ^{35}S-labeled sodium sulfate. Before aerosolization, the bacteria were repeatedly washed and centrifuged for removal of unattached radiolabel. Bacterial counts were determined in a Petroff-Hauser counting chamber by dark field microscopy and also by culture plate technique. Radioactive counts were measured with a Mark III Liquid Scintillation System Model 6880 (Searle Analytic Inc.).

Health-effect assays. Groups of 4- to 6-week-old female CD_1 mice (Charles River Laboratories) were exposed to either aerosols of the test particles at 2 mg/m^3 mass concentration or to filtered air for 3 h/day, 5 days/week, for one, two, or four weeks. Health-effect assays followed within one hour of the last exposure.

Pulmonary free cells were obtained from mice by tracheo-bronchial lavage. Total cell counts were made in a hemocytometer, and differential counts were made of smears of cells fixed in methanol and stained with Wright's stain. Viability was determined

by dye exclusion. Cellular ATP concentration in the lavaged cells
was monitored, using a Dupont 760 Luminescence Biometer.

Pulmonary bacterial activity was determined in the lungs of
individual animals through a modification of the method of Green
and Goldstein (1966), whereby mice exposed to particles and control
mice exposed to filtered air are challenged with radiolabeled live
bacteria. The ratio of the viable bacterial count to the
radioactive count in each animal's lung gives the rate at which
bacteria are destroyed by the lung in a given time after infection.
The streptococcus infectivity model (Ehrlich, 1966; Gardner, 1979)
was used to determine the effect of particle exposure on
susceptibility to respiratory infection. Groups of exposed and
control mice were simultaneously challenged with streptococcus
aerosol. After the challenge, the mice were removed to a clean-air
isolation room, and mortality rate and survival time were recorded
over a 14-day holding period.

RESULTS

In Vitro Tests

The effects of in vitro incubation of the various particulate
samples on rabbit AM were monitored in dose-response experiments,
in which cell viability (percent), total protein (micrograms,
percent of control), and ATP (femtograms ATP per microgram protein,
percent of control) were measured. The means and standard errors
for these parameters were calculated from six or nine replicates at
each concentration, with triplicate assay determinations of each
replicate. When regression analysis was applied to these data,
highly significant negative linear dose-response relationships were
observed for each parameter in all samples (P was generally
< 0.001 and occasionally < 0.01).

The estimated concentrations that were required to reduce the
experimental parameters to 50% of the control responses were
calculated from the linear regressions. From these data, the
samples could be ranked by relative toxicity, as shown in Table 1.
Two of the samples examined, those collected from the copper and
aluminum smelters, were highly cytotoxic; the copper smelter sample
was the more toxic of the two. Three of the power plant coal fly
ashes (nos. 4, 3, and 2) showed intermediate-to-low cytotoxicity,
and the last two samples had very little effect on AM. In fact,
the EC_{50} values for fly ash no. 1 and the steel foundry particles
could be obtained only by extrapolation above the tested
concentration range for all three experimental parameters; in the
cases of fly ashes nos. 2 and 3, this was necessary only for
viability.

Table 1. Concentration of Particles Required to Reduce Alveolar
Macrophage Viability, Total Protein Content, and
ATP Levels to 50% (EC_{50})[a]

	EC_{50} (µg/ml)		
Sample	Viability	Total Protein	ATP/Protein
Steel foundry particles	>1000[b]	>1000[b]	>1000[b]
Coal fly ash no. 1 (conventional)	>1000[b]	>1000[b]	>1000[b]
Coal fly ash no. 2 (fluidized-bed)	>1000[b]	952	537
Coal fly ash no. 3 (conventional)	>1000[b]	930	553
Coal fly ash no. 4 (conventional)	949	856	445
Aluminum smelter dust	114	139	60
Copper smelter dust	11	6	5

[a]Estimated by linear regression analysis for experimental
parameters expressed as: viability, %; total protein (µg), % of
control; and ATP/protein (fg/µg), % of control.
[b]Highest concentration tested.

The samples collected from the copper and the aluminum
smelters were not only much more toxic than the other samples, but
also behaved differently. At exposure concentrations between 250
and 1000 µg/ml (used for all other test particles), they produced
initial large decreases in the experimental parameters that
remained fairly constant over this entire range. Only at
concentrations below 250 µg/ml could a monotonic decrease in the
parameters be observed with increasing exposure concentration.

The high cytotoxicity of the copper and aluminum smelter
samples at low concentrations and the absence of a dose-dependent
response at higher concentrations, suggested that soluble compounds
released continuously into the medium during the incubation period
produced these results, in addition to the particles per se. To
substantiate this hypothesis, the test particles were preincubated
without AM in the maintenance medium at the highest exposure
concentration (1000 µg/ml), under similar conditions (20 h at 37°C)
to those used in the cytotoxicity experiments. After incubation,
the particles were removed from the media by ultracentrifugation

and Millipore filtration (0.22-μm pore size). These filtered
culture media were subsequently incubated for 20 h at 37°C with AM,
and the viability of the cells was compared with that of the
unexposed control AM. The viability of the cells exposed to the
supernatant fractions from the copper and aluminum smelter dust
samples was substantially lower than that of the unexposed control
cells, demonstrating that soluble cytotoxic components were
released from these samples during incubation (see Table 2). No
such difference in viability was found for coal fly ash and steel
foundry particles, indicating that if any compounds were
solubilized from the particles of these samples, the amounts were
not toxic to the AM.

Table 2. Examination of Test Particles for Soluble Cytotoxic
Components by Alveolar Macrophage Viability[a]

Sample	Viability (%)
Control	96.0 to 98.0
Steel foundry particles	97.5
Coal fly ash no. 1 (conventional)	98.3
Coal fly ash no. 2 (fluidized-bed)	97.3
Coal fly ash no. 3 (conventional)	96.7
Coal fly ash no. 4 (conventional)	97.0
Aluminum smelter dust	46.8
Copper smelter dust	65.1

[a]Determined after exposure at 37°C for 20 h to the particle-free
medium separated from the preincubated particles. For details,
see text.

Spark-source mass spectroscopic analysis by EPA of the copper
smelter particles showed that major trace metal·constituents with
concentrations ranging up to 20% by weight were lead, arsenic,
copper, iron, antimony, and zinc, with the arsenic as high as 13%,
present mostly in water-soluble form. Thus, soluble arsenic could
be one of the components responsible for the cytotoxicity of the
copper smelter particles. No information is now available on the
solubility of the other trace metals in this sample, nor on similar
properties of the aluminum smelter dust.

Thus, the in vitro RAM test has enabled us to evaluate the
relative cytotoxicity of seven industrial particulate samples. The
data have demonstrated that five of these--a foundry particulate

and fly ash samples from three conventional combustion processes
and one fluidized-bed process--had low to intermediate effects on
AM. The samples collected from a copper and an aluminum smelter,
however, were much more toxic than all others, and the copper
smelter sample was the more toxic of the two. In the case of the
two smelter samples, we found that in addition to the particles per
se, cytotoxic soluble components released during incubation
contributed to the total toxicity to AM.

In Vivo Tests

The next step, and a major objective of these studies, was to
confirm the in vitro evaluation by demonstrating parallel effects
in in vivo assessments of the inhalation hazard of these samples in
the intact animal. Female CD_1 mice were exposed, as described, to
aerosols of the copper smelter dust and the fluidized-bed coal fly
ash (fly ash no. 2), particles that had shown very high and low
cytotoxicity, respectively, in vitro. The means and standard
deviations of aerosol mass concentration were 2033 ± 153 $\mu g/m^3$ for
the copper smelter dust and 2043 ± 308 $\mu g/m^3$ for the fly ash. Both
aerosols had log-normal size distributions, with count median
diameters and σ_g's of 0.225 μm and 2.9 for the copper smelter dust
and 0.134 μm and 2.1 for the fly ash.

The effects of inhalation of the particles were evaluated
after 5, 10, and 20 exposures by examining changes in pulmonary
cellular lavage, in susceptibility to respiratory streptococcal
infection, and in pulmonary bactericidal activity to inhaled
radiolabeled K. pneumoniae. Results are summarized in Figures 1,
2, and 3.

Total cell counts and ATP levels (expressed as percent of the
control responses; see Figure 1) generally did not change
significantly or exhibit any trend related to the number of aerosol
exposures. Similarly, differential cell counts and percent
viability of the lavaged cells (not shown) were not affected by the
exposures. However, as seen in Figure 2, the percent mortality
after streptococcus inhalation challenge was greater in aerosol-
exposed than in control mice, for 5, 10, and 20 daily 3-h exposures
to copper smelter dust aerosols (2 mg/m^3). No significant changes
were observed for any of the exposures to the coal fly ash. Mean
survival time (not shown in the figures) was significantly lower
than that of the control groups only following inhalation of the
copper smelter dust (i.e., treatments that also significantly
increased the mortality rate). The percent of bactericidal
activity in response to inhaled radiolabeled K. pneumoniae was
significantly less for exposed than for control mice (Figure 3),
for 5, 10, or 20 daily 3-h exposures to the copper smelter dust
aerosol. Similar doses of the coal fly ash aerosol had no effect.

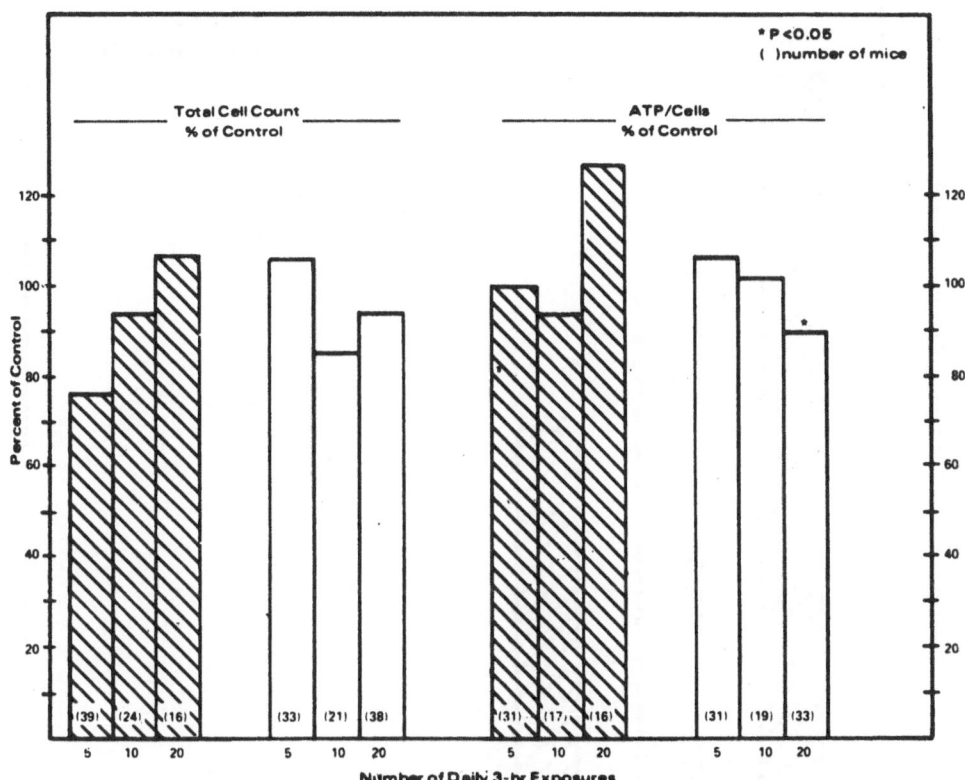

Figure 1. Changes in pulmonary cellular lavage following multiple
           daily 3-h aerosol exposures of mice to 2 mg/m^3 of copper
           smelter dust (shaded bars) or fluidized-bed coal fly ash
           (unshaded bars).

Thus, these data clearly demonstrate that the copper smelter dust
was not only more cytotoxic in vitro to AM but also more
deleterious in inhalation exposures in vivo than was the
fluidized-bed coal fly ash.

CONCLUSIONS

     In assays of the < 3 µm size fraction of a series of
stationary point-source samples by the RAM test, generally
low-to-intermediate cytotoxicity was found for samples collected
from a foundry and from several coal-fired power plants. However,
particulate samples from an aluminum and a copper smelter were
highly toxic to AM, as measured by viability and total cellular
protein and ATP levels. In contrast to all others, the two smelter
samples also contained soluble components that contributed
substantially to their overall in vitro cytotoxicity.

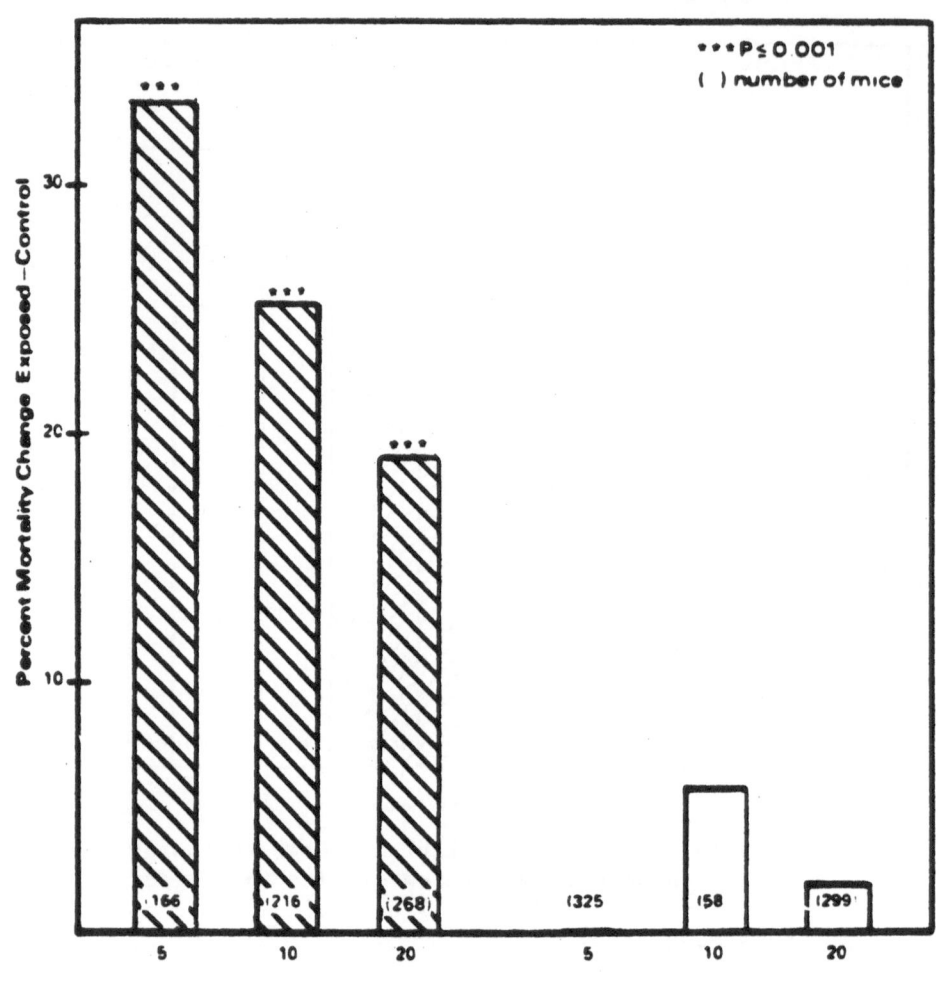

Figure 2.   Excess mortality from streptococcus aerosol infection in
            exposed mice following multiple daily 3-h aerosol
            exposures to 2 mg/m^3 of copper smelter dust (shaded
            bars) or fluidized-bed coal fly ash (unshaded bars).

        The copper smelter particles and the fluidized-bed coal fly
ash, chosen on the basis of their respectively high and low in
vitro cytotoxicity, were used in aerosol exposures to examine their
effects in vivo on pulmonary free cells, bactericidal activity, and
resistance to respiratory infection in mice. The results obtained
after multiple daily 3-h exposures to 2 mg/m^3 of these aerosols
correlated well with the in vitro data; inhalation of the aerosols

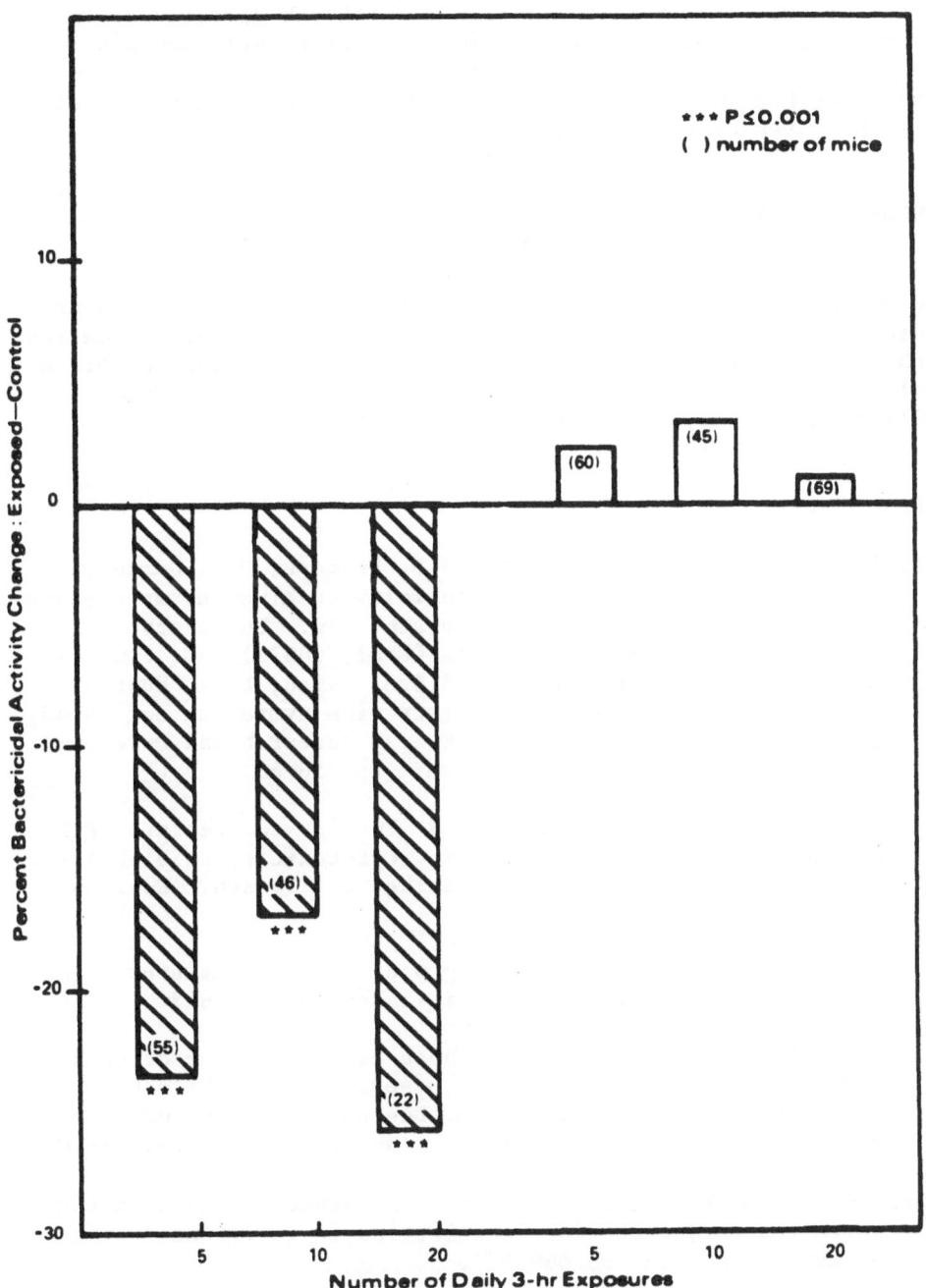

Figure 3. Percent change in bactericidal activity in response to inhaled K. pneumoniae in exposed mice following multiple daily 3-h aerosol exposures to 2 mg/m³ of copper smelter dust (shaded bars) or fluidized-bed coal fly ash (unshaded area).

of the copper smelter dust produced significant differences from
the controls in more of the experimental parameters than did
aerosols of the coal fly ash sample. Thus, the validity of
inhalation hazard prediction on the basis of an in vitro screening
assay has been demonstrated.

ACKNOWLEDGMENTS

These studies were supported by the U.S. Environmental
Protection Agency under grant no. R80514101/02. The authors are
indebted to Mr. J. Hingeveld and Mr. W. O'Shea for their excellent
technical assistance. The critical review and comments of Dr. R.
Ehrlich and Dr. J. Fenters of IITRI and Dr. R. Merrill from
EPA/IERL are highly appreciated.

REFERENCES

Aranyi, C., S. Andres, R. Ehrlich, J.D. Fenters, D.E. Gardner,
    and M.D. Waters. 1977. Cytotoxicity to alveolar macrophages
    of metal oxides absorbed on fly ash. In: Pulmonary
    Macrophage and Epithelial Cells (Conf-760972). C. Sanders,
    R.P. Schneider, G.E. Dagle, and H.A. Ragan, eds. Energy
    Research and Development Administration Symposium Series 43,
    Technical Information Center, Energy Research and Development
    Administration: Springfield, VA. pp. 58-65.

Aranyi, C., F.J. Miller, S. Andres, R. Ehrlich, J. Fenters, D.E.
    Gardner, and M.D. Waters. 1979. Cytotoxicity to alveolar
    macrophages of trace metals absorbed on fly ash. Environ.
    Res. 20:14-23.

Ehrlich, R. 1966. Effect of nitrogen dioxide on resistance to
    respiratory infection. Bacteriol. Rev. 30:604-613.

Gardner, D.E. 1979. Alteration in host-bacteria interaction by
    environmental chemicals. In: Assessing Toxic Effects of
    Environmental Pollutants. S.D. Lee and J. Bryan, eds. Ann
    Arbor Science Publishers, Inc.: Ann Arbor, MI. pp. 87-103.

Green, G.M., and E. Goldstein. 1966. A method for quantitating
    intrapulmonary bacterial inactivation in individual animals.
    J. Lab. Clin. Med. 68:669-677.

Griest, W.H., and M.R. Guerin. 1979. Identification and
    quantification of polynuclear organic matter on particulates
    from coal-fired power plant. EPRI EA-1092/DOE No. RTS 77-58
    (Interim Report), Dept. of Energy and Electric Power Res.:
    Palo Alto, CA.

Mahar, H. 1976. Evaluation of Selected Methods for Chemical and
    Biological Testing of Industrial Particulate Emissions.
    EPA 600/2-76-137. U.S. Environmental Protection Agency:
    Research Triangle Park, NC.

Waters, M.D., D.E. Gardner, C. Aranyi, and D.L. Coffin. 1975a.
    Metal toxicity for rabbit alveolar macrophages in vitro.
    Environ. Res. 9:32-47.

Waters, M.D., D.E. Gardner, and D.L. Coffin. 1974a. Cytotoxic
    effects of vanadium on rabbit alveolar macrophages in vitro.
    Toxicol. Appl. Pharmacol. 28:253-263.

Waters, M.D., J.L. Huisingh, and N.E. Garrett. 1978. The cellular
    toxicity of complex environmental mixtures. In: Application
    of Short-term Bioassays in the Fractionation and Analysis of
    Complex Environmental Mixtures. M.D. Waters, S. Nesnow, J.L.
    Huisingh, S. Sandhu, and L. Claxton, eds. Plenum Press: New
    York. pp. 125-167.

Waters, M.D., T.O. Vaughan, J.A. Campbell, F.J. Miller, and D.L.
    Coffin. 1974b. Screening studies on metallic salts using
    the rabbit alveolar macrophage. In Vitro 10:342-343.

Waters, M.D., T.O. Vaughan, J.A. Campbell, A.G. Stead, and D.L.
    Coffin. 1975b. Adenosine triphosphate concentration and
    phagocytic activity in rabbit alveolar macrophages exposed to
    divalent cations. J. Reticuoloendothel. Soc. 18:296.

MUTAGENICITY AND CARCINOGENICITY OF A RECENTLY CHARACTERIZED CARBON BLACK ADSORBATE: CYCLOPENTA(CD) PYRENE

Avram Gold
Department of Environmental Sciences Engineering
University of North Carolina
Chapel Hill, North Carolina

Eric Eisenstadt   Stephen Nesnow, Martha M. Moore, Helen Garland,
School of Public Health   Gaynelle Curtis, Barry Howard, and Deloris Graham
Harvard University   Health Effects Research Laboratory
Boston, Massachusetts   U.S. Environmental Protection Agency
                          Research Triangle Park, North Carolina

INTRODUCTION

Polycyclic aromatic hydrocarbons (PAH) are widespread environmental contaminants that may be metabolically activated to mutagenic or carcinogenic derivatives (Particulate Polycyclic Organic Matter, 1972; Gelboin and Ts'o, 1978). Intensive research indicates that PAH are promutagens or procarcinogens containing the bay region geometric feature (Jerina et al., 1978). Studies show that the biological activity of these compounds results from metabolism to bay region diol-epoxides, which are capable of forming covalent adducts at nucleophilic sites within DNA.

Cyclopenta(cd)pyrene (CPP, I)(see Figure 1), a non-bay region PAH, was recently characterized and shown to be highly mutagenic in the Salmonella typhimurium assay (Eisenstadt and Gold, 1978). CPP was initially identified in extracts of furnace black as the major contributor to the high mutagenic activity of the extracts. Because of its unique structure and wide environmental distribution as a component of soots (Lee et al., 1977; Grimmer, 1977; Kaden et al., 1979), carbon blacks (Wallcave et al., 1975; Gold, 1975), and cigarette smoke (Snook et al., 1977), CPP has evoked considerable interest (Ittah and Jerina, 1978; Konieczny and Harvey, 1979; Ruehle et al., 1979). The 3,4-oxide of CPP, predicted as an ultimate mutagenic metabolite, has been synthesized and shown to be a powerful direct-acting mutagen to S. typhimurium. Confirmation of the 3,4-oxide as a primary metabolite and ultimate mutagen requires that the expected enzymatic hydration product, trans CCP 3,4-dihydrodiol, be identified as a metabolic product.

445

Figure 1.   Structure of cyclopenta(cd)pyrene (CPP, I).

     Response of different assay systems to treatment with specific
mutagens may vary (Maher et al., 1978), and for this reason, the
mutagenicity of CPP and its 3,4-oxide in mammalian cells as well
as in microbial systems is of interest.

     This study reports identification of trans-CPP 3,4-dihydrodiol
as the major CPP metabolite of both 3-methylcholanthrene- (3-MC)
and Aroclor-1254-induced rat liver microsomes. CPP and its
3,4-oxide were tested for mutagenicity in the L5178Y mouse lymphoma
system and for cell transformation in the C3H10T1/2CL8 mouse embryo
fibroblast system. The results are discussed in relation to the
proposal that CPP 3,4-oxide may be an ultimate mutagenic or
carcinogenic metabolite of CPP.

MATERIALS AND METHODS

Chemicals

     CPP, CPP 3,4-oxide, and 4-oxo-CPP were obtained as previously
described (Gold et al., 1978, 1979). Tritiated CPP ([H^3]CCP) was
prepared by catalytic exchange labeling (New England Nuclear,
Boston, MA) and was purified prior to use by chromatography over
silica.

Carbon Black Analysis

     Semireinforcing furnace black was extracted with methylene
chloride (CH_2Cl_2) and fractionated by standard procedures (Gold,
1975; Rosen and Middleton, 1955). A 120-mg aliquot of the PAH

fraction was further separated on a size B Lo Bar silica column (Merck) with an initial eluent of 4% CH_2Cl_2 in hexane changed to 5% CH_2Cl_2 in hexane after 500 ml. The separation was followed by ultraviolet (UV) detection, and individual peaks were collected.

Ames assays were performed with strain TA100 as described by Ames et al. (1975). Dose-response curves were obtained from 1, 2, 4, 10, and 20 µg of test mixture per plate, and mutagenic potency (revertants per microgram) was determined from the slope of the linear portion of the curve. Metabolic activation was supplied by 0.5 ml S-9 from Aroclor-treated rats.

Metabolism Studies

Rat liver microsomes and S-9 were prepared from Aroclor-1254-treated male Sprague-Dawley rats weighing 150 to 200 g, according to the procedures of Ames et al. (1975), except that the microsomal pellet was resuspended in buffer and centrifuged a second time at 100,000 x g for 60 min. Microsomes were stored frozen at -80°C until used.

The concentrations of ingredients in the 1-ml microsomal metabolism mixture were: 50 mM Tris-hydrochloride, pH 7.4; 3 mM magnesium chloride ($MgCl_2$); 0.8 mM NADP; 5 mM glucose-6-phosphate; 0.4 units glucose-6-phosphate dehydrogenase; 0.89 mM [G-^{3}H]CPP (specific activity, 5 x 10^4 dpm/nmol) and 0.2 mg microsomal protein. The reaction was started by first adding [^{3}H]CPP to the microsomal mixture at 4°C and then shaking the mixture at 37°C for the indicated times. One volume of acetone was added to stop the reaction. Two volumes of ethyl acetate were added and the mixture was shaken or vortexed vigorously. The ethyl acetate-acetone phase was removed, dried with anhydrous sodium sulfate (Na_2SO_4) and evaporated to dryness under nitrogen gas. The residue was dissolved in a small volume of methanol and subjected to high performance liquid chromatography (HPLC) analysis. Ninety to ninety-five percent of the input radioactivity was recovered in the organic extract.

Aroclor-1254- and 3-MC-induced enzymes produced similar metabolite profiles (Figure 2), and preparative work was done with 3-MC-induced S-9 activation using 50- to 100-ml reaction volumes.

HPLC separations were performed on a Perkin-Elmer Series II liquid chromatograph with a Perkin-Elmer 4.6-mm x 25-cm ODS-SilX-I column or 2.6-mm x 25-cm ODS SilX-II column. A UV detector at 254 nm was used to monitor fractions for preparative work. Fractions were collected (0.5 ml) and counted by liquid scintillation for quantitative analysis of [^{3}H]CPP metabolites.

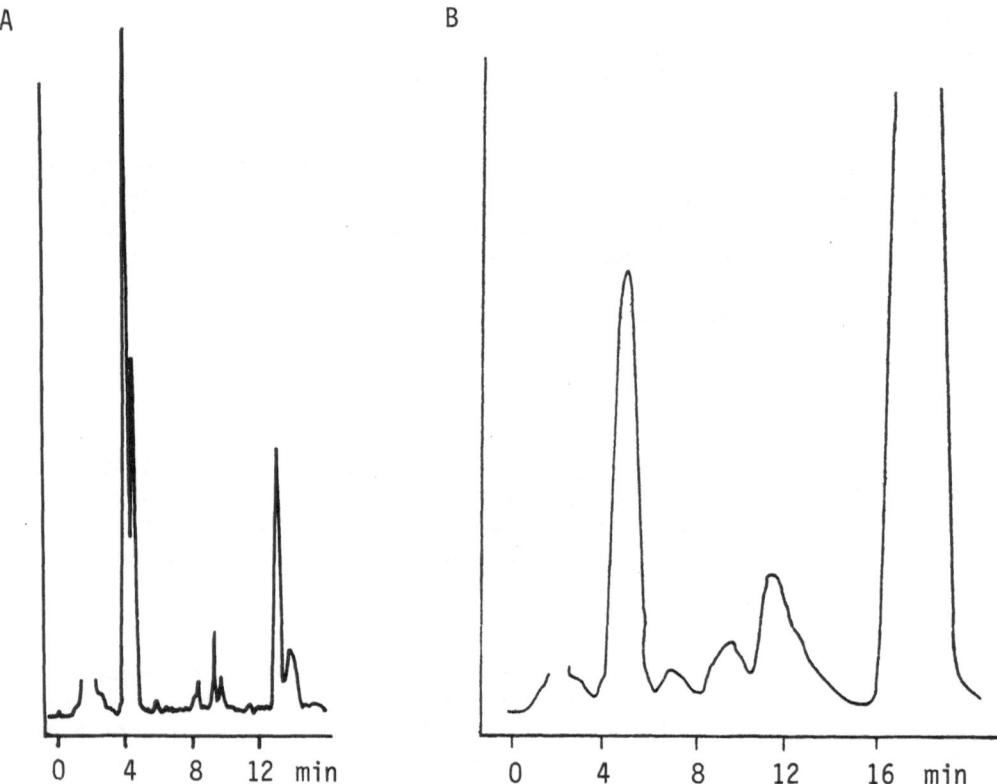

Figure 2.   HPLC trace at 254 nm of CPP metabolites from (A)
            Aroclor-1254-induced rat liver microsomes (ODS SilX-II
            column, 2.6 mm x 25 cm) and (B) 3-MC-induced rat liver
            microsomes (ODS SilX-I column, 4.6 mm x 25 cm).

The same linear gradient was used for all separations: a 1%/min
gradient from 35:65 H_2O:methanol to 100% methanol, at a flow rate
of 1.5 ml/min.

In Vitro Mammalian Mutagenesis Assay

     The $TK^{+/-}$L5178Y mouse lymphoma mutagen assay, developed by
Clive and co-workers to identify mutagens that induce genetic
damage at the thymidine kinase (TK) locus, was performed according
to published methods (Clive et al., 1979; Clive and Spector, 1975),
using Fischer's medium.   Trifluorothymidine (1 µg/ml) was used to
select for thymidine-kinase-deficient mutants.   CPP, dissolved in

to published methods (Clive et al., 1979; Clive and Spector, 1975), using Fischer's medium. Trifluorothymidine (1 µg/ml) was used to select for thymidine-kinase-deficient mutants. CPP, dissolved in dimethylsulfoxide (DMSO), was tested in 3% horse serum at a cell concentration of 0.6 x 10^6 cells/ml, with and without an Aroclor-1254-induced rat hepatic S-9 activation system. The CPP 3,4-oxide (dissolved in DMSO) was tested in 10% horse serum at a cell concentration of 0.6 x 10^6 cells/ml without metabolic activation. The positive controls ethyl methanesulfonate (EMS) and 2-acetylaminofluorene (AAF) were dissolved in saline and DMSO, respectively. Two days were allowed for the expression of newly induced mutants.

Oncogenic Transformation Assay

The mouse embryo fibroblast cell line C3H10T1/2CL8 was derived by Reznikoff et al. (1973; Nesnow and Heidelberger, 1976; Gehly et al., 1979) and donated by Dr. Charles Heidelberger for use in these experiments. Cell cultures were incubated at 37°C in humidified incubators with an atmosphere of 5% carbon dioxide in air. All cultures were grown in Eagle's Basal Medium with Earle's salts and L-glutamine supplemented with 10% heat-inactivated fetal calf serum (Grand Island Biological Co.). The cells were routinely checked for Mycoplasma contamination and found to be Mycoplasma-free.

For transformation, cells were seeded onto 60-mm petri dishes (1000/dish) in 5 ml of medium (12 replicates/treatment) and 24 h later treated with the hydrocarbon dissolved in acetone (25 µl). After an additional 24 h, the medium was removed, and the cells received fresh complete medium containing penicillin (100 units/ml) and streptomycin (50 µg/ml). Medium was changed weekly until the cells reached confluency, whereupon the fetal calf serum concentration was reduced to 5%.

At the end of six weeks, the dishes were washed with 0.9% sodium chloride solution, fixed with methanol, stained with Giemsa, and scored for oncogenic transformation. Three different types of foci have been described after treatment of C3H10T1/2CL8 cells with polycyclic hydrocarbons. Only Type II and Type III foci were scored in this assay, since it has been demonstrated that these foci produce fibrosarcomas upon injection into irradiated C3H mice, 33% and 67% of the time, respectively.

Control cultures received the appropriate solvent and were treated the same way as the exposed cells. Cytotoxicity assays were performed concurrently with the transformation assays, and using the same protocol, except that the dishes were plated with 200 cells (6 replicates/treatment) and stained 10 to 12 days later.

RESULTS AND DISCUSSION

A fractionation scheme was applied to the organic extract of semireinforcing furnace black used in the rubber industry. The Ames test was used in conjuction with the chemical separation to identify mutagens. The data shown in Table 1 indicate that the PAH fraction of the CH_2Cl_2 extract of the semireinforcing furnace black accounted for essentially all of the mutagenicity in the total extract. Table 2 shows the mutagenic activity of the fractions obtained on further chromatographic resolution of the PAH fraction. Specific activity of the total PAH fraction calculated from the activities of the fractions resolved in Table 2 appeared to be about the same as that observed for the total PAH fraction (see Table 1). A similar observation has been reported for kerosene soot (Kaden et al., 1979). It is apparent that benzo(a)pyrene (B[a]P) contributed only slightly to the mutagenic activity of the PAH fraction and therefore to the mutagenic activity of the total organic extract, while CPP was the compound principally responsible for the activity. An estimate of mutagenicity based on B(a)P content would have seriously underestimated the mutagenicity of the sample. This result underscores the inherent danger in using a marker compound such as B(a)P as an index of hazard even for samples with similar pyrogenic origins like soots and carbon blacks.

Table 1. Mutagenic Activity of Fractionated Carbon Black Extracts in Salmonella Strain TA100

Fraction	Weight (mg)	Activity (rev/µg)
Neutral		
saturated hydrocarbon	9	not active
polycyclic aromatic	135	38
polar	27	4
Acidic	49	0.5[a]
Basic	6	not active
Total Extract[b]	346	14

[a]Nonlinear dose-response curve.
[b]From 314 g furnace black.

Table 2. Distribution of Mutagenic Activity in Aromatic Fraction
of Furnace Black Extract with Ames Strain TA100

Fraction Number	Composition	Fraction Weight (mg)	Activity (rev/μg)	Contribution to Total Polycyclic Aromatic Activity (% wt. of Polycyclic Aromatic Fraction x Activity)
1	naphthalene acenaphthylene phenanthrene	10.4	not active	
2	pyrene	70	not active	
3	fluoranthene benzo(ghi)fluoranthene[a] cyclopenta(cd)pyrene	30	93	21
4	benzo(ghi)fluoranthene cyclopenta(cd)pyrene[a]	8.4	380	24
5	cyclopenta(cd)pyrene isomers of molec. wt. 252	2	90	1
6	benzo(a)pyrene benzo(e)pyrene benzo(ghi)perylene isomers of molec. wt. 276 and 300	4.6	29	1

Calculated activity of total polycyclic aromatic fraction (rev/μg)   47

[a]Major component.

Structure-reactivity relationships for CPP were of considerable interest because of its high mutagenicity and unusual structure--containing a fused cyclopenteno ring and lacking a bay region. Hence, an investigation of the metabolism, mutagenicity, and carcinogenicity of CPP was undertaken.

To identify the structure of the metabolic products of CPP, synthetic metabolites were produced by acid-catalyzed decomposition of the 3,4-oxide. This procedure cleanly yielded three products (Figure 3). The two most polar peaks were identified as the 3,4-dihydrodiol isomers based on the mass spectrum and UV spectrum of the mixture. As a result of the rigid planarity of the fused five-membered ring in CPP, the trans diol was expected to be the less polar isomer, because the dipole moments of the hydroxy substituents are largely opposing. The rigid geometry also led to the prediction of greater shielding for the C^3 and C^4 protons of the trans isomer in the nuclear magnetic resonance (NMR) spectrum. The major diol resulting from hydration was the less polar isomer (peak B, Figure 3) and had the more shielded C^3 and C^4 protons in the NMR spectrum (Figure 4a). On this basis, it was assigned the trans configuration. This assignment was consistant with the expectation that the major isomer of acid-catalyzed epoxide hydration would be the trans isomer (Bruice et al., 1976; Keller and Heidelberger, 1976).

Based on its physical-chemical properties, the third and major product of the acid catalyzed reaction was identified as 4-oxo CPP. As illustrated in Figure 5, it was readily distinguished by UV spectroscopy from the known 3-oxo compound (Gold et al., 1978).

The HPLC chromatograms (Figure 2) of the metabolic mixtures produced by 3-MC- and Aroclor-1254-induced rat liver microsomes indicated a single major metabolite accounting for 50% of the products. This metabolite was identified as trans-CPP 3,4-dihydrodiol. Its chromatographic retention time and NMR spectrum (Figure 4b) corresponded to those of the trans isomer from the hydration of the 3,4-oxide. Also, its UV and mass spectrum were consistent with the assigned structure.

The $TK^{+/-}$ L5178Y mouse lymphoma assay has been used to measure the mutagenicity of diverse chemical agents (Clive et al., 1979). Since these cells lack the enzymes necessary to activate promutagens, CPP was tested both with and without S-9 activation, to confirm the activation-dependence of mutagenicity typical of PAH. As expected, CPP was not mutagenic without activation over the concentrations tested (0.75 to 30 µg/ml)(unpublished data). On activation by Aroclor-1254-induced hepatic S-9, CPP was mutagenic, with twice the mutation frequency of the control (Table 3). The 3,4-oxide was tested without activation to determine

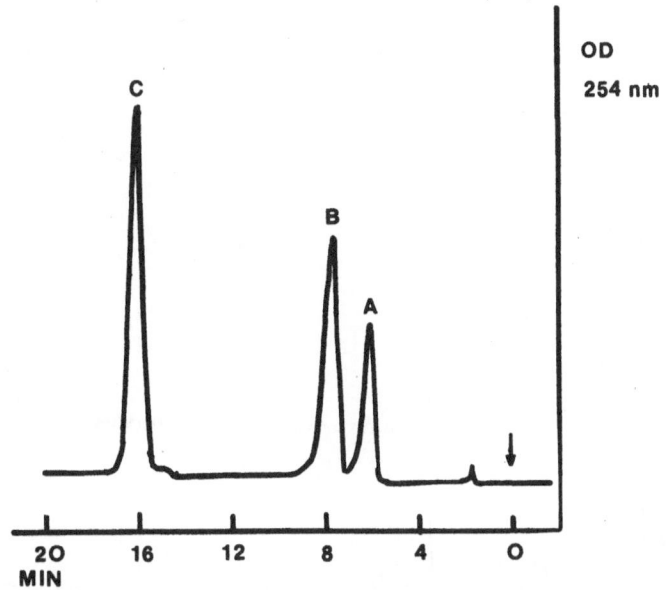

Figure 3.   HPLC trace at 254 nm of acid catalyzed decomposition
            products of CPP 3,4-oxide, peak A, cis-CPP
            3,4-dihydrodiol; peak B, trans-CPP 3,4-dihydrodiol;
            peak C, 4-oxo CPP.

whether it was a direct-acting (and a possible ultimate) mutagen.
It was found to be mutagenic to L5178Y cells over a dose range
similar to that of CPP. Over the 70 to 20% survival range, the
oxide was two- to six-fold more mutagenic than the parent
hydrocarbons.

    The C3H10T1/2CL8 transformation assay responds to a variety of
carcinogens, including PAH (Nesnow and Heidelberger, 1976).
C3H10T1/2CL8 mouse embryo cells contain cytochrome P-450 mixed
function oxidase, epoxide hydrase, and conjugating enzymes
necessary to metabolize and activate or detoxify chemical
carcinogens, especially PAH (Gehly et al, 1979; Nesnow and
Heidelberger, 1976). As shown in Table 2, CPP produced a dose-
related response in the formation of both Type II and Type III
transformed foci. At the highest dose used, every plate contained
at least one Type III focus. In accord with the recent report
(Wood et al., 1980) that CPP is a weaker tumor initiator than
B(a)P, the data in Table 4 indicate that CPP was also less active
than B(a)P in the C3H10T1/2CL8 transformation assay.

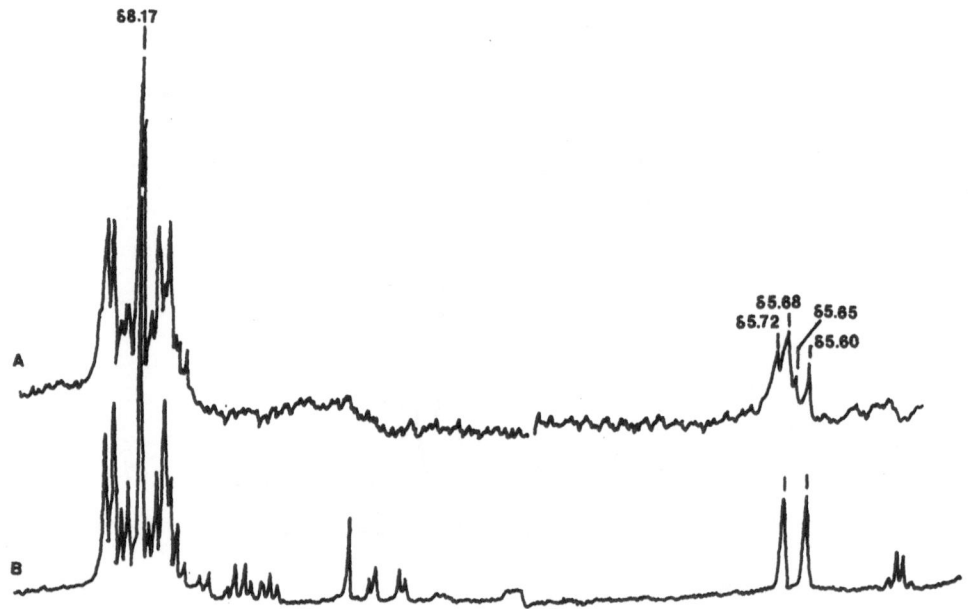

Figure 4. In part a, 270 MHz NMR (acetone-d$_6$) of <u>cis-trans</u>-CPP
3,4-dihydrodiol mixture. In part b, 270 MHz <u>NMR</u>
(acetone-d$_6$) of major CPP metabolite. Underlined
resonances are consistent with CPP 9.0-dihydrodiol
cochromatographing with the 3,4-dihydrodiol.

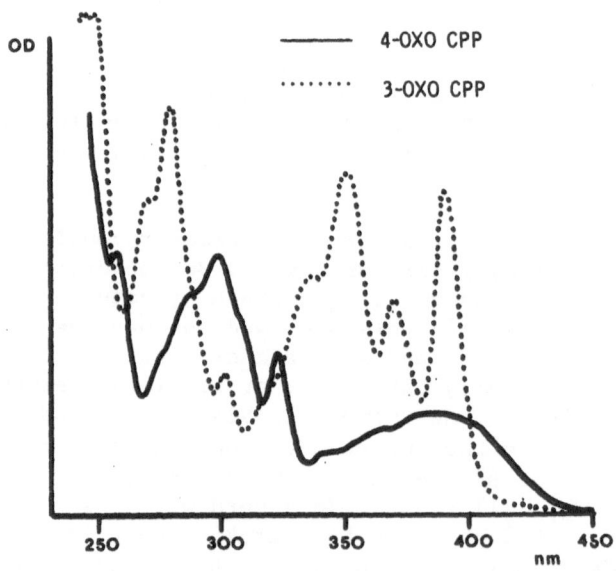

Figure 5. UV-VIS spectrum (CH$_2$Cl$_2$) of 4-oxo and 3-oxo CPP.

Table 3. Mutagenesis of TK$^{+/-}$ L5178Y Mouse Lymphoma Cells
by CPP and CPP 3,4-Oxide

Concentration (µg/ml)	Total Viable Clones	Total Mutant Clones	Total Survival[a] (% of Control)	Mutant Frequency (x 10^6)
CPP (with S-9)				
1.20	439	263	110	120
1.30	365	317	74	172
1.40	436	300	56	138
1.50	407	284	44	139
1.60	399	375	20	188
2-AAF				
30	270	785	31	581
DMSO (1%)	453	178	100	79
CPP 3,4-oxide				
0.60	536	783	71	292
0.70	472	803	63	340
0.84	536	912	72	340
0.96	463	776	58	335
1.08	475	1234	52	519
1.20	476	1065	44	448
1.32	406	1174	32	578
1.44	341	1090	22	640
1.56	360	1423	23	790
1.68	459	1118	36	487
1.80	273	1204	10	882
2.04	333[b]	1214	6	1460
EMS				
500	259	1638	25	1267
DMSO (1%)	623	297	100	95
Untreated Control	558	220	100	79

[a]Survival calculations described by Clive and Spector (1975)
combine both relative growth in suspension and relative plating
efficiency.
[b]Cells cloned at a density of 12 cells/ml; all other cultures
cloned at 6 cells/ml to determine viability.

Table 4. Morphological Transformation of C3H10T1/2CL8 Cells by CPP and CPP 3,4-Oxide

Concentration (μg/ml)	Plating Efficiency (%)	No. Type II Foci /Total Dishes	No. Type III Foci /Total Dishes	Dishes with Type II and III Foci (%)
CPP				
0.01	27	0/12	2/12	17
0.03	28	1/12	1/12	17
0.1	26	0/12	1/12	8
0.3	27	3/12	2/12	25
1.0	29	4/11	7/11	73
3.0	29	6/12	7/12	75
10.0	22	13/12	12/12	100
CPP 3,4-oxide				
0.001	32	2/12	0/12	8
0.003	31	1/11	0/11	9
0.01	30	0/12	0/12	0
0.03	32	1/11	0/11	9
0.1	32	1/12	0/12	8
0.3	32	0/12	2/12	17
1.0	31	0/12	2/12	17
3.0	27	1/12	2/12	25
Acetone (0.5%)[a]	30	2/24	0/24	8
B(a)P[a] (1)	18	38/24	59/24	100

[a]Results from two separate experiments combined.

CPP 3,4-oxide transformed C3H10T1/2CL8 cells at concentrations of 0.3 μg/ml. At 3 μg/ml, it produced both Type II and Type III foci, with virtually no toxicity. The lack of toxicity is remarkable, because structurally related K-region oxides of B(a)P and 3-MC were highly cytotoxic (unpublished data). Although the direct-acting oxide was a more potent mutagen than CPP, it was less active than the parent hydrocarbon in the C3H10T1/2CL8 transformation assay. This was probably due to the C3H10T1/2CL8 cells' ability to detoxify arene oxides to dihydrodiols (Nesnow and Heidelberger, 1976; Gehly et al., 1979). L5178Y mouse lymphoma cells lack this ability (Clive et al., 1979).

The potent direct-acting mutagenicity of the 3,4-oxide in both bacterial and mammalian assays, its ability to transform C3H10T1/2CL8 cells, and the identification of CPP 3,4-dihydrodiol as a major metabolite (Gold et al., 1979) in 3-MC- and Aroclor-induced metabolism of CPP are strong evidence that CPP 3,4-oxide is both a primary product of enzymatic oxidation and an ultimate mutagen or carcinogen. Further support for this conclusion is found in the report that the addition of epoxide hydrase to a purified, reconstituted monooxygenase activating system drastically reduces the mutagenicity of CPP (Wood et al., 1980). The effect of epoxide hydrase can readily be explained if CPP 3,4-oxide is the ultimate mutagen and, like other arene oxides (Gelboin and Ts'o, 1978), is a good substrate for epoxide hydrase. Since the B(a)P bay region diol-epoxides are poor substrates for epoxide hydrase (Gelboin and Ts'o, 1978), CPP would be less mutagenic than B(a)P in the C3H10T1/2CL8 system due to the more rapid deactivation of metabolically-generated CPP 3,4-oxide.

An important distinction between B(a)P diol-epoxides and CPP 3,4-oxide is that the latter is an arene oxide: the epoxidized bond is adjacent to the aromatic nucleus at both termini. The arene oxides tested, 3-MC 11,12-oxide and B(a)P 4,5-oxide, both failed to transform C3H10T1/2CL8 cells (unpublished data). CPP is the first arene oxide reported to transform this cell type.

REFERENCES

Ames, B.N., J. McCann, and E. Yamasaki. 1975. Methods of detecting carcinogens and mutagens with the Salmonella/ mammalian-microsome mutagenicity test. Mutation Res. 31:347-364.

Bruice, P.Y., T.C. Bruice, P.M. Dansette, H.G. Selander, H. Yagi, and D.M. Jerina. 1976. Comparison of the mechanisms of solvolysis and rearrangement of K-region vs. non-K-region arene oxides of phenanthrene; comparison solvolytic rate constants of K-region and non-K-region arene oxides. J. Am. Chem. Soc. 98:2965-2973.

Clive, D., K.O. Johnson, J.F.S. Spector, A.G. Batson, and M.M.M. Brown. 1979. Validation and characterization of the L5178/TK$^{+/-}$ mouse lymphoma mutagen assay system. Mutation Res. 59:61-108.

Clive, D., and J.F.S. Spector. 1975. Laboratory procedures for assessing specific locus mutations at the TK locus in cultures L5178Y mouse lymphoma cells. Mutation Res. 31:17-29.

Eisenstadt, E., and A. Gold. 1978. Cyclopenta(cd)pyrene: a highly mutagenic polycyclic aromatic hydrocarbon. Proc. Nat. Acad. Sci. USA 75:1667-1669.

Gehly, E.B., W.E. Fahl, C.R. Jefcoate, and C. Heidelberger. 1979. The metabolism of benzo(a)pyrene by cytochrome P450 in transformable and non-transformable C3H mouse fibroblasts. J. Biol. Chem. 254:5041-5048.

Gelboin, H.V., and P.O.P. Ts'o, eds. Polycyclic Hydrocarbons and Cancer, Vols. 1 and 2. Academic Press: New York.

Gold, A. 1975. Carbon black adsorbate: separation and identification of a carcinogen and some oxygenated polyaromatics. Anal. Chem. 47:1469-1472.

Gold, A., J. Brewster, and E. Eisenstadt. 1979. Synthesis of cyclopenta(cd)pyrene-3,4-epoxide, the ultimate mutagenic metabolite of the environmental carcinogen cyclopenta(cd)pyrene. J. Chem. Soc. Chem. Commun. 903-904.

Gold, A., J. Schultz, and E. Eisenstadt. 1978. Relative reactivities of pyrene ring positions: cyclopenta(cd)pyrene via an intramolecular Friedel-Crafts acylation. Tetrahedron Lett. 4491-4494.

Gold, A., J. Schultz, and E. Eisenstadt. 1979. Synthesis and metabolism of cyclopenta(cd)pyrene. In: Polynuclear Aromatic Hydrocarbons. P.W. Jones and P. Leber, eds. Ann Arbor Science: Ann Arbor, MI. pp. 695-704.

Grimmer, G. 1977. IARC monographs on the evaluation of carcinogenic risk of chemicals to man, Vol. 16. International Agency for Research on Cancer, Lyon, France. p. 29.

Ittah, Y., and D.M. Jerina. 1978. Synthesis of cyclopenta(cd)pyrene. Tetrahedron Lett. 4495-4498.

Jerina, D.M., H. Yagi, R.E. Lehr, D.R. Thakker, M. Schaeffer-Ridder, J.M. Karle, W. Levin, A.W. Wood, R.L. Chang, and A.H. Conney. 1978. Bay region theory of carcinogenesis by polycyclic aromatic hydrocarbons. In: Polycyclic Hydrocarbons and Cancer, Vol. 1. H.V. Gelboin and P.O.P. Ts'o, eds. Academic Press: New York. pp. 173-188.

Kaden, D.A., R.A. Hites, and W.G. Thilly. 1979. Mutagenicity of soot and associated polycyclic aromatic hydrocarbons to Salmonella typhimurium. Cancer Res. 39:4152-4159.

Keller, J.W., and C. Heidelberger. 1976. Polycyclic K-region arene oxides; products and kinetics of solvolysis. J. Am. Chem. Soc. 98:2328-2336.

Konieczny, M., and R.G. Harvey. 1979. Synthesis of cyclopenta(cd)pyrene. J. Org. Chem. 44:2158-2160.

Lee, M.L., P.G. Prado, J.B. Howard, and R.A. Hites. 1977. Source identification of urban airborne polycyclic aromatic hydrocarbons by gas chromatographic mass spectrometry and high resolution mass spectrometry. Biomed. Mass Spectrom. 4:182-186.

Maher, V.M., and J.M. McCormick. 1978. Mammalian cell mutagenesis by polycyclic aromatic hydrocarbons and their derivatives. In: Polycyclic Hydrocarbons and Cancer, Vol. 2. H.V. Gelboin and P.O.P. Ts'o, eds. Academic Press: New York. pp. 137-160.

Nesnow, S., and C. Heidelberger. 1976. The effect of modifiers of microsomal enzymes on chemical oncogenesis in cultures of C3H mouse cell lines. Cancer Res. 36:1801-1080.

Particulate Polycyclic Organic Matter. 1972. National Academy of Sciences: Washington, DC.

Reznikoff, C.A., J.S. Bertram, D.W. Brankow, and C. Heidelberger. 1973. Quantitative and qualitative studies of chemical transformation of cloned C3H mouse embryo cells sensitive to postconfluence inhibition of cell division. Cancer Res. 33:3239-3249.

Rosen, A.A., and F.M. Middleton. 1955. Identification of petroleum refinery wastes in surface waters. Anal. Chem. 27:790-794.

Ruehle, P.H., D.L. Fischer, and J.C. Wiley. 1979. Synthesis of cyclopenta(cd)pyrene, a ubiquitous environmental carcinogen. J. Chem. Soc. Chem. Commun. 302-303.

Snook, M.E., R.E. Severson, R.F. Arrendale, H.C. Highman, and O.T. Chortyk. 1977. The identification of high molecular weight polynuclear aromatic hydrocarbons in a biologically active fraction of cigarette smoke condensate. Beitrage zur Tabakforsch. 9:97-101.

Wallcave, L., D.L. Nagel, J.W. Smith, and R.D. Waniska. 1975. Two pyrene derivatives of widespread environmental distribution: cyclopenta(cd)pyrene and acepyrene. Environ. Sci. Technol. 9:143-145.

Wood, A.N., W. Levin, R.L. Chang, M. Huang, D.E. Ryan, P.E. Thomas, R.E. Lehy, S. Kumer, M. Koreeda, H. Akagi, Y. Ittah, P.M. Dansette, H. Yagi, D.M. Jerina, and A.H. Conney. 1980. Mutagenicity and tumor-initiating activity of cyclopenta(cd)pyrene and structurally related compounds. Cancer Res. 40:642-649.

MUTAGENICITY OF COAL GASIFICATION AND LIQUEFACTION PRODUCTS

Rita Schoeny, David Warshawsky, Lois Hollingsworth, and
Mary Hund
Department of Environmental Health
University of Cincinnati College of Medicine
Cincinnati, Ohio

George Moore
Pittsburgh Energy Technology Center
U.S. Department of Energy
Pittsburgh, Pennyslvania

INTRODUCTION

As it becomes evident that the shortage of oil resources is
not a transient phenomenon, emphasis is being directed to the use
of domestic coal. Increased use of coal, however, adds coal
combustion products to the pollution burden. Although various
technologies are being developed to produce cleaner-burning fuels
from coal, such as gaseous fuels, de-ashed low-sulfur boiler fuels,
and synthetic crude oils, problems remain in the production of
these materials.

The production of liquid fuels from coal has been associated
with an increased risk of cancer, and certain of these coal-derived
liquids have been shown to be carcinogenic in experimental animals
(Bingham, 1975; Ketcham et al., 1960; Sexton, 1960a, b; Weil et
al., 1960). Heavy exposure to coal hydrogenation materials causes
both benign and malignant skin tumors. Composition data
(Swansiger, 1974; Battelle, 1974; ORNL, 1975; Electric Power Res.
Inst., 1975; ERDA, 1976) suggest that various liquefaction products
and by-products are likely to contain polycyclic substances of
considerable carcinogenic potential; these compounds are most
likely to be found in the high-boiling-point aromatic fractions of
the product liquids (TWR, 1976). The total products of
hydrogenation, high-boiling-point distillates, centrifuged oils,
char, residues, recycled solvent oil, recycled solvent, and liquid
coal are all potentially hazardous materials (Freudenthal et al.,
1975).

Synthetic natural gas is not expected to pose a carcinogenic risk, as any trace elements, organic carcinogens, or cocarcinogens present in the raw product will have been removed during clean-up and scrubbing operations. Rather than the fuel produced, it is the coal gasification process itself that should be the primary concern (Kornreich, 1976). The greatest potential hazard in coal gasification is in the early stages of the processes, in which coal goes through a series of structural degradations of complex organic compounds. During these early stages, leaks and spills should contain hazardous material (Freudenthal et al., 1975). Potentially carcinogenic polycyclic organic material is likely to concentrate in the tars, oils, and char (Kornreich, 1976; TRW, 1976). The crude gas, if it contains tars of high-boiling-point oils, must also be considered a potential hazard.

Hazardous chemicals may be synthesized whenever coal is subjected to severe conditions such of those of pyrolysis, hydrogenation, or gasification. As use of synthetic fuels will probably increase, sensitive and rapid in vitro studies should be carried out on products, intermediate streams, and wastes of coal conversion processes to determine potential hazards. A variety of techniques are available for assessing the environmental risks and potential health effects of coal processing technologies. These techniques include chemical and physical characterizations, microbial assays, and tests for acute toxicity and irritation, subchronic toxicity and teratology, chronic toxicity, and carcinogenesis, among others. To conduct so many tests on all new technological developments would not be appropriate, as each change of experimental condition or operation would lead to new health study requirements. The cost would be prohibitive, and much of the resulting data useless, as it would apply to defunct processes. Delaying biological assessment until the process is ready for production would likewise be inappropriate. An acceptable position on testing new fossil energy processes must be found. A suggested compromise would involve chemical characterizations and rapid bioassay studies in small-scale developmental programs followed by detailed characterization and short-term and long-term toxicological testing programs on pilot processes.

For this project, materials ranging from solid residue to liquid products and waters, produced through advanced coal-conversion technologies (including gasification and liquefaction), have been selected and screened using the Salmonella/microsomal mutagenesis assay. This assay, developed by B.N. Ames et al. (1975), is recognized as one of the most useful short-term assays for mutagenesis, based on the number of compounds it detects as mutagens and on its high correlative of positive responses with known carcinogens (Bridges, 1976; McCann et al., 1975; McCann and Ames, 1976; Purchase et al., 1976; Simmon 1979; Committee 17, 1975). It is being used to investigate the health effects of

compounds already in the environment and of materials under
development. It is particularly useful in the evaluation of
mixtures of substances, such as cigarette smoke concentrate (Kier
et al., 1974; Sato et al., 1977), synthetic crude oil (Epler, 1978;
Epler et al., in press), and organic extracts of drinking water
(Loper et al., 1978).

MATERIALS AND METHODS

Sample Preparation

     Coal-related materials were provided by the Department of
Energy. These samples are not considered to be process discharges
and may not be identical to materials eventually generated from
advanced coal processes developed for commercial use. Although the
samples are not representative of all materials derived from
advanced coal processes, they were selected for study because of
their immediate availability and the desirability of testing
materials of widely differing properties.

     All samples were stored at 5°C. Materials ETTM-01, ETTM-02,
ETTM-08, and ETTM-09, which were tars or viscous liquid, were
prepared for testing by weighing a small amount (20 to 70 mg) and
adding dimethylsulfoxide (DMSO) to obtain a presumptive
concentration of 10 mg/ml. In no case did all the material
dissolve. The amount of insoluble sample was subtracted from the
total to give the concentration used in calculating the mutagenic
doses. The sample solutions were filter-sterilized prior to
testing. All samples were applied in 0.1-ml aliquots. The sample
solutions were further diluted in DMSO, so that the following
percentages of the sample solution were assayed: 100%, 50%, and
0.5%, or 100%, 50%, 10%, 5%, and 1%. For ETTM-02, the liquid
fraction was assayed by applying 0.1 ml of the undiluted substance,
as well as the concentrations listed above.

     The powdered samples ETTM-03 and ETTM-04 did not dissolve in
DMSO, nor were mutagenic substances in detectable quantities washed
from them into the DMSO. Aqueous leachates were made from these
samples. Five grams of the powder were added to 45 g distilled
water at pH 3.0, 5.0, 6.0, 7.0, and 10.0. These suspensions were
stoppered and shaken at room temperature overnight. Each
suspension was centrifuged, and the supernatant fraction was
aspirated and filter-sterilized for use in the mutagenic assays.

     Liquids ETTM-06 and ETTM-07 changed from yellow to turquoise
when mixed with DMSO, and ETTM-06 underwent an exothermic reaction.
These two samples were also diluted in double-distilled water. The
black particulate matter floating on sample ETTM-06 was removed

during filter-sterilization, and the solvent and sample phases were mixed vigorously immediately prior to assay.

Organic extracts of ETTM-01, ETTM-02, ETTM-08, and ETTM-09 were prepared as follows: Approximately 5 g or less of each sample was weighed, and a volume (in ml) of solvent equal to five times the sample weight (in g) was added. This mixture was agitated vigorously in the dark at room temperature for two hours. After centrifugation to settle particulates, the solvent was removed, and an equal amount of fresh solvent was added. The two extracts were pooled and evaporated under nitrogen gas. This procedure was carried out sequentially with hexane, toluene, methylene chloride, and acetonitrile. This simple organic extraction procedure was chosen for its ease and appropriateness for these samples. The type of organic extracts produced, moreover, is suitable for analysis by high performance liquid chromatography.

Mutagenicity Assays

Salmonella/microsomal assays were carried out according to the methods described by Ames et al. (1975). Microsomal extracts for routine assays (S-9) were made from livers of male Sprague-Dawley rats (150 to 200 g body weight) that had been administered 500 mg/kg Aroclor 1254 (PCB) on day one and killed on day six. On the day of assay, S-9 from four animals was pooled. Assays were also done using S-9 from rats treated with 3-methylcholanthrene (3-MC, 40 mg/kg) and from rats given corn oil. Plates were scored using an automatic colony counter. Only counts of at least twice the spontaneous values were considered to indicate mutagenicity. Colony counts below the range observed for spontaneous revertants, clearing of bacterial lawns, or the appearance of pinpoint his⁻ colonies was scored as a toxic response.

RESULTS AND DISCUSSION

Table 1 summarizes the mutagenic activity of the samples. No sample was direct acting (i.e., mutagenic in the absence of S-9). The liquefaction products, distillate oils, heavy liquid, and residue were mutagenic. Of the gasification products, only the tar was active in these assays. Neither DMSO extracts nor aqueous leachates of gasification particulate and ash were mutagenic.

Figure 1 is a plot of dose-response data for ETTM-01. This plot, typical of those for the other active samples, indicates a linear dose response at the lowest concentrations tested, with decreases in the slopes of the curves at the higher concentrations.

Table 1. Mutagenicity of Dimethylsulfoxide-soluble Components
of Coal-related Materials

	Sample	Mutagenicity[a]			
		TA1535	TA1538	TA98	TA100
ETTM-01	Liquefaction product	–	+	+	+
ETTM-02	Gasification tar	–	+	+	+
ETTM-03	Gasification particulate	ND	ND	–	–
ETTM-04	Gasification ash	ND	ND	–	–
ETTM-06	Liquefaction untreated water	ND	ND	–	–
ETTM-07	Liquefaction light oils	ND	ND	–	–
ETTM-08	Liquefaction heavy liquid (with solids)	ND	ND	+	+
ETTM-09	Liquefaction product (filtered)	ND	ND	+	+
ETTM-10	Liquefaction distillate oils	–	+	+	+
ETTM-11	Liquefaction residual	–	+	+	+

[a]+ = mutagenic; – = not mutagenic; ND = not determined.

As these samples are mixtures, there are several possible explanations for this pattern:

1) The samples may be toxic to the organisms at the higher doses. When these concentrations of sample were tested without S-9, there was apparent toxicity.

2) At the high concentrations, activation enzymes may be saturated, or metabolism channelled into detoxification pathways.

3) Components in the mixture may interact, contributing to nonlinear kinetics.

Revertant colonies per microgram of sample, derived through regression analysis of the linear portion of the dose response, are given in Table 2. Strains TA98 and TA1538 (frameshift mutants with and without a misrepair-enhancing plasmid) were the most sensitive to the mutagenic action of the samples. The amount of Aroclor-induced S-9 routinely used in the assay was 50 µl/plate. Figure 2 is a representative plot of mutagnenicity in response to varying

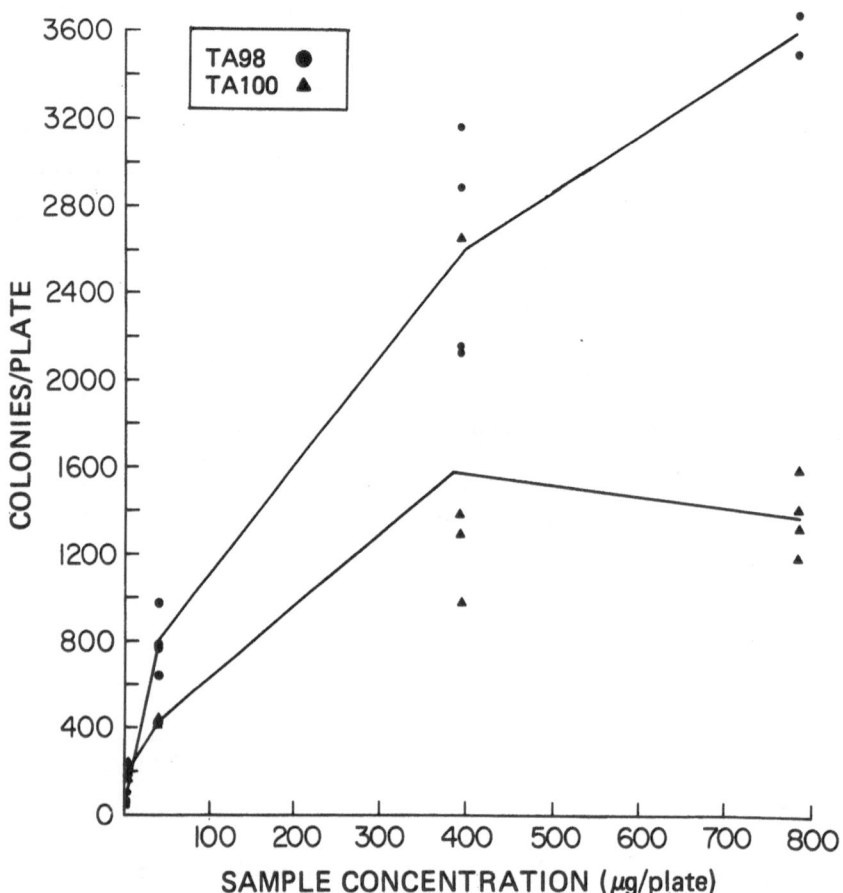

Figure 1.   Mutagenicity of ETTM-01 in <u>Salmonella</u> with Aroclor-
            induced S-9 (50 μl/plate).   Correlation coefficients (r)
            for the linear regression lines up to 39.3 μg of sample
            are as follows: strain TA98, 0.982; TA100, 0.971.
            For concentrations above 39.3 μg/plate, lines are drawn
            through the means of the data points.

amounts of S-9. For all the mutagenic samples, 50 μl of S-9
provided optimal or nearly optimal activation for mutagenesis. The
active samples were also assayed for mutagenicity in the presence
of 50 μl/plate uninduced S-9 and 50 μl/plate 3-MC-induced S-9.
Uninduced S-9 was uniformly poor for sample activation. For all
but one sample (ETTM-01; see Figure 3), Aroclor-induced S-9 was
most effective in generating metabolites mutagenic for TA98 and
TA100.

Table 2. Relative Mutagenic Activities of Coal-related Materials

A.	Revertant Colonies/μg[a]			
Sample	TA1535	TA1538	TA98	TA100
ETTM-01	–	30.12	18.54	6.89
ETTM-02	–	11.16	6.75	6.49
ETTM-08	ND	ND	16.66	8.38
ETTM-09	ND	ND	11.55	6.76
ETTM-10	–	7.30	10.88	2.10
ETTM-11	–	27.92	26.01	11.28

B.	Coefficients of Linear Correlation (R)					
	TA1358		TA98		TA100	
Sample	N[b]	R	N	R	N	R
ETTM-01	16	0.921	32	0.888	59	0.918
ETTM-02	6	0.999	46	0.918	45	0.857
ETTM-08		–	60	0.960	57	0.843
ETTM-09		–	39	0.948	54	0.887
ETTM-10	18	0.987	64	0.905	78	0.911
ETTM-11	28	0.912	64	0.933	65	0.950

[a]Calculated from linear postions of dose-response curves; – = no dose response; ND = not determined.
[b]N = number of data points.

Sequential organic extracts were prepared on two occasions from each of the six mutagenic samples. Whenever possible, five-point dose-response assays were done with strains TA98 and TA100. For comparison, a dose-response assay of the unfractionated whole sample was run in tandem with these assays. Representative results are presented in Tables 3 and 4 and Figures 4 and 5. No extracted material was mutagenic in the absence of S-9. The samples varied widely in the percentage extractable by the solvents and in the mutagenicity of the extracted fractions. For the majority of samples, the toluene-extractable components were most mutagenic, although they were not always the largest fractions by

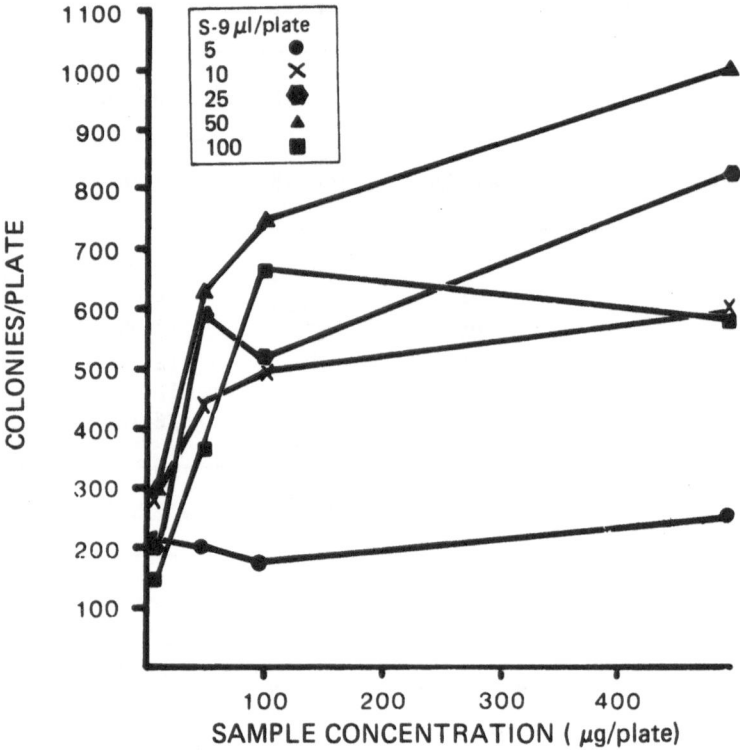

Figure 2.   Effect of varying S-9 concentration on TA98 mutagenesis
             by ETTM-10 with Aroclor (PCB)-induced S-9.

weight.   Tables 3 and 4 show the mutagenic contribution of each
fraction to the whole sample (i.e., percent of whole sample present
in fraction x revertant colonies per milligram).   Comparing the
sums of the fractional mutagenic contributions with the activity of
the unfractionated sample reveals two patterns.   In the first,
typified by ETTM-01, the sum of the activities of the fractions was
less than the activity of the whole sample.   Synergistic actions
among components of the sample could contribute to the greater
mutagenicity of the unfractionated sample.   It is also possible
that material was lost or altered during extraction.   ETTM-10
represents the second pattern, in which the sum of fraction
activities was greater than the activity of the parent mixture.
This probably indicates that antagonistic components in the whole
sample inhibit total mutagenicity.   It is also conceivable that

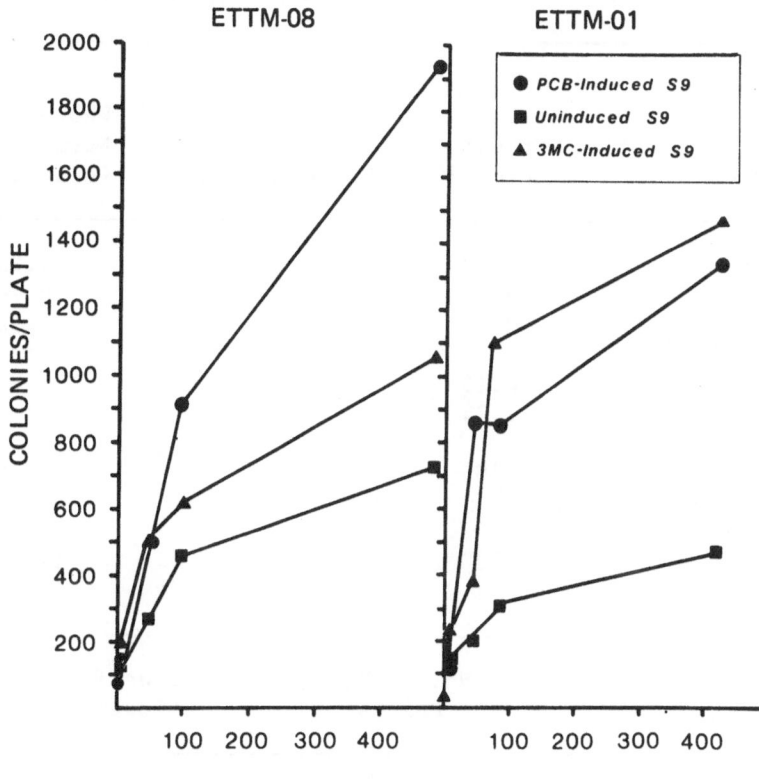

Figure 3.   Effect of induction of S-9 (50 µl/plate) enzymes on
            sample mutagenicity for TA98.

mutagenic forms were generated by the extraction process, although
this appears unlikely, as no direct-acting forms were produced.

     While mutagenicity in this assay is not proof of a compound's
carcinogenic potential, it does indicate an urgent need for further
study. If the types of material represented by these samples are
to be produced in large amounts or are found to be widespread in
the environment, they may pose a significant health problem.

     It is likely that production of synthetic fuels will increase,
necessitating the use of in vitro assays to assess the health
hazards of products, intermediate streams, and wastes of coal
processes. Future work should include samples from alternative
processes, feedstocks, and varying process conditions. Bioassay

Table 3.   Liquefaction Product (ETTM-01)

A.

Extract	I Percent Extracted[a]	II TA98 (colonies/mg)[b]	I x II Whole Sample Normalization (colonies/mg)
Hexane	29.09	7,640	2,222
Toluene	65.26	16,490	10,761
Methylene chloride	3.84	5,220	200
Acetonitrile	0.16	–	–
Residue	1.74	–	–
Sum of fractions	100		13,183 (74.5%)
Whole sample	100		17,690 (100%)

B.

	No. of Data Points	Coefficient of Linear Regression
Whole	30	0.902
Hexane	34	0.975
Toluene	28	0.923
Methylene chloride	28	0.905

[a]Initial weight prior to extraction: ETTM-01A, 5.2948 g; ETTM-01B, 2.9320 g. Sum of organic extract weights: ETTM-01A, 5.4655 g (a gain of 3.22%); ETTM-01B, 3.2128 g (a gain of 9.60%).
[b]Calculated from linear portions of dose response curves: – = no dose response.

data from both complete and fractionated samples will be correlated with chemical characterizations to identify specific compounds or functionality effects. Such results will be compared with those from toxicology testing on larger-scale advanced coal processes. Assessing hazards early in the process-development cycle will facilitate development of technology to minimize health risks.

Table 4. Liquefaction Distillate Oils (ETTM-10)

A.

Extract	I Percent Extracted[a]	II TA98 (colonies/mg)[b]	I x II Whole Sample Normalization (colonies/mg)
Hexane	97.47	9,500	9,260
Toluene	0.86	271,370	2,334
Methylene chloride	1.21	63,890	774
Acetonitrile	0.05	63,510	32
Residue	0.41	13,260	54
Sum of fractions	100		12,453 (134%)
Whole sample	100		9,280 (100%)

B.

	No. of Data Points	Coefficient of Linear Regression
Whole	24	0.815
Hexane	31	0.937
Toluene	24	0.777
Methylene chloride	24	0.934
Acetonitrile	18	0.958
Residue	38	0.868

[a]Initial weight prior to extraction: ETTM-10A, 5.0212 g; ETTM-10B, 4.7879 g. Sum of organic extract weights: ETTM-10A, 4.2488 g (a loss of 15.4%); ETTM-10B, 4.5672 g (a loss of 4.61%).
[b]Calculated from linear portions of dose response curves.

ACKNOWLEDGMENT

This work was supported by U.S. Department of Energy contract No. AS-22-78ET0022.

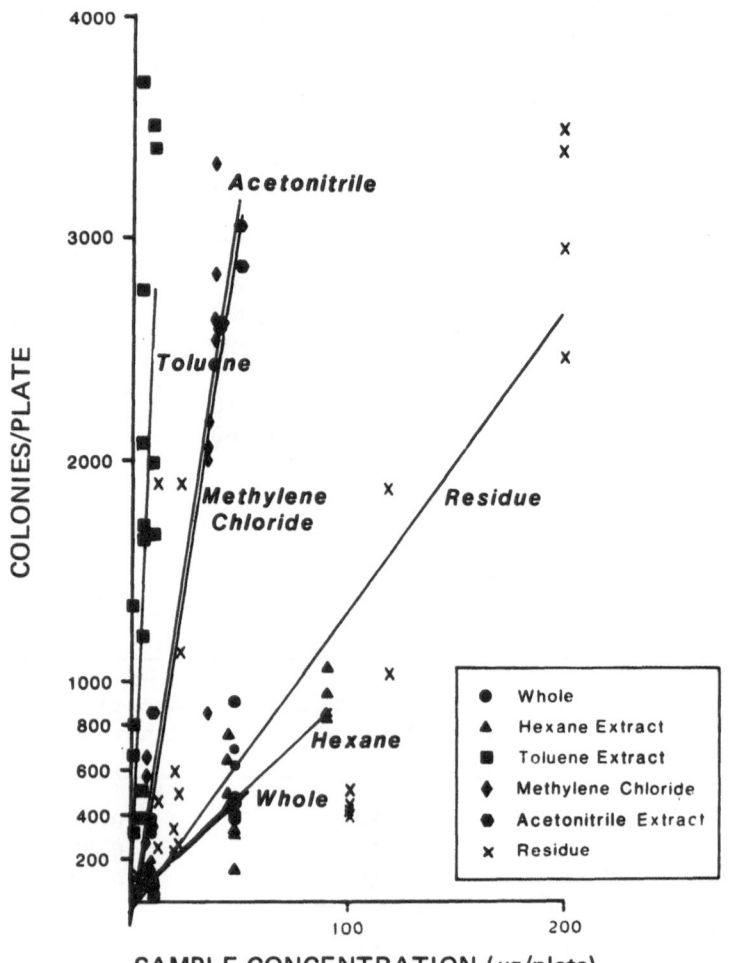

Figure 4.   Mutagenicity of organic solvent extracts of ETTM-10 for
strain TA98 with Aroclor-induced S-9 (50 µl/plate).

REFERENCES

Ames, B.N., J. McCann, and E. Yamasaki. 1975. Methods for
    detecting carcinogens and mutagens with the Salmonella/
    mammalian microsome mutagenicity test. Mutation. Res.
    31:347-364.

Battelle. 1974. Liquefaction and chemical refining of coal.
    Battelle Columbus Laboratories, Columbus, OH. pp. 52-54.

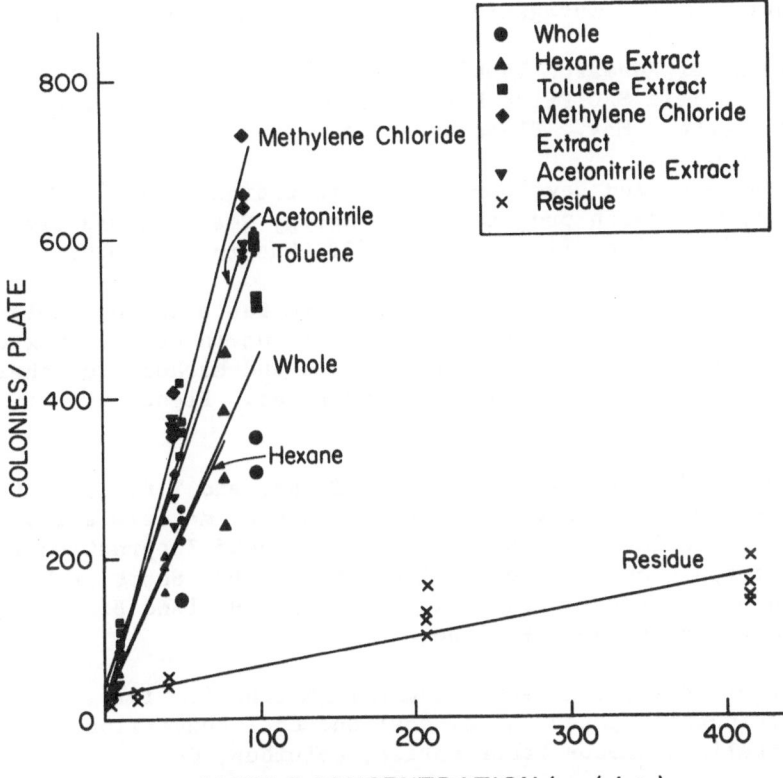

Figure 5.   Mutagenicity of organic solvent extracts of ETTM-02B for
strain TA98 with Aroclor-induced S-9 (50 μl/plate).
Correlation coefficients (r) for the linear regressions
are as follows: whole sample, 0.924; hexane extract,
0.940; toluene extract, 0.980; methylene chloride
extract, 0.994; acetonitrile extract, 0.989; residue,
0.965.

Bingham, E. 1975. Carcinogenic potency of oil fractions derived
from fossil fuels. Presented at Workshop on Health Effects of
Coal and Oil Shale Mining Conversion Utilization, Department
of Environmental Health, Kettering Laboratory, University of
Cincinnati, Cincinnati, OH.

Bridges, B.A. 1976. Short-term screening tests for carcinogens.
Nature 261:195-200.

Committee 17 Report, Environmental Mutagen Society USA. 1975.
    Environmental mutagenic hazards. Science 187:503-514.

Electric Power Research Institute. 1975. Status Report of
    Wilsonville SRC Pilot Plant. Electric Power Research
    Institute. May. Pleasanton, CA. p. 32.

Energy Research and Development Administration. 1976. Fossil
    energy research program of the Energy Research and Development
    Administration, fiscal year 1977. Washington, D.C.

Epler, J.L. 1978. The use of short-term tests in the isolation
    and identification of chemical mutagens in complex mixtures.
    In: Chemical Mutagens: Principles and Methods for their
    Detection, Vol. VI. A. Hollaender, ed. Plenum Press: New
    York. pp. 1-54.

Epler, J.L., F.W. Larimer, C.E. Nix, T. Ho, and T.K. Rao. (in
    press). Comparative mutagenesis of test materials from
    synthetic fuel technologies. In: Second International
    Conference on Environmental Mutagens. D. Scott, F.H. Sobels,
    and B.A. Bridges, eds. Elsevier/North Holland Biomedical
    Press: Amsterdam, Holland.

Freudenthal, R.I., G.A. Lutz, and R.I. Mitchell. 1975.
    Carcinogenic potential of coal and coal conversion products.
    Battelle Columbus Laboratories, Columbus, OH.

Ketcham, N., and R.W. Norton. 1960. The hazards to health in the
    hydrogenation of coal. III. The industrial hygiene studies.
    Arch. Environ. Hlth. 1:194-207.

Kier, L.D., E. Yamasaki, and B.N. Ames. 1974. Detection of
    mutagenic activity in cigarette smoke condensates. Proc.
    Natl. Acad. Sci. USA 71:4159-4163.

Kornreich, M.R. 1976. Coal conversion processes: potential
    carcinogenic risk. MTR-7155, MITRE Technical Report, 4-14 to
    4-16.

Loper, J.C., D.R. Lang, R.S. Schoeny, B.B. Richmond, P.M.
    Gallegher, and C.C. Smith. 1978. Residue organic mixtures
    from drinking water show in vitro mutagenic and transforming
    activity. J. Toxicol. Environ. Hlth. 4:919-938.

McCann, J., and B.N. Ames. 1976. Detection of carcinogens as
    mutagens in the Salmonella/microsome test: assay of 300
    chemicals: discussion. Proc. Natl. Acad. Sci. USA
    73:950-954.

McCann, J., E. Choi, E. Yamasaki, and B.N. Ames. 1975. Detection of carcinogens as mutagens in the Salmonella/microsomal test: assay of 300 chemicals. Proc. Natl. Acad. Sci. USA 72:5135-5139.

Oak Ridge National Laboratory Coal Technology Program Annual Report. 1975. ORNL-5069. Oak Ridge, TN. p. 52.

Purchase, I.F.H., E. Longstaff, J. Ashby, J.A. Styles, D. Anderson, P.A. Lefevre, and F.R. Westwood. 1976. Evaluation of six short-term tests for detecting organic chemical carcinogens and recommendation for their use. Nature 264:624-627.

Sato, S., Y. Seino, T. Ohka, T. Yahagi, M. Nagao, T. Matsushima, and T. Sugimura. 1977. Mutagenicity of smoke condensates from cigarettes, cigars, and pipe tobacco. Cancer Lett. 3:1-8.

Sexton, R.J. 1960a. The hazards to health in the hydrogenation of coal. I. An introductory statement on general information, process description, and a definition of the problem. Arch. Environ. Hlth. 1:181-186.

Sexton, R.J. 1960b. The hazards to health in the hydrogenation of coal. IV. The control program and its effects. Arch. Environ. Hlth. 1:208-231.

Simmon, V.F. 1979. In vitro mutagenicity assays of chemical carcinogens and related compounds with Salmonella typhimurium. J. Natl. Cancer Inst. 62:893-899.

Swansiger, J.T. 1974. Liquid coal composition analysis by mass spectrometry. Anal. Chem. 46:730-734.

TRW Systems and Energy. 1976. Carcinogens relating to coal conversion processes. ERDA Contract E(49-18)-2213. Washington, DC. pp. 26-28.

Weil, C.S. and N.I. Condra. 1960. The hazards to health in the hydrogenation of coal. II. Carcinogenic effect of materials on the skin of mice. Arch. Environ. Hlth. 1:187-193.

SESSION 6

HAZARD ASSESSMENT

THE ROLE OF SHORT-TERM TESTS IN ASSESSING THE HUMAN HEALTH HAZARDS OF ENVIRONMENTAL CHEMICALS: AN OVERVIEW

Michael D. Waters
Health Effects Research Laboratory
U.S. Environmental Protection Agency
Research Triangle Park, North Carolina

The problems to be addressed in this paper and in this session are twofold: first, to identify and confirm mutagenic and presumptive carcinogenic chemicals through short-term tests (based on the use of plant as well as animal materials); and second, to determine the relationship of demonstrated effects in short-term tests to actual hazards and risks to human health.

When we consider the complexity of environmental samples, it becomes clear that removing all chemical hazards is not feasible. Increased exposure to man-made carcinogens and mutagens in the environment must be presumed to entail additional human health risk. Therefore, our research objective and our regulatory objective must be to identify the potential hazards of such compounds and to minimize the risk to human health.

Even the task of identifying those chemicals that pose mutagenic or carcinogenic hazards is complex, because of the diversity of such agents and the multiplicity of their interactions with biological systems. Mutagens can induce heritable changes in the genetic material of either somatic or germinal cells of living organisms. Alterations can occur at the single gene and the chromosomal level. Many carcinogens act through mutational mechanisms, but some may act through other physiological changes that may alter, for example, the individual sensitivity or pattern of tumor expression. Eventually, we shall have to develop hazard assessment procedures that take into account the range of genotoxic and physiological properties of substances evaluated as carcinogens.

Our understanding of the mechanistic basis of genetic toxic
effects is undoubtedly best developed in the field of mutagenesis.
Much of this information was obtained from the kinds of short-term
systems that we are discussing. We know much less about the
measurement of human genetic effects. Cytogenetic techniques have
demonstrated that changes in structure and number of chromosomes
are associated with a variety of human diseases. The cri-du-chat
syndrome and one form of Down's syndrome are associated with
structural alterations (a chromosomal deficiency and translocation,
respectively). Kleinfelter's and Turner's syndromes are
attributable to alterations in chromosome number. It is estimated
that chromosomal abnormalities occur in the human population at a
combined frequency of around 0.5% of all live births. Some 50% of
all spontaneous abortions involve chromosomal defects. The
relationship between heritable chromosomal alterations in man and
exposure to environmental mutagens remains uncertain. But exposure
of various experimental organisms to such agents definitely
increases the frequency of chromosomal abnormalities.

Similarly, the induction of gene mutations by environmental
chemicals is well known in experimental organisms. Such mutations
give rise to new alleles that can have a variety of effects,
depending on the mode of gene expression. In man, dominant alleles
may be responsible for physical defects, such as dwarfism. It is
estimated that dominant mutations may cause adverse effects in up
to 0.5% of the human population. Perhaps of greater concern are
the recessive mutations. Well over 1000 disease states display
inheritance patterns characteristic of recessive mutations,
including phenylketonuria (PKU), Tay-Sachs, and cystic fibrosis.
The association of such genetic anomalies with chemically-induced
mutation in man has not been established. However, since induced
recessive mutations can remain unexpressed for many generations,
exposures to environmental mutagens could be covertly increasing
the genetic load of the human gene pool. Certainly human exposure
to man-made mutagens has increased, and the genetic risk associated
with such exposure must be evaluated.

Heritable genetic damage is not our only concern. The somatic
mutation theory of carcinogenesis extends the concern to
carcinogenesis as well. Fortunately, microbial mutagenesis tests,
when coupled with mammalian metabolic activation, detect a major
proportion of the chemicals that have been shown to be carcinogenic
in animals. The overall qualitative correlation between the
mutagenicity of chemicals in microbial systems and the
carcinogenicity of the same chemicals in experimental animals is
quite good. Limited data suggests that the quantitative
correlation between mutagenic potency in mammalian cell systems and
carcinogenic potency in animals may be somewhat better than in the
case of the microbial systems, but fewer chemicals have been
tested. Dr. de Serres discusses in his paper (1980) results of

various mutagenicity and related tests conducted on a series of carcinogens and noncarcinogens.

To the extent that such correlations depend on more accurate measurement of metabolism, we have made several recent advances. Perfusion techniques and cell dissociation methods have made it possible to prepare metabolically active primary cells for co-cultivation with mutagenesis indicator systems or with cells that are subject to transformation in vitro. While these technical innovations may improve our ability to detect genetically toxic agents missed with microsomal activation methods, there is no substitute for the intact mammal, when mutagenic or carcinogenic potency must be quantified directly. In addition to pharmacokinetic considerations, species, strain, and sex differences must be considered, as well as organ specificity and differences in sensitivity of various cell types and cell stages. Ultimately, potency is a question of dose-response relationships and the probability of effects at environmental dose levels. Short-term tests that can be applied directly to man are highly desirable in this regard. Dr. Wyrobeck (1980) specifically addresses such tests using sperm. Similar procedures are being developed with laboratory animals. The ability to identify mutations in single cells of exposed animals will have significant advantages over presently available in vitro and in vivo mutagenesis bioassays.

Finally, I would like to discuss two major approaches to assessing genetic risk to humans resulting from exposure to chemical mutagens. One uses experimental data obtained directly from induced germinal mutations in intact animals, and the other uses data on chemical dose to the germ cells of animals together with mutagenesis results obtained with other-than-germinal cells in various short-term tests.

Tests that directly provide information on induced germinal mutations include the specific locus test, the X-Y chromosome loss test, and the heritable translocation test. These tests are complete, in that they measure genetic damage in germ cells which is expressed in the subsequent generation. This information on induced mutation frequency may be combined with information on levels of human exposure. To estimate human risk, the induced mutation frequency is extrapolated downward to the estimated level of human exposure. Theory suggests the use of a linear or no-threshold model for point- or gene-mutational effects. Chromosomal alterations, on the other hand, are thought to proceed by multi-hit mechanisms. Thus, linear extrapolation of translocation data is likely to overestimate risk at lower levels of exposure. Other models may be more appropriate when supported by sufficient data. As was mentioned earlier, gene mutations usually occur at lower doses than do chromosomal mutations;

therefore, gene mutation data would be expected to be more
sensitive for human risk assessment.

The issue of sensitivity tends to favor the second approach to
genetic risk assessment. Because short-term test systems,
particularly the in vitro systems, are usually more powerful, it is
possible to detect more-readily-induced mutants and to relate this
number to the chemical estimation of mutagen-DNA interactions or
binding to DNA. A similar determination can be made of mutagen-DNA
interactions in the germinal cells of the intact animal that has
been exposed via an appropriate route. With knowledge of
mutagen-DNA interactions in the intact animal and in the short-term
mutagenesis test, a relationship can be constructed between
exposure in the one system and induced mutation frequency in the
other. To the extent that binding of the test mutagen to DNA can
be measured at anticipated human exposure levels, it may not be
necessary to perform a high-to-low dose extrapolation. If
extrapolation is required, it may be assumed that DNA binding is
directly proportional to the exposure level, unless there is
evidence to the contrary. It is extremely important to understand
the relationship between exposure, DNA binding, and mutation in all
of our short-term mutagenesis tests in order to make maximum use of
resulting data for relative chemical potency evaluation and for
hazard assessment. These same considerations of exposure,
effective dose, and response apply as well to other genetically
mediated effects.

As far as carcinogenesis is concerned, we are in a better
position to address the issue of human risk assessment because we
have epidemiological evidence of cancer in man. This evidence
provides us with the necessary information on human exposure and
response to validate our long-term whole animal models. On the
strength of correlations with whole animal data, short-term test
results provide evidence of the carinogenic potential of previously
untested pure chemicals and complex mixtures. We have seen in this
symposium that comparative studies of the relative biological
activity of complex mixtures can be used to provide early
information on the carcinogenic potential of such materials.
However, at this stage in the development of short-term tests,
their results are best considered suggestive rather than
conclusive. Given the evidence from these tests and human exposure
considerations, it is reasonable to require the performance of
long-term whole animal tests for carcinogenesis and related effects
to define more precisely the extent of human health risk. Dr.
Albert, who is directly involved in the process of health risk
assessment, elaborates on these points in his paper (1980).

REFERENCES

Albert, Roy E. 1980. Assessing carcinogenic risk resulting from
     complex mixtures. Presented at U.S. Environmental Protection
     Agency Second Symposium on the Application of Short-term
     Bioassays in the Fractionation and Analysis of Complex
     Environmental Mixtures, Williamsburg, VA.

de Serres, Frederick J. 1980. International program for the
     evaluation of short-term tests for carcinogenicity. Presented
     at the U.S. Environmental Protection Agency Second Symposium
     on the Application of Short-term Bioassays in the
     Fractionation and Analysis of Complex Environmental Mixtures,
     Williamsburg, VA.

THE INTERNATIONAL PROGRAM FOR THE EVALUATION OF SHORT-TERM TESTS FOR CARCINOGENICITY (IPESTTC)

Frederick J. de Serres
National Institute of Environmental Health Sciences
Research Triangle Park, North Carolina

INTRODUCTION

The major impetus for developing the International Program for the Evaluation of Short-term Tests for Carcinogenicity (IPESTTC) was that our need for rapid identification and control of carcinogens is not satisfied by traditional rodent bioassays. Because of resource limitations, rodent studies cannot be carried out on a large enough scale to identify all carcinogenic chemicals in the environment within a reasonable period of time. This need places tremendous pressure on the scientific community to develop test systems for identifying chemical carcinogens in the environment at a lower cost and on a shorter time scale.

A major problem in selecting short-term tests for carcinogenicity has been that the mechanisms of action of chemical carcinogens have not been well understood. However, during the past decade, developments in genetic toxicology have allowed progress towards a unifying theory of the action of many carcinogens. This theory is based on the hypothesis that chemical carcinogens induce mutations in somatic cells, and that these mutations change the behavior of the cells so that cancer develops. It is a logical step from accepting the somatic mutation theory of cancer to using short-term mutation tests for detecting chemicals that potentially produce mutations and cancer. However, other theories of the induction of cancer exist, and many other highly accurate tests have been developed to identify carcinogens.

The availability of a large number of short-term tests has created a problem for both the scientist and administrator who must select the most appropriate and accurate test systems for

carcinogenicity screening. During the past few years, various
laboratories have tackled this problem using well-known validation
studies. The purpose of these studies is to assess how effectively
the short-term test can distinguish between carcinogens and
noncarcinogens that have been classified as such by their activity
in whole-animal systems. However, the problem remains of how to
compare the performance of different test systems when the data
describing the performance has been developed in different
laboratories using tests with different protocols and different
criteria.

IPESTTC was designed specifically to examine the abilities of
various test systems to distinguish between known chemical
carcinogens and known noncarcinogens. Although many of the test
systems may also detect mutagenic events, this was not the primary
purpose of this program. As the program developed, it became
apparent that a secondary aim could be achieved, namely, providing
a body of data on the mutagenicity of various chemicals. This body
of data would not only describe the effects of the chemicals on
various test systems, but would help to make clear what additional
information was required to describe completely the biological
activity of these chemicals.

HISTORICAL DEVELOPMENT OF IPESTTC

This program emerged from research on short-term tests
supported by the United Kingdom Medical Research Council (MRC)
under commission from the United Kingdom Health and Safety
Executive (HSE). Early discussions on how best to carry out this
study led to a proposal by scientists at the Imperial Chemicals
Industries, Ltd. (ICI) to expand the study to include a larger
number of chemicals, which would be tested blind as coded samples.
ICI scientists took on the responsibility for selecting the
chemicals and preparing them in a high state of purity and in large
enough quantity that all investigators could work with samples
taken from the same batch. As it has evolved, the program is a
unique attempt to gather a large set of test results with which to
objectively evaluate the ability of short-term tests to correctly
distinguish between the carcinogens and noncarcinogens.

The initial selection of test systems included a wide variety
of assays, but none involved inducing cancer in animals. In each
case a correlation with carcinogenic activity in laboratory animals
was sought. It was agreed that the most useful method for
assessing the performance of short-term tests would be to compare
the results for pairs of structually-related chemicals where one
was a known carcinogen and the other a known noncarcinogen. The
investigators would not know the identity of the chemicals tested;
in other words, the test chemical would be evaluated as coded

samples in a blind trial. Scientists of Imperial Chemicals
Industries, Limited, agreed to supply the chemicals for this
testing program and selected the 25 carcinogens and 17
noncarcinogens (including 14 paired compounds). All chemicals were
synthesized and prepared for the study in as pure a state as
possible and in large enough quantity that all investigators could
use samples from the same batch. Since the 50-g quantities
prepared exceeded the requirements of the original HSE/MRC program,
the National Institute of Environmental Health Sciences (NIEHS)
developed a plan to include a larger variety of test systems. This
expansion made it possible to look for interlaboratory variation in
test performance. In other cases, where the time required to test
the set of 42 chemicals exceeded that allowed in the program plan,
samples were divided among laboratories performing the same test,
in an effort to make the final test data as complete as possible.

Many investigators financed their own testing. In other
cases, the work was financed by the HSE/MRC science in the United
Kingdom, NIEHS or the U.S. Environmental Protection Agency in North
America, or various other mechanisms in other parts of the world
such as Japan and the Soviet Union. By the time of the final
meeting at St. Simons Island, GA, in October, 1979, 30 different
assay systems were part of the program and data from over 50
laboratories were considered.

SELECTION OF TEST CHEMICALS

The three main criteria for selecting the test chemicals were
1) to have as large a range of chemical types and chemical classes
as possible in a group of 42; 2) to ensure the highest possible
purity and to ensure that all samples of each chemical under test
would be taken from the same batch; and 3) to obtain a balance
among different chemical classes, not including too many chemicals
from any particular class. In addition, 11 of the 25 chemical
carcinogens selected were included, because they were known to be
difficult to detect as mutagens in assays for point mutation in
either <u>Salmonella typhimurium</u> or <u>Escherichia coli</u>.

In the selection of noncarcinogens, an attempt was made to
find structural analogs of the chosen carcinogens. When a test
system gives a positive response to a carcinogen, one cannot
determine whether the assay is responding to some chemical property
other than the carcinogenicity inherent in the chemical structure.
Therefore, the most meaningful assay systems for screening tests
would be those that gave a positive response with a given
carcinogen and a negative result with its noncarcinogenic
structural analog. This is another reason for stressing purity of
the test chemicals: to be sure that the noncarcinogens (for
example) are not contaminated with trace levels of carcinogenic

structural analogs. Such contamination would elicit a positive
response for noncarcinogenic chemicals that should give negative
results. All chemicals (except auramine) were more than 99% pure,
and at least six different criteria of purity were evaluated.

The chemicals selected for testing are given in Table 1. This
table lists the 25 carcinogens and the 11 of these 25 that are
difficult to detect in bacteria. Also listed are the 17
noncarcinogens arranged in order of the strength of the data
available in the literature for making this evaluation. Selection
of the noncarcinogens proved exceptionally difficult and somewhat
disappointing, since much attention has been given in the past to
the development of criteria for carcinogenicity, but not
noncarcinogenicity. In Table 2, the list of chemicals are
reorganized to show the 14 pairs of carcinogens and noncarcinogenic
analogs.

SELECTION OF ASSAY SYSTEMS

An effort was made to include representatives of all available
types of short-term assays thought to be potentially useful
carcinogenicity pre-screening tests. Several of the assays
included in the study were part of the original HSE/MRC study in
the United Kingdom. As the program grew, subsequent sponsors
selected other assays to fill in gaps in the program. Many assays
were included because of their sponsors' interest in participation
and willingness to do so without financial support. The
developmental status of the assays ranges from those considered to
be the best-standardized and -validated (such as the Salmonella/
microsome assay) to those in the earliest stages of development
(such as the inductest and the diptheria toxin resistance system in
human fibroblasts).

No effort was made to standardize protocols. Without
knowledge of optimum protocols, standardization would only insure
that each investigator using a given assay would make the same
mistakes. As a result, the program allowed a comparison of results
from different protocols. Among the investigators using
established assays, general agreement on a protocol was reached for
maximum comparability of the results (as with the sex-linked
recessive lethal test in Drosophila and the micronucleus test in
mice).

The assay systems may be divided into five groups as shown in
Table 3. In group 1 are listed the two inductests used in the
study; one assays the lysis of bacteria (E. coli and B. subtilis),
the other, the expression of genes linked to prophage lambda. The
other two tests are the degranulation test (which assays the
dissociation of ribosomes or polysomes from rat liver endoplasmic

Table 1. Chemicals Selected for Testing in the International Program for the Evaluation of Short-term Tests for Carcinogenicity

---

Chemicals Classified as Carcinogens

4-Nitroquinoline-N-oxide
Benzo(a)pyrene
9,10-Dimethylanthracene[a]
2-Acetylaminofluorene
N-nitrosomorpholine
2-Naphthylamine
Hydrazine sulfate
Diethylstilbestrol[b]
Cyclophosphamide
3-Aminotriazole[b]
4-Dimethylaminoazobenzene
  (butter yellow)[b]
4,4'-Methylenebis-
  (2-chloroanilin) (MOCA)
Hexamethylphosphoramide (HMPA)[b]

Methylazoxymethanolacetate
Auramine (technical grade)[b]
Benzidine
β-Propiolactone
Chloroform[b]
Dimethylcarbamoyl chloride
Urethane[b]
DL-Ethionine[b]
Ethylenethiourea[b]
Safrole[b]
Epichlorohydrin
O-Toluidine hydrochloride[b]

Chemicals Classified as Noncarcinogens

Pyrene[a]
Anthracene[a]
4-Acetylaminofluorene[a]
1-Naphthylamine[d]
Azoxybenzene[e]
Sugar (sucrose)[e]
Dinitrosopentamethylene
  tetramine[e]
Isopropyl N(3-chlorophenyl)
  carbamate[c]
3-Methyl-4-nitroquinoline-
  N-oxide[e]

3,3',5,5'-Tetramethylbenzidine[e]
γ-Butyrolactone
1,1,1-Trichloroethane[d]
Dimethylformamide[d]
Diphenylnitrosamine[2]
Methionine[a]
Ascorbic acid[a]
4-Dimethylaminoazobenzene-4-
  sulfonic acid Na salt[d]

---

[a]Data not convincing.
[b]Carcinogen difficult to detect in bacterial assays.
[c]Best evidence for noncarcinogenicity.
[d]Intermediate evidence.
[e]Poorest evidence for noncarcinogenicity.

Table 2.   Pairs of Structural Analogs Among the Chemicals Tested

Carcinogen	Noncarcinogenic Analog
4-Nitroquinoline-N-oxide	3-Methyl-4-nitroquinoline-N-oxide
Benzidine	3,3',5,5'-Tetramethylbenzidine
4-Dimethylaminoazobenzene (butter yellow)	4-Dimethylaminoazobenzene-4-sulfonic acid Na salt
Benzo(a)pyrene	Pyrene
β-Propiolactone	γ-Butyrolactone
9,10-Dimethylanthracene	Anthracene
Chloroform	1,1,1-Trichloroethane
2-Acetylaminofluorene	4-Acetylaminofluorene
Dimethylcarbamoyl chloride	Dimethylformamide
2-Naphthylamine	1-Naphthylamine
N-Nitrosomorpholine	Dinitrosopentamethylene tetramine
Urethane	Isopropyl N(3-chlorophenyl)carbamate
Methylazoxymethanol acetate	Azoxybenzene
DL-Ethionine	Methionine

reticulum) and the nuclear enlargement assay (in which a positive result is indicated by an increase in the size of the nuclei in both HeLa cells and fibroblasts in culture).

Group II includes assays for the induction of point mutations in bacteria, including assays for forward and reverse mutation in both Salmonella and E. coli. The Salmonella/microsome reverse-mutation assay was conducted in 13 separate laboratories using a total of seven strains. Several procedures were used for metabolic activation, and the variations here included 1) source of S-9, 2) use of different chemicals as inducers, 3) variation in the amount of S-9, 4) plate incorporation versus pre-incubation, 5) the use of hepatocytes from rat liver, and 6) the addition of the comutagen norharman. Data were also obtained for a Salmonella assay that measured forward mutations to azaguanine resistance. In E. coli, two systems were included: one in which both forward and reverse mutation at four loci are screened simultaneously and another in which tryptophan reversion is measured in different strains of E. coli strain WP2.

The tests in group III measure various types of genetic damage in yeast, including forward mutation in Saccharomyces pombe and the following endpoints in S. cerevisiae: reverse mutation, mitotic crossing over in five different strains, induction of aneuploidy in strain D6, and differential survival of wild type and a multiply repair-deficient strain.

Table 3. Assay Systems Used in the International Program for the
Evaluation of Short-term Tests for Carcinogenicity

I. Prokaryotic Repair, Phage Induction, Nongenetic Assays

Inductests:
  Bacillus subtilis rec assay   Escherichia coli pol assay
  Escherichia coli   Nuclear enlargement
                                       Degranulation test

II. Prokaryotic Mutation Assays

Salmonella/microsome fluctuation assay   Escherichia coli 343
Salmonella 8-AZA resistance

III. Yeast Assays

Forward mutation--S. pombe   Aneuploidy - D6
Reverse mutation--XV185-14C   Repair assay - RAD, URA
Mitotic recombination - PG-154, PG-155,
  D4, D7, JD1

IV. Mammalian In Vitro Assays

Unscheduled DNA synthesis (WI-38, HeLa)   CHO-HGPRT, APRT, TK, OUS
Sister-chromatid exchange (CHO)   V79-HGPRT
Chromosome aberrations (CHO, RL_1)   Human fibroblasts--
BHK21--transformation   diphtheria toxin
                                              resistance

V. In Vivo Assays

Sex-linked recessive lethal--   Micronucleus--mouse
  Drosophila   Sperm morphology--mouse
Sister-chromatid exchange--mouse

The assays in group IV are mammalian in vitro systems, including assays for gene mutation in Chinese hamster, V79, mouse lymphoma, and human fibroblast cells. Chromosome effects in vitro include sister-chromatid exchange, chromosome aberrations, and unscheduled DNA synthesis. The BHK21 cell transformation assay based on growth in soft agar is also in this group of assays.

The Group V assays are whole-animal systems, including sex-linked recessive lethals in Drosophila, the micronucleus test in mice, and sister-chromatid exchange in liver cells and bone marrow. The only plant system included in the program is the induction of gene mutations resulting in color changes in Tradescantia stamen hairs.

EVALUATION OF THE DATA BASE

The data base that has come out of this program is broad and complex. Eventually, it can be used not only to compare the performances of the individual assay systems, but also to evaluate the effects of the chemicals both qualitatively and quantitatively. One of the difficulties in comparing the performances of the various assay systems is that due to resource limitations, not all 42 chemicals were tested in each assay. Since the samples were coded, the chemicals tested should be a random sample of the 42. However, the chemicals were usually tested in order of their receipt, so that the gaps in the data base involve the same group or groups of chemicals. Resource limitations also precluded interlaboratory comparisons of each assay system.

Each assay system's performance has been evaluated (Table 4) by its ability to correctly classify as positive the carcinogenic chemicals (sensitivity) and as negative the noncarcinogenic chemicals (specificity) and by its overall ability to correctly classify both carcinogens and noncarcinogens (accuracy). It should

Table 4. Terms Used in Describing Assay Performances

$$\text{Sensitivity (true positive fraction)} = \frac{\text{\# positive results with carcinogens}}{\text{\# carcinogens tested}}$$

$$\text{Specificity (true negative fraction)} = \frac{\text{\# negative results with noncarcinogens}}{\text{\# noncarcinogens tested}}$$

$$\text{Accuracy} = \frac{\text{\# correct results}}{\text{\# chemicals tested}}$$

---

be remembered that two assay systems can have the same sensitivity, specificity, and accuracy, and yet differ in the sets of chemicals that they correctly classify.

In the present study, 11 of the 25 carcinogens were selected because they are difficult to detect as mutagens in the bacterial assays for point mutation. Because of this, the sensitivity, specificity, and accuracy figures will differ markedly from previous studies. The data from the present program are being selectively analyzed so that these comparisons can be made for selected portions of the data base, with groups of chemicals omitted or with their classifications changed.

The calculations for the 42 chemicals show that no single assay system is sufficiently sensitive, specific, and accurate to be used in isolation. The data clearly indicate the need for a battery of tests.

EVALUATION OF TEST DATA ON THE 42 CHEMICALS

The main problem with this evaluation was that six of the noncarcinogens may not have been correctly classified. The following six chemicals gave a high frequency of positive results over a wide range of assay systems: 3-methyl-4-nitroquinoline-N-oxide, 4-acetylaminofluorene, 1-naphthylamine, azoxybenzene, diphenylnitrosamine, and 4-dimethylaminoazobenzene sulfonic acid (methyl orange). Those results indicate that they are probably carcinogens and are thus misclassified for the purpose of the present program. An advantage of computerization of the IPESTTC data is that the classification of these six .chemicals can be changed and new estimates of sensitivity, specificity, and accuracy obtained.

PROGRAM COORDINATION AND COMPLETION

IPESTTC was managed by a coordinating committee consisting of F.J. de Serres, Chairman, J. Ashby, P. Brookes, B. Bridges, I. Purchase, M. Shelby, and T. Sugimura. This group was responsible for the initial selection of assay systems and investigators, distribution of test samples, collection of data, and organization of the various workshops held during the course of the study. In addition, this group was responsible for the follow-up work required to complete the present data base, as well as its final analysis.

An interim report of this program will be published as a book (de Serres and Ashby, in press) containing reports from all of the investigators. It will also contain summary reports from the various test-system and test-chemical work groups that met to evaluate the data base after decoding of the test chemicals. The data, along with more detailed discussion, will be published early in 1981.

REFERENCES

de Serres, F.J., and J. Ashby. (in press). International Program for the Evaluation of Short-Term Tests for Carcinogenicity. Elsevier/North Holland: Amsterdam.

SPERM ASSAYS IN MAN AND OTHER MAMMALS AS INDICATORS OF CHEMICALLY INDUCED TESTICULAR DYSFUNCTION

Andrew J. Wyrobek
Lawrence Livermore Laboratory
University of California
Livermore, California

INTRODUCTION

Concern about human exposure to chemical agents has led to the development of numerous bioassays to detect mutagens and carcinogens rapidly and inexpensively. Recent attempts to compare and evaluate the efficacy of these bioassays have shown clearly that no single assay is sufficient (Coordinating Committee, 1978). A number of sperm assays have been developed and evaluated (Wyrobek, in press). Although the mutagenic basis of chemically induced sperm anomalies is generally not well understood, these assays play an important role in mutagenesis and carcinogenesis testing. First, sperm assays can measure chemical damage to the germ cells occurring during spermatogenesis or transit through the efferent ducts. Agents that are found to be mutagenic or carcinogenic in other bioassays can be tested directly for their spermatotoxic effects. This possibility is of major importance because the activity of an agent in bacteria or mammalian somatic cells is often a poor predictor of its activity in the testes after exposure in vivo (Coordinating Committee, 1978). Second, since animal sperm assays are as inexpensive as other short-term tests, many agents can be tested. Third, sperm assays have not only been developed and applied to mice, other rodents, and a variety of domestic animals, but they are also applicable to men exposed to chemicals (Wyrobek and Glendhill, in press).

This paper describes the application and methodology of sperm assays in men and laboratory animals and discusses the predictive value of induced sperm changes, correlations of results to carcinogenicity and mutagenicity, and the relative strengths and weaknesses of sperm assays (for more detailed accounts with

specific agents, see Topham, 1980b; Wyrobek and Bruce, 1975;
Wyrobek et al., in press b, c).

METHODOLOGY

Human Sperm Assays

Human semen assays have a long history in the diagnosis of
infertility. Thus, it is not surprising that early attempts to
assess altered testicular function in men exposed to chemicals have
involved measuring changes in the sperm parameters commonly used in
fertility diagnosis, such as sperm density (counts), motility, and
morphology.

Of these parameters, morphology is the most constant in an
unexposed man and is statistically highly sensitive to small
changes (Wyrobek et al., in press c). In the past, this assay has
not been widely used, because the scoring criteria were generally
difficult to standardize, and interlaboratory discrepancies were
unavoidable. We have improved the human morphology assay by
describing 10 classes of spermhead shapes and categorizing at least
500 sperm per individual into one of these classes. Through the
intermittent use of coded standard slides, we have shown constancy
in scoring the same set of standard slides for up to three years
and have demonstrated the objective nature of the scoring criteria
(Wyrobek, in press). We have adapted this method to men
occupationally exposed to carbaryl (Wyrobek et al., in press e),
anesthetic gases (Wyrobek et al., in press a), dibromochloropropane
(DBCP), and mercury (Wyrobek, et al., in preparation). Men exposed
to carbaryl and DBCP showed marked sperm changes when compared with
unexposed men. Men exposed to cancer-chemotherapeutic agents
showed drug-related decreases in sperm counts and increases in
sperm shape abnormalities (Wyrobek et al., 1980).

Among the more recently developed human sperm assays is the
YFF test, which is based on the unique fluorescence of the human Y
chromosome when stained with quinacrine (Kapp, 1979). Men exposed
to adriamycin and DBCP, for example, showed exposure-related
increases in the proportion of sperm having two fluorescent spots,
which are presumably due to the presence of two Y chromosomes in
one sperm because of errors in meiotic disjunction (Kapp, 1979).

A recent survey of the literature (Wyrobek et al., in press c)
showed that human sperm assays have been more widely used than was
generally suspected: more than 80 papers were found on the use of
semen assays in assessing testicular function. About 75% of the
exposures involved experimental or therapeutic drugs; about 15%
were occupational exposures; and about 10% involved personal drug
use. The studies cover 37 single agents, 10 complex mixtures, and

12 sets of multiple agents. Tables 1 and 2 list the single agents and complex mixtures categorized by those agents that 1) were found to induce significant changes in the sperm parameters of exposed men, 2) gave suggestive but inconclusive evidence of change, and 3) had no effect. Of the 37 single agents, 21 were positive, 9 suggestive, and 6 negative. Almost all of the agents showing some effect had a reduction in semen quality (e.g., reduction in sperm counts, decrease in sperm motility, decrease in the proportion of sperm with normal shapes, or increased frequencies of YFF sperm). However, five agents (chlomiphene citrate, coenzyme Q-7, fluoxymestrone, kallikrein, and methadone) were found to increase sperm counts and/or motility in some of the infertile patients studied.

Table 1. Single Agents Studied with Human Sperm Assays[a]

Positive Effects	Suggestive but Inconclusive Effects	No Effects Observed
Aspartic acid	Centchroman	CIBA-32644 Ba
Chlorambucil	Colchicine	Lysine
Chlomiphene citrate	Methadone	Methyl testosterone
Cyclophosphamide	Methotrexate	Ornithine
Cyproterone acetate	Metronidazol	Testosterone
Doxorubin hydrochloride	Nitrofurantoin	WIN 59491
Enovid	Trimeprimine	
Gossypol		
6 Medroxyprogesterone		
Metandienone		
Nilevar		
Norlutin		
Prednisone		
Progesterone		
Salicylazosulfapyride		
Testosterone enanthate		
Testosterone propionate		
WIN 13099		
WIN 17416		
WIN 18446		

[a]For data on specific agents and their chemical names, see Wyrobek et al. (in press c). Table entries are based on studies of sperm counts, motility, morphology, and YFF. The assignment of individual agents to columns is based on the data provided in the papers reviewed by the U.S. Environmental Protection Agency (EPA) Gene-Tox panel (see Waters, 1979) and may change as more data becomes available.

Table 2.   Complex Mixtures Studied with Human Sperm Assays[a]

Positive Effects	Suggestive but Inconclusive Effects	No Effects Observed
Alchoholic beverages	Carbaryl[b]	Epichlorohydrin[b]
Carbon disulfide[b]	Diethylstilbestrol	Glycerine workers[b]
Dibromochloropropane[b]	Tobacco smoke	Polybrominated
Lead[b]		biphenyls[b]
Marijuana		

[a]For data on specific agents and their chemical names, see Wyrobek et al. (in press c). Table entries are based on studies of sperm counts, motility, and YFF. The assignment of individual mixtures to columns is based on the data provided in the papers reviewed by the EPA Gene-Tox panel (Waters, 1979) and may change as more data becomes available.
[b]Occupational exposures.

Eleven complex mixtures were studied using the sperm assays. As shown in Table 2, five showed positive effects, three showed suggestive but inconclusive effects, and three showed no effect. Most of these studies involved occupational exposures in which single active agents were implicated (e.g., DBCP or lead).

Sperm assays have also been used in men exposed to at least 12 sets of two or more agents in consort (Wyrobek et al., in press c). Ten combinations caused reductions in semen quality (e.g., cyclophosphamide plus prednisone, Danazol plus testosterone enanthate, and MVPP cancer therapy).

Of the 60 different human exposures evaluated, including all single agents, multiple agents, and complex mixtures, 97% used sperm counts as one of the parameters measured, of which 25% used counts as the only parameter measured. Furthermore, 58% of all the exposures evaluated studied motility, 42% studied morphology, and only 7% used YFF.

Because of our poor understanding of the genetic mechanisms underlying various induced sperm anomalies, the only information that can be gained from these assays at present is whether human spermatogenesis is affected by exposure to a chemical agent or mixture of agents. These data, together with results of other short-term assays for mutagenicity, may indicate which of these agents are potential germ cell mutagens. Clearly, more research is needed to develop sensitive sperm assays with defined mutational end points, so that risks of heritable damage may be assessed directly using sperm.

Animal Sperm Assays

Sperm assays have also been used to assess chemically induced changes in testicular function in a variety of animal species. At least 13 agents have been studied in rabbits, 14 in rats, 2 each in sheep, cattle, dogs, hamsters, and monkeys, and 1 in pigs (for review, see Wyrobek et al., in press b). Most of the studies (covering approximately 159 agents) used the mouse sperm morphology assay (Wyrobek and Bruce, 1975). Approximately 40 agents produced dose-dependent increases in sperm shape abnormalities, 105 were negative, and 5 showed marginal responses. However, of the negatives, only about 50 were known to have been tested to lethal doses, and the remainder should be retested at higher doses. Positive results came from a wide variety of chemical classes, including antimetabolites, alkylating agents, spindle poisons, polyaromatic hydrocarbons, aromatic amines, estrogens, and others. Negative responses will occur with any compound that rapidly kills the animal or whose active form does not reach the testes either because of the route of exposure or the metabolism required to activate the agent. Because only seven of the chemical agents reviewed were tested in two or more species, meaningful interspecies comparisons of results of sperm assays are not yet possible.

DISCUSSION

The Genetic Implications of Chemically Induced Sperm Anomalies

Evidence from human studies. Although it is generally agreed that major reductions in sperm counts and motility are linked to reduced fertility, it remains unclear which sperm parameter(s) indicates embryonic failure or heritable genetic abnormalities. Human data on this question is very limited. In one study, fathers of 201 spontaneous abortions showed significantly higher sperm abnormalities and lower sperm counts than 116 fathers of normal pregnancies (Furuhjelm et al., 1962), suggesting a link between poor semen quality and frequency of spontaneous abortions. Although several studies support this observation (Czeizel et al., 1976; Joel, 1966; Takala, 1957), some studies found no such correlation (Kneer, 1957). Clearly, more human studies are needed to compare exposure of the male parent, induced sperm defects, and reproductive outcome.

Evidence from animal studies. Most of the studies on genetic validation of induced sperm abnormalities have been conducted with mice. Three lines of evidence link induction of abnormal sperm and heritable genetic abnormalities. First, it is clear that sperm shaping and the production of abnormal sperm is polygenically controlled by autosomal as well as sex-linked genes (Beatty, 1972; Brozek, 1970; Hugenholtz and Bruce, 1979; Krzanowska, 1972; Topham, 1980a; Wyrobek and Bruce, 1978; Wyrobek, 1979). Second, in at least three independent studies with numerous mutagens and nonmutagens, germ cell mutagens generally induced sperm abnormalities, while nonmutagens generally had no effect (Bruce and Heddle, 1979; Topham, 1980b; Wyrobek and Bruce, 1978). Third, in several studies using agents that induce sperm abnormalities, sperm abnormalities were transmitted to the male offspring of the exposed mice (Hugenholtz and Bruce, 1979; Sotomayor, 1979; Staub and Matter, 1976; Topham, 1980a; Wyrobek and Bruce, 1978).

A brief survey of the literature indicates that many of the compounds that are active in the mouse sperm morphology test are also active in the heritable specific locus, F_1 sperm morphology, heritable translocation, and/or dominant lethal tests in mice (for review, see Wyrobek et al., in press b). Therefore, the mouse sperm morphology test may be a useful screening test for compounds that constitute a potential genetic hazard for mammals. Spindle poisons that may cause nondisjunction in germ cells can also be identified. Further murine studies are needed to understand the quantitative relationships among dosage regime, appearance of abnormal sperm shapes in the semen, time between exposure and conception, fertility of the exposed male, frequency of genetically abnormal offspring, and fertility of the abnormal offspring.

Correlations of Sperm Abnormality Results with Results of Short-term Assays for Carcinogenesis

Several attempts have been made to compare results of the murine sperm morphology assay with other short-term tests for carcinogenesis. As part of the International Program for the Evaluation of Short-term Tests for Carcinogenesis, six pairs of carcinogens and noncarcinogens and five unpaired carcinogens were surveyed as unknowns, using the mouse sperm abnormality assay (Wyrobek et al., in press d). No false positive responses were found, suggesting that the sperm assay has a high specificity for carcinogens. However, several false negatives were obtained, indicating that not all carcinogens induce sperm abnormalities in mice. This data may be very important in assessing which carcinogens may also be active in the testes. The detailed comparison of the sperm abnormality assay and the other assays surveyed in this program is still in progress.

A comparison of the potency (dose required to double background frequencies) of some 30 agents studied in both the Salmonella/microsome and sperm abnormality assays showed no apparent correlation, suggesting that the assays are measuring different biological phenomena (unpublished results). In a different study of 61 agents (Bruce and Heddle, 1979) with the mouse sperm morphology and Salmonella/microsome assays, each assay was found to correctly identify approximately 60% of the carcinogens and noncarcinognens tested. The assays together identified nearly 90% of the agents, suggesting that the two assays measure different end points. These authors recommended that the number of false negatives, which is relatively high with each of the two assays, may be reduced by using a battery of both assays for the identification of potential carcinogens.

CONCLUSIONS

Advantages

The major advantages of sperm assays are that the cells examined are readily available in both animals and man, and that sperm carry the paternal genome in the form that will be ultimately involved in fertilization. Other advantages are the following:

1) Sperm are examined after exposure of a whole mammal. This helps ensure that artifacts (false positives and false negatives) due to problems of tissue penetration, metabolism, pharmacokinetics, and dosage encountered in non-gonadal, cultured-cell, or nonmammalian systems are minimized.

2) The changes in sperm parameters probably arise from interference by the test substance with the differentiation of the sperm cell. Thus, these changes are intrinsically relevant to safety evaluation and assessment of potential effects of the agent on male fertility.

3) The laboratory methods are generally rapid, inexpensive, and quantitative.

4) Sperm assays have major advantages over other approaches for assessing induced changes in testicular function. Testicular biopsies are impractical, traumatic, invasive, and may themselves affect testicular function. Epidemiological surveys of reproductive function using questionnaires exclusively require large sample sizes and are generally expensive. Analyses of blood levels of gonadotrophins are expensive and generally insensitive to

small changes in testicular function. Compared with these
methods, sperm assays are noninvasive, inexpensive,
require small sample sizes for effective analyses, and are
sensitive to small changes.

Disadvantages

The major disadvantages of sperm assays are the following:

1) Heritability of the induced damage is not yet clearly
    demonstrated.

2) Limited sperm sampling times and dosage regimens may
    reduce the sensitivity of the assay (e.g., agents that
    only exert transient effects may be missed by using single
    sampling times).

3) Other factors such as ischemia, infection, and starvation
    may produce spurious false positive responses.

Applications

The availability of animal and human sperm assays suggests
several applications in the assessment of chemically induced
spermatotoxicity (antifertility effects) and heritable genetic
abnormalities. Animal sperm assays (such as the mouse morphology
assay) may be used to screen large numbers of agents to establish a
ranking that sets priorities for sperm studies in exposed men.
This approach would minimize the use of human studies that
generally have complex requirements for epidemiological and
statistical input and often require lengthy interactions with union
officials, industry representatives, employees, physicians, and
patient-donors. Furthermore, animal sperm studies also may be
useful in evaluating the relative effects of the components of a
complex mixture that is suspected of affecting human sperm (such as
an occupational exposure).

Since little is known of the quantitative relationships
between induced sperm abnormalities and heritable genetic damage,
indirect methods are needed to assess the genetic risk to offspring
of men who show induced sperm anomalies. By combining data from
short-term mutagen bioassays (e.g., Salmonella/microsome assay,
mammalian somatic cell mutation assays), which may demonstrate
mutagenic potential, with data from animal and human sperm assays,
which may demonstrate activity in the testes, we may be able to
evaluate whether a mutagen (or carcinogen) is active in the testes.
For select agents, a more objective assessment of germ cell
mutagenicity may be required. This could be done using various

murine F_1 generation bioassays (e.g., heritable chromosomal translocation, dominant skeletal mutations, and heritable sperm abnormalities) to quantify the relationships between heritable consequences and chemically induced sperm anomalies. The combined use of data from animal and human sperm assays, short-term in vitro bioassays, and murine F_1 generation mutational bioassays may provide a feasible approach to genetic risk assessment in men exposed to agents that cause sperm anomalies.

ACKNOWLEDGMENTS

I wish to thank G. Watchmaker and L. Gordon for their technical input and M. Mendelsohn and B. Ishida for their incisive suggestions in the preparation of this manuscript. I am especially thankful to the members of the EPA Gene-Tox panel on human and sperm assays for access to some of their data and results presented in this manuscript. This work was performed under the auspices of the U.S. Department of Energy, under contract no. W-7405-ENG-48, and the U.S. Environmental Protection Agency, under a Pass-Through Agreement and under contract no. 79-D-X0826.

REFERENCES

Beatty, R.A. 1972. The genetics of size and shape of spermatozoon organelles. In: The Genetics of the Spermatozoon: Proceedings of an International Symposium. R.A. Beatty and S. Gleucksohn-Waelsch, eds. University of Edinburgh: Edinburgh. pp. 97-115.

Brozek, C. 1970. Proportion of morphologically abnormal spermatozoa in two inbred strains of mice, their reciprocal F_1 and F_2 crosses and backcrosses. Acta Biol. Cracov. (Ser. Zool.) 13:189-198.

Bruce, W.R., and J.A. Heddle. 1979. The mutagenic activity of 61 agents as determined by the micronucleus, Salmonella, and sperm abnormality assays. Can. J. Genet. Cytol. 21:319-334.

Coordinating Committee. 1978. International program for the evaluation of short-term tests for carcinogenicity. Mutation Res. 54:203-206.

Czeizel, E., M. Hancsok, and M. Viczian. 1976. Examination of the semen of husbands of habitually aborting women. Orvosi Hetilap 108:1591-1595.

Furuhjelm, J., B. Jonson, and C.G. Lagergren. 1962. The quality of human semen in spontaneous abortion. Int. J. Fertil. 7:17-21.

Hugenholtz, A.P., and W.R. Bruce. 1979. Radiation-induced heritable sperm abnormalities in mice. Environ. Mutagen. 1:127-128.

Joel, C.A. 1966. New etiologic aspects of habitual abortion and infertility, with special reference to the male factor. Fertil. Steril. 3:374-380.

Kapp, R.W. 1979. Detection of aneuploidy in human sperm. Environ. Hlth. Perspect. 31:27-31.

Kneer, M. 1957. Der habituelle Abort. Dtsch. Med. Wochenschr. 82:1059-1061.

Krzanowska, H. 1972. Influence of Y chromosome on fertility in mice. In: The Genetics of the Spermatozoon: Proceedings of an International Symposium. R.A. Beatty and S. Gleucksohn-Waelsch, eds. University of Edinburgh: Edinburgh. pp. 370-386.

Sotomayor, R.E. 1979. Spermatid head abnormalities in translocation heterozygotes from EMS- or CPA-treated sires. Environ. Mutagen. 1:129. (abstr.)

Staub, J.E., and B.E. Matter. 1976. Heritable reciprocal translocations and sperm abnormalities in the F_1 offspring male mice treated with triethylenemelamine (TEM). Arch. fuer Genet. 49:29-41.

Takala, M.E. 1957. Studies on the seminal fluid of fathers of congentially malformed children (199 sperm analyses). Acta Obst. et Gynec. Scandinav. 36:29-41.

Topham, J.C. 1980a. Chemically-induced transmissible abnormalities in sperm head shape. Mutation Res. 70:109-114.

Topham, J.C. 1980b. The detection of carcinogen-induced sperm head abnormalities in mice. Mutation Res. 69:149-155.

Waters, M.D. 1979. The Gene-Tox program. In: Mammalian Cell Mutagenesis: The Maturation of Test Systems, Banbury Report 2. A.W. Hsie, J.P. O'Neill, and V.K. McElheny, eds. Cold Spring Harbor Laboratory: Cold Spring Harbor, NY. pp. 449-457.

Wyrobek, A.J. 1979. Changes in mammalian sperm morphology after x-ray and chemical exposures. Genetics (Suppl.) 92:s105-s119.

Wyrobek, A.J. (in press). Methods for human and murine sperm assays. In: Short-Term Tests for Chemical Carcinogens. H.F. Stich and R.H.C. San, eds. Springer Verlag: New York.

Wyrobek, A.J., J. Brodsky, L. Gordon, G. Watchmaker, and E. Cohen. (in press a). Sperm studies in anesthesiologists.

Wyrobek, A.J., and W.R. Bruce. 1975. Chemical induction of sperm abnormalities in mice. Proc. Natl. Acad. Sci. USA 72:4425-4429.

Wyrobek, A.J., and W.R. Bruce. 1978. The induction of sperm-shape abnormalities in mice and humans. In: Chemical Mutagens: Principles and Methods for Their Detection, Vol. 5. A. Hollaender and F.J. de Serres, eds. Plenum Press: New York. pp. 257-285.

Wyrobek, A.J., J.G. Burkhart, M.C. Francis, L.A. Gordon, R.W. Kapp, G. Letz, H.V. Malling, J.C. Topham, and M.D. Wharton. (in press b). An evaluation of the mouse sperm morphology assay and sperm assays in other animals: a report for the Gene-Tox program. Mutation Res., Rev. Genet. Toxicol.

Wyrobek, A.J., J.G. Burkhart, M.C. Francis, L.A. Gordon, R.W. Kapp, G. Letz, H.V. Malling, J.C. Topham, and M.D. Wharton. (in press c). Chemically induced alterations of spermatogenic function in man as measured by semen analysis parameters: a report for the Gene-Tox program. Mutation Res., Rev. Genet. Toxicol.

Wyrobek, A.J., M. DaCunha, L. Gordon, G. Watchmaker, B. Gledhill, B. Mayall, J. Gamble, and M. Meistrich. 1980. Sperm abnormalities in cancer patients. Cancer Res. 21:196.

Wyrobek, A.J., and B.L. Gledhill. (in press). Human semen assays for workplace monitoring. In: Proceedings of the Workshop on Methodology for Assessing Reproductive Hazards in the Workplace, NIOSH.

Wyrobek, A.J., L. Gordon, and G. Watchmaker. (in press d). The effects of 17 chemical agents including 6 carcinogen/non-carcinogen pairs on sperm shape abnormalities in mice. In: Short-term Tests for Carcinogens: Report of The International Collaborative Program. F.J. de Serres and J. Ashby, eds. Elsevier/North Holland: Amsterdam.

Wyrobek, A.J., G. Watchmaker, K. Wong, and D. Moore II.
    (in press e). Sperm abnormalities in carbaryl exposed
    workers. Environ. Hlth. Perspect.

ASSESSING CARCINOGENIC RISK RESULTING FROM COMPLEX MIXTURES

Roy E. Albert
Institute of Environmental Medicine
New York University Medical Center
New York, New York

The evaluation of carcinogenic risks from complex mixtures, in contrast to pure substances, adds a dimension of uncertainty to a situation already frought with uncertainties and complexities.s. The uncertainties of evaluating complex mixtures are likely to be lost in the overall uncertainties of the risk assessment process. There is considerable controversy about the risk assessment of carcinogenic substances these days. Some liken the assessment of carcinogenic risks to the theater, in that a willing suspension of disbelief is required. To others, like myself, the assessment of carcinogenic risks reflects Mark Twain's definition of work: "It is something which a body is obliged to do." I believe that the risk assessment of carcinogens is something that one is obliged to do in a regulatory setting in order to make regulatory judgements in as rational a manner as possible. The responsibility of those doing the risk assessment is to make the most sensible use of current science, paying proper regard to cautioning those who are making the decisions about the uncertainties in the process.

One of the problems with the assessment of carcinogenic risks either of pure substances or complex mixtures is that it is a relatively new field. Before 1970, the only well-established risk assessment area involved ionizing radiation. It is worth noting that the standards for permissible exposure to ionizing radiation were set not on the basis of risk assessment but on the traditional basis of applying safety factors to observed levels of carcinogenic responses, particularly to cancer induction in the bone, lung, and bone marrow. It was in the field of ionizing radiation that the dominant concept of risk assessment emerged, namely, the linear non-threshold extrapolation model. This model was based on the correlation between carcinogenesis and mutagenesis and on the recognition that the linearity of dose response is applicable to mutagenesis and consistent with linearity in the dose response for leukemia, particularly among Japanese atom-bomb survivors.

The linear non-threshold concept has proven to be a powerful tool for predicting small excess risks of cancer at levels that can't be confirmed or denied by direct observation either in animals or in epidemiologic follow-up studies of exposed populations. Both animal studies and the epidemiologic follow-up methods are too insensitive to detect levels of risk below a few percent, levels which are far too high to be tolerated willingly. Consequently, the linear non-threshold concept of dose response can be considered a two-edged sword. It provides the impetus to regulate, because we are essentially accepting the position that any exposure, however small, will produce an excess of cancer, but it leaves us in a quandary as how to regulate, because it puts us in a position of trying to figure out which excess risks are tolerable under given circumstances. The non-threshold concept was first incorporated into the regulation of carcinogens in the Delaney amendment, which bans food additives that show evidence of carcinogenic action. The Delaney amendment reflects the scientific concept that there is no safe level of carcinogen exposure and the political judgement that no excess risk of cancer from food additives is tolerable.

In the 1970's the regulatory movement was accelerated with the formation of the U.S. Environmental Protection Agency (EPA), the Occupational Safety and Health Administration (OSHA), and the Consumer Protection and Safety Commission (CPSC). Many laws dealing with the regulation of carcinogens were formulated using different approaches to the control of carcinogens: the Clean Air Act calls for protecting everyone with a margin of safety—clearly impossible under a non-threshold concept; the Federal Insecticide, Fungicide, and Rodenticide Act calls for weighing risks and benefits; the Toxic Substances Control Act refers to making regulatory judgments based on reasonable risk; the Occupational Safety and Health Act calls for using "best available technology" combined with economic considerations. A number of other laws call for the use of "best available technology."

These different regulatory approaches use risk assessment in different ways and to different degrees. Some of them simply call for the characterization of an agent as a carcinogen with the application of "best available technology," thereby applying considerable economic and technical pressure. Others call for weighing risks and benefits, whereby an agent is not only characterized as a carcinogen, but its public health impact is estimated as a basis for evaluating the benefits and costs of regulation.

In 1976, EPA developed its own guidelines for the risk
assessment of carcinogens in response to a great deal of criticism
of its approach to regulating pesticides. The EPA guidelines for
risk assessment are based on the evidence approach. Risk
assessment was viewed as an exercise trying to answer two
questions: First, is the agent likely to be a human carcinogen?
Second, if the agent is a carcinogen, how much cancer is likely to
be produced? The former question requires a qualitative judgment,
the latter a quantitative one.

The qualitative judgement is based on the sum of evidence
concerning the quality, scope, and kinds of tumorigenic responses.
The judgements can range from the characterization of an agent
with strong evidence, based on good epidemiologic data backed by
animal data, to the opposite end of the spectrum where only a
single test, a single strain, a single sex, or a single species
may have a marginal response. The guidelines regard short-term
tests as being in a suggestive category when they stand alone, yet
recognize their great value in providing support for animal
bioassays or epidemiologic evidence of carcinogenicity. It is
difficult to know when a scientific approach achieves sufficient
consensus to provide the basis for regulatory action. So far,
short-term bioassays haven't achieved this stature. I am not sure
when they will. A tremendous amount of work is certainly aimed in
that direction. It may very well be that within a forseeable
time, short-term bioassays will have sufficient stature to provide
a very strong impetus toward regulation on their own merits.

The quantitative assessment looks at how much cancer is
likely to be produced, and assumes some background knowledge about
exposure. From our experience in the Carcinogen Assessment Group,
I think that exposure assessment is one of the weakest areas in
EPA. I am sure that this situation also exists in other agencies.
A quantitative assessment is based on an estimate of the exposure
as well as the use of models for extrapolation from high doses to
low doses; in many cases it also involves extrapolation from
animals to humans. The U.S. Environmental Protection Agency
started off in its guidelines calling for the use of more than one
extrapolation model. Over the last four years, the assessments
have focussed on the linear non-threshold extrapolation model.
While no one model has a resoundingly solid scientific foundation,
the linear non-threshold model has more scientific plausibility
than other models and also tends to provide conservative estimates
of risk. This model has been used almost exclusively in EPA's
Carcinogen Assessment Group over the past four years. Given the
uncertainties in the quantitative assessment of risk, the linear
non-threshold model can be regarded as providing a plausible upper
limit of estimated risk.

This assessment approach has been used in EPA for the last
four years on about 100 agents. More recently, an Interagency
Regulatory Liaison Group looked at the assessment of carcinogenic
risks and formulated a position document. The agencies
represented were EPA, OSHA, CPSC, and the Food and Drug
Administration. The position taken in this interagency document
can be read to be essentially consistent with the approach that I
already mentioned. Realistically, however, it reflects
considerable resistance to quantitative risk assessment. The
document, to a considerable extent, gives only qualified support
to quantitative risk assessment and can be cited as support by
agencies that wish to do or not do quantitative assessment,
depending on their point of view. More recently the Federal
Regulatory Council published a cancer policy: its approach to the
assessment of carcinogens supports the approach that has been
taken by the EPA and IRLG, but with more qualification of the
quantitative aspects of risk assessment than has been given by
EPA.

The assessment of complex mixtures is no different, at least
in principle, to the pattern that I have described already. There
is certainly nothing unusual about complex mixtures of
carcinogens. After all, the first demonstration of chemical
carcinogenesis in animals was in the 1920's by the Japanese
Yamagiwa and Ichikawa. They painted coal tar on rabbits' ears and
first displayed the action of chemical carcinogens. One hundred
and fifty years earlier the first epidemiologic observations on
cancer induced by environmental chemicals dealt with scrotal
cancer caused by soot in chimney sweeps. Soot is a complex
mixture. From the beginning, the field of chemical carcinogenesis
has been firmly embedded in the problem of complex mixtures.

Probably the most important exercise that has been undertaken
by the EPA Carcinogen Assessment Group involving complex mixtures
has concerned diesel exhaust particulates. The approach here has
been to peg the evaluation of diesel particulates to some fairly
solidly established epidemiologic dose-response data for agents
that are chemically similar to diesel exhaust, namely, lung cancer
among coke-oven workers, cigarette-smoke-induced lung cancer, and
lung cancer in workers who have used coal tar and asphalt roofing
materials. The approach has been to develop a cross comparison of
the potency of diesel particulate exhaust and the other three
materials on the basis of short-term in vitro assays, mouse skin
tumorigenesis studies, and intratracheal intubation studies in
hamsters. If we can bracket the carcinogenic potency of diesel
exhaust with respect to the materials that have been observed to
produce lung cancer in humans, we can use the human epidemiology
data as surrogates for diesel particulates data in quantitative
risk assessment. I think that this approach covers about as much
as can be done in terms of dealing with complex mixtures of this

sort. The same principle could be applied to the comparison of complex mixtures with pure materials using animal studies; certainly the use of short-term assays, both in vitro and in vivo, is particularly important when complex mixtures have a variable composition.

The diesel story is certainly not yet finished. The research program is still in the stage of producing data. In addition to the efforts involving cross comparisons of different materials with diesel particulates for carcinogenic potencies, there are also studies dealing with exposure to the diesel exhaust per se. One should wait until much of the experimental data is in hand before making any firm quantitative risk assessments.

The risk assessment process is receiving its major challenge in the quantitative area, although the publicity over the recent saccharin epidemiology studies is probably going to provide ammunition to those who are opposed to the use of assessment even in the qualitative area, namely, the applicability of rodent bioassays to human cancers. I must express my dismay at the way in which the publicity about the saccharin studies failed to highlight the fact that the duration of exposure to saccharin is completely inadequate to permit a proper characterization of its carcinogenicity in humans. It is perfectly clear that even if one were dealing with an agent that produced cancer, if the effects on human populations were evaluated before the end of the latent period, no effects would be found.

But the most serious challenges to risk assessment come in the quantitative area. I have mentioned the kind of objections that have been made by some agencies within the IRLG; also, the Federal Regulatory Council tended to downplay quantitative risk assessment on the basis of its uncertainty. Industry doesn't particularly like the EPA brand of quantitative risk assessment because of the use of the conservative linear non-threshold dose-response model. I think the position of industry is that if one is going to do quantitative risk assessment, which they think is not a bad idea, it would be better to use models that are less conservative, namely that show a curvilinear dose response and yield much lower risk estimates. The Occupational Safety and Health Administration is against quantitative risk assessment, because it feels that the law doesn't require it and it would only interfere with its regulatory program. There will be a decision made on this question by the Supreme Court in the case of benzene. The National Academy of Science, in a yet-to-be-released document on how to regulate pesticides, also takes a crack at quantitative risk assessment. This particular committee would go so far as to agree that one could extrapolate from animals to humans, as a basis for characterizing carcinogens in terms of relative potency, but they balk at the use of extrapolation models to predict

responses at low levels of exposure. It has been pointed out, to
no effect, that several other committees in the National Academy
of Science have used the linear non-threshold extrapolation model
for estimating risks. Probably the most vocal group to express
opposition at the present time to quantitative risk assessment is
the environmentalists. In a joint comment by the Environmental
Defense Fund and the Natural Resources Defense Council to EPA on
its newly proposed Air Cancer Policy, they say that "EPA must
abandon its arbitrary and unlawful reliance on quantitative risk
assessment methods. Quantitative risk estimates can be wrong by a
factor of five million times or more and thus are too unreliable
and imprecise to play a rational role in determining the levels of
control applied to a hazardous pollutant. Such estimates may have
a role in the grossest form of priority setting, but that will be
valid only if the agency much more explicitly recognizes the
uncertainties of the estimates and commits itself to not using
them in any way in subsequent standard setting."

     That quote is quite a forceful expression of opinion. The
notion that quantitative risk estimates can be wrong by a factor
of five million stems from the National Academy of Sciences
Saccharin Report, which listed the risk estimates from saccharin
using a variety of extrapolation models. One can pick the
extrapolation model that yields the kind of result that one wants.
It is quite easy to pick a set of models that give estimates that
vary by many orders of magnitude. Although I do believe that it
is fair to say that the use of the linear non-threshold
extrapolation model may somewhat overestimate risk, it is not
likely to produce much of an underestimate of risk. Certainly,
for example, the estimation that one percent of cancer is due to
background ionizing radiation can't be too small by five million
times. That wouldn't be very likely.

     Clearly, these expressed reservations about the use of
quantitative risk assessment reflect the very real weakness in the
scientific foundation underlying quantitative risk assessment.
This problem should be one of the major priority areas for
research related to carcinogen assessment. The regulation of
carcinogens is a public health program, and it is difficult to
mount a public health program without having some idea of the
likely benefits in relation to costs. Hopefully, we will make
rapid progress in gaining the knowledge necessary for doing risk
assessments with greater confidence.

INDEX